Adapting Information and Communication Technologies for Effective Education

Lawrence Tomei
Robert Morris University, USA

Information Science REFERENCE

INFORMATION SCIENCE REFERENCE
Hershey · New York

Acquisitions Editor:	Kristin Klinger
Development Editor:	Kristin Roth
Senior Managing Editor:	Jennifer Neidig
Managing Editor:	Sara Reed
Assistant Managing Editor:	Carole Coulson
Copy Editor:	Katie Smalley
Typesetter:	Michael Brehm
Cover Design:	Lisa Tosheff
Printed at:	Yurchak Printing Inc.

Published in the United States of America by
Information Science Reference (an imprint of IGI Global)
701 E. Chocolate Avenue, Suite 200
Hershey PA 17033
Tel: 717-533-8845
Fax: 717-533-8661
E-mail: cust@igi-global.com
Web site: http://www.igi-global.com/reference

and in the United Kingdom by
Information Science Reference (an imprint of IGI Global)
3 Henrietta Street
Covent Garden
London WC2E 8LU
Tel: 44 20 7240 0856
Fax: 44 20 7379 0609
Web site: http://www.eurospanonline.com

Library of Congress Cataloging-in-Publication Data

Adapting information and communication technologies for effective education / Lawrence Tomei, editor.

p. cm.

Summary: "This book addresses ICT assessment in universities, student satisfaction in management information system programs, factors that impact the successful implementation of a laptop program, student learning and electronic portfolios, and strategic planning for e-learning. It provides innovative research on several fundamental technology-based initiatives, and will make a valuable addition to every reference library"--Provided by publisher.

Includes bibliographical references and index.

ISBN-13: 978-1-59904-922-9 (hardcover)

ISBN-13: 978-1-59904-925-0 (ebook)

1. Educational technology. 2. Education, Higher--Effect of technological innovations on. 3. Information technology. I. Tomei, Lawrence A.

LB1028.3.A333 2008

371.33'4--dc22

2007024487

Adapting Information and Communication Technologies for Effective Education is part of the IGI Global series named *Advances in Information and Communication Technology Education Series (AICTE)* (ISSN: 1935-3340).

British Cataloguing in Publication Data
A Cataloguing in Publication record for this book is available from the British Library.

Table of Contents

Section III
Assessment

Detailed Table of Contents

Section II
Educational Initiatives

This chapter presents the development of technology-assisted problem solving (TAPS) packages at the University Tenaga Nasional (UNITEN). This project is the further work of the development of interactive multimedia based packages targeted for students having problems in understanding the subject of engineering mechanics dynamics. One facet of the project is the development of engineering mechanics dynamics problems for core undergraduate engineering courses. This chatper discusses the development of an interactive multimedia environment for solving relative motion of a rigid body using rotating axes, and more specifically outlines the framework used to develop the multimedia package, highlighting our multimedia design process and philosophy.

Parallel to advancements in information technology usage, there are increasing demands for basic computer skills at minimum from today's college graduates. As a consequence, many colleges and universities have chosen to stimulate campus laptop initiatives as a way to provide their students opportunities to grow their computer skills and experiences. However, the success of laptop programs is very much dependent on the degree to which students and faculty are accepting a laptop environment and are willing to implement such programs. Defining which conception factors are necessary is essential for successful implementation. This study examines such factors by focusing on university student perceptions of required laptop programs in order to distinguish which factors they perceive as important. In understanding what factors encourage student support of laptop initiatives, such programs can be made more useful to students as well as more beneficial to universities.

Schools of business can benefit from the adoption of geographic information gystems (GIS). A brief overview of GIS is presented with an example showcasing how it can be presented in a business school, the benefits for business schools, their students, and faculty, and a comprehensive approach for promoting such spatial thinking. The goal is to empower faculty to adopt GIS for their research and teaching, producing a large number of business school graduates that can promote spatial thinking in their organizations.

Decision trees are one of the most successful machine learning paradigms. This chapter presents a library of decision tree algorithms in Java that was eventually used as a programming laboratory workbench. The initial design focus was, in regards to the non-expert user, to conduct experiments with decision trees using components and visual tools that facilitate tree construction and manipulation, and in regards to the expert user, to be able to focus on algorithm design and comparison with few implementation details. The system was built over a number of years and various development contexts and has been successfully used as a workbench in a programming laboratory for junior computer science students. The underlying philosophy was to achieve a solid introduction to object-oriented concepts and practices based on a fundamental machine learning paradigm.

This chapter develops an educational strategy to foster the interest and persistence of middle school girls in science, technology, engineering, and mathematics (STEM) careers using existing Websites. Criteria are specified that enable middle school teachers to evaluate Websites as supplemental learning activities within prescribed curricula. In particular, the evaluative criteria help assess sites that provide materials appealing to boys and girls, allowing teachers to adopt them without concern that they are providing an unfair advantage to girls.

A challenge in teaching and providing any type of instruction in the online learning environment is to ensure that participants are engaged in the process and find meaning in their learning. This case study investigated the use of vignettes as a teaching strategy and learning activity of the generative learning model in a hybrid online course. Vignettes are short and realistic stories that may help bridge participants' previous experiences to applying course material in relevant situations. The generative learning model, consisting of five main components: attention, motivation, knowledge, generation, and metacognition (Wittrock, 2000), was incorporated when requiring students to answer teacher-generated vignettes and to generate their own vignettes. Two outcomes were anticipated using vignettes within the generative learning model in a hybrid online course: (1) enhancement of academic achievement, and (2) higher order thinking. This study considered data from student work collected from the instructional techniques course, GITED 631, taught in the graduate school of education at Duquesne University, in Pittsburgh, Pennsylvania, in the fall of 2003. Eight participants responded to teacher-generated vignettes, created diagrams and rubrics, created their own vignettes, and recorded their observations concerning vignettes in reflective learning logs. The adult online learners in this study professionally focused on teaching

Universities teach project management to information technology (IT) students. The project management principles that students have previously learned are often put into practice in a project course, intended to give final year students the experience of applying their knowledge to real or simulated projects. This chapter reports on research that investigated the use of, and usefulness of, project management in student IT projects. The results show that there was a wide range in the application of project management practices, with students being more likely to produce the initial documentation associated with some of the project management knowledge areas than to make use of it throughout the project to monitor the project's progress. The results also showed that the number of project management guidelines applied in student projects was not linked with IT project success. However, there was a strong relationship between project management plan quality and obtaining a good software product.

Motivating students to learn TCP/IP network fundamentals is often difficult because students find the subject rather technical when it is presented using a lecture format. To overcome this problem we have prepared some hands-on exercises (practicals) that give students a practical learning experience in TCP/IP networking. The practicals are designed around a multi-user, multi-tasking operating system and are suitable for classroom use in undergraduate TCP/IP networking courses. The effectiveness of these practicals has been evaluated both formally by students and informally in discussion within the teaching team. The implementation of the practicals was judged to be successful because of the positive student feedback and that students improved their test results. This chapter describes the practicals and their impact on student learning and comprehension, based on the author's experiences in undergraduate computer networking courses.

The aim of this study is to investigate the influence of faculty affiliation and teaching experience on the use of the Internet by faculty members in six first generation universities in Southern Nigeria. A total of 476 faculty members from nine faculties across the six universities participated in the study. The data for the study was collected by means of a questionnaire survey and this was deemed appropriate as it allowed the views of all the participants to be sought on a Likert-type scale options. The results of this study provide a number of insights: (a) the faculties of engineering, science and arts in that order were the foremost users of the Internet for instructional purposes; (b) the faculties of education and agriculture

Preface

MODELS, INITIATIVES, AND ASSESSMENTS

Introduction

During the publication year 2006, the International Journal of Information Communication and Technology Education (IJICTE) evidenced a focus on models, educational initiatives, and assessment of instructional technology. Many of the articles shared with our readership throughout the year concentrated on the results of investigations on these three educational apparatus.

By definition, a model is a pattern, plan, representation, or description designed to show the structure or workings of an object, system, or concept. (Wikipedia, 2007). They may refer to abstractions, concepts, and theories used to estimate, predict, or forecast events. In his paper, Gerald Grow (1996) offers a cognitive model of learning that begins with comprehension to predict what prior knowledge will be relevant and which strategies might prove to be useful in teaching. Next, learning occurs when this new information becomes a part of a learner's existing knowledge network. Finally, recall comes into play to call up stored information in response to some cue for use in a process or activity. Memory is reconstructive. Grow's final comment is worth remembering here, "In a nutshell: cognition is an active, recursive, integrated process by which we continuously model the world and continuously modify the model." These working definitions of a model will serve us well.

Adapting Information and Communication Technologies for Effective Education re-introduces a series of models for consideration that include the TUI model for faculty development, blended ICT models for higher education, the KARPE model for differentiating teaching and learning with technology, the ADDIE model applied to online instruction, and the TRAKS model for IT training in organizations.

Educational initiatives are nothing new to education. In point of fact, the discipline is replete with examples of initiatives started and dissolved, tried and abandoned, successes and failures. Educational initiatives attempt to introduce or promote a culture of quality within education by raising concerns related to student learning, providing services related to assessment, professional development of teachers, curriculum and pedagogy, and influencing educational policy, for our purposes, in the realm of technology.

In this text, the reader is provided updated investigations into several important technology-based initiatives. They include technology-assisted problem packages for engineering, incorporating geographic information systems, programming with decision trees, a scheme for increasing female interest in science curriculum, using vignettes to exapnd higher order thinking, anchoring e-commerce courses with business plans, supporting special needs learners in cyber schools, game modding and customized learning opportunities, infusing project management in student technology projects, and teaching networking using practical laboratory exercises.

Evaluating educational technology programs is challenging. Thankfully, the research and the literature are beginning to fill the void in what we know about the successful integration of technology. As more and more universities, schools, and corporate training organizations develop technology plans to ensure technology will directly benefit learning and achievement, the demand for more investigations into an understanding of how technology impacts learner achievement becomes even greater. The question, thus, becomes how do you evaluate educational technology programs that vary in the students they serve, the curriculum they teach, and the technologies employed?

Adapting Information and Communication Technologies for Effective Education offers four revised articles from 2006 that address ICT assessment in universities, student satisfaction in management information system programs, factors that impact the successful implementation of a laptop program, student learning and electronic portfolios, and strategic planning for e-learning.

A recap of the chapters, by category (models, educational initiatives, and assessment) follows.

Models

In their chapter, Graham and Semich introduced a three-step staff development program for linking technology training with theory to transform pedagogy. The model proposed three key phases: training, application, and integration. Their updated research, seen in Chapter I, on the three-phase model entitled, "Integrating Technology to Transform Pedagogy: Revisiting the Progress of the Three Phase TUI Model for Faculty Development," highlights the progress that one university has made to transform the teacher-centered classroom into a technology rich, learner-centered environment.

Information transfer is a tradition in higher education. In the information transfer model, knowledge is passed from the experts (tutors) to the learners (students) by means of lectures and textbooks. Increased costs often dash any hopes of increasing the educational impact of these traditional resources by augmenting them with more advanced technology-enhanced ICT tools. Drossos, Vassiliadis, Stefani, and Xenos argue that new, low-cost educational models based on constructivism can be used in parallel with traditional learning to introduce a blended (or enhanced) learning approach. In such a blended environment, organizational, educational and technological issues need to be considered as a whole. Their initial manuscript introduced a light-weight blended educational model based on cooperation and experimentation. Chapter II, "Blended ICT Models for Use in Higher Education," adds a developmental framework and discuss its quality aspects based on the ISO standard.

The knowledge, application, research, practice, and evaluation (K-A-RPE) model was initially offered as a benchmark for differentiating technology-oriented teaching and learning. The K-A-RPE model was added to the progressive, hierarchical classification systems of other taxonomies. Additional undergraduate, masters, and doctoral programs in instructional technology were added to the original data presented in the 2 2006 article. The findings shared by Tomei in Chapter III titled, "The KAR-P-E Model Revisited: An Updated Investigation for Differentiating Teaching and Learning with Technology in Higher Education," now include some 87 programs, 1542 courses, and over 14,000 learning objectives.

Online education has quickly become a widespread and accepted mode of instruction among higher education institutions throughout the world. Although many faculties who teach traditional courses now embrace teaching online, others still feel intimidated when asked to develop a course using technology. The ADDIE model, first presented in the July-September 2006 issues of the *International Journal of Information and Communication Technology Education*, is a five-step process that has proven equally adept at designing both traditional and online instruction. The five steps, analysis, design, develop, implement, and evaluate, provide the framework for solid instructional design techniques. In Chapter IV, "Applying the ADDIE Model to Online Instruction," Shelton and Saltsman add to their assembled best practices and augment their initial findings with new suggested strategies for online class design, syllabus development, and online class facilitation. Both novice and experienced online instructors alike will benefit from the ideas, tips, and tricks published in this chapter.

Introduction of new information technology (IT) in organizations is a necessary, but not sufficient, condition for organizational success. Effective adoption and use is fostered by the integration of IT into an organization's strategic planning in areas of technology use, planning, and training. Despite the strategic nature of technology training in organizations, most existing studies on technology training address only operational issues (e.g., training needs assessment, learning, delivery methods, etc.). More strategic concerns (i.e., enhancing business productivity) are largely not addressed by the current literature. To address this gap, Srivastava and Teo explored the role of IT training in hierarchical organizations, in Chapter V, entitled "TRAKS Model: A Strategic Framework for IT Training in Hierarchical Organizations," the authors synthesize various ideas related to change management, training needs analysis and IT adoption to evolve a strategic IT training framework for hierarchical organizations; namely, the TRAKS model. The first contribution presented in volume 2, number 4, of the *International Journal of Information and Communication Technology Education* offered framework for recognizing the differences in IT training requirements at various levels of employees. The model suggested tracking training requirements based on attitudes, knowledge, and skills for different segments of employees. The revised manuscript augments the original study with discussions of actionable and comprehensive tools that can be used for systematically planning IT training. The result: enhanced productivity and a more complete and robust training itinerary.

Educational Initiatives

Sidhu and Ramesh present their work in Chapter VI, entitled "Technology Assisted Problem Packages for Engineering," on the development of technology-assisted problem solving (TAPS) packages at the University Tenaga Nasional (Nigeria) that began with an investigation into the development of interactive multimedia based packages targeted for engineering. Their original work was shared in volume 2, number 1 issue of the *International Journal of Information and Communication Technology Education*. This chapter continues the research into the philosophy, design, and development of interactive multimedia for solving engineering dynamics problems.

Increasing demands for basic computer skills at today's colleges parallel advancements in overall information technology use. As a consequence, many colleges and universities have initiated campus laptop programs to provide their students opportunities to grow their computer skills and experiences. However, the success of laptop programs is very much dependent on the degree to which students and faculty are accepting a laptop environment and are willing to implement such programs. Defining which conception factors are necessary is essential for successful implementation. In their initial investigations reported in the *International Journal of Information and Communication Technology Education*, Changchit, Cutshall, and Elwood examined student perceptions of the required laptop programs in order to distinguish which factors they perceive as important. In Chapter VII, "Perceptions of Laptop Initiatives: Examining Determinant Factors of University Students for Laptop Successful Implementation," the authors add to our understanding of the factors that encourage student support of laptop initiatives and how such programs can be made more useful to students as well as more beneficial to universities.

Schools of business can benefit from adoption of geographic information systems (GIS). In Chapter VIII, "Incorporating Geographic Information Systems for Business in Higher Education," Gadish presented a brief overview of this technology along with an example of how it can be incorporated into a business school curriculum. Benefits for business schools, their students, and faculty are discussed and a comprehensive approach for promoting such spatial thinking is presented. The goal of the research was to determine ways to empower faculty to adopt GIS-based research and teaching while producing business school graduates who can promote spatial thinking in their own organizations. The follow-on manuscript validates the findings and uncovered that, with time and effort, an increased awareness of spatial thinking and use of GIS technology benefits business school administrators, faculty and students.

The design for this chapter focused on a library of decision tree algorithms in Java that were eventually used as a programming laboratory workbench. Kalles and Papagelis found decision trees to be one of the most successful machine learning paradigms. Chapter IX's experiments with decision trees found that using components and visual tools facilitate decision tree construction. The resulting system has been built over a number of years and has been successfully used in a programming laboratory for junior computer science students. The underlying philosophy, expanded in this follow-on study of "Programming Drills with a Decision Trees Workbench," was to achieve a solid introduction to object-oriented concepts and practices based on a fundamental machine learning paradigm.

Chapter X, "CareerQuesting Revisited: A Protocol for Increasing Girls' Interest in STEM Careers," by White and Wasburn, introduces an educational strategy to foster the interest and persistence of middle school girls in science, technology, engineering, and mathematics (STEM) careers. In the chapter, criteria are offered that would assist middle school teachers in the evaluation of Websites to serve as supplemental learning activities within prescribed curricula. As the authors' investigations continued, new evaluative criteria distinguished successful factors between boys and girls, allowing teachers to adopt them reducing the concern that they are providing an unfair advantage to either sex.

A challenge in teaching and providing any type of instruction in the online learning environment is to ensure that participants are engaged in the process and find meaning in their learning. Kish's previous case study, "Overview of Using Vignettes to Develop Higher Order Thinking and Academic Achievement in Adult Learners in an Online Learning Environment," investigated the use of vignettes as a teaching strategy and learning activity in a hybrid online course. The generative learning model was explored and two outcomes were anticipated: (1) enhancement of academic achievement and (2) higher order thinking. The modified study in Chapter XI discusses the methods used to teach adult learners how to respond to and create vignettes for their own teaching and presentation purposes. Participants responded to teacher-generated vignettes, created diagrams and rubrics, created their own vignettes, and recorded their observations concerning vignettes in reflective learning logs. The research findings indicate that the use of teacher-generated vignettes can increase academic achievement, and that learner-generated vignettes can help students achieve higher order thinking; a most appropriate example of a technology-based initiative.

Graduate-level educators are challenged by the diversity and currency of subjects covered in e-commerce courses. In Chapter XII, "Business-Plan Anchored E-Commerce Courses at the MBA-Level," Huang found the use of the business plan model a viable means for addressing those challenges. The apparatus of a business plan links subjects together while tendering students with real-life experiences. Learning, with proper curriculum design and delivery, gives students an opportunity to be reflective practitioners. Results from Huang's initial study are sustained in this revised paper as he continues to show one successful methodology for learning.

In Chapter XIII, "Cyber Schools and Special Needs: Making the Connection," Hipsky and Adams introduced the concepts of a new educational delivery network formally coined "cyber schools." For targeted K-12 students, the cyber school strategies have become particularly successful, especially to meet the needs of students with exceptionalities. In the *International Journal of Information and*

Communication Technology Education (volume 2, no. 4), Hipsky and Adams studied 468 students of the Pennsylvania Cyber Charter School and six dominant themes including: communication, interests, focus, less-stigma from the special education label, education differences in comparison to other methods, and cyber school shortcomings. The study employed the action research model to uncover the techniques and strategies at work in today's cyber schools. The revised investigation and latest results presented in this text augmented the teacher-tested documents from the original study and enhanced the cyber learning model for special needs strategies established through this research.

A game mod(ification) describes an adaptation for another purpose of an existing commercially available computer-based game originally created for an entirely different intention. Using game modding, a user can participate in the creative process by taking the setting of their favorite game and customizing it for entertainment (or educational) purposes or to deliver new knowledge and fresh information. For years, commercial computer-based game developers committed considerable resources towards preventing users from "hacking" into or "hijacking" their games. Now several computer-based game developers actually encourage partner-users to build additional content and seek the advantages of producing quality commercial computer games and customized instruction. Chapter XIV, "Game Mods: Customizable Learning in a K16 Setting," focuses on mainstream, accessible games with straightforward tools that are easily integrated into a learning environment. Read the author's updated version of these interesting instructional technologies and how they might be applied to today's classrooms.

Chapter XV, entitled "Project Management in Student Information Technology Projects," by Rojas, McGill, and Depickere reports on their investigations into the use and usefulness of project management in student IT projects. The results show there was a wide range in the application of project management practices with students more likely to produce the initial documentation associated with some of the project management knowledge areas than to make use of it throughout the project. The results also show that the number of project management guidelines applied in student projects is not linked to project success. The revised chapter continues to show the strong relationship between project management plan quality and a good software product discovered in the initial study, and goes further in exploring this aspect of how universities teach project management to information technology students. The project management principles that students have previously learned remain applicable to experiential learning in a project-based course; the experience of applying knowledge to real or simulated projects makes an important contribution to this text.

Finally, motivating students to learn TCP/IP network fundamentals is often difficult because students find the subject rather technical when presented via the lecture format. To overcome this problem we have prepared some hands-on exercises (practicals) that give students a practical learning experience in TCP/IP networking. The practicals are designed around a multi-user, multi-tasking operating system and are suitable for classroom use in undergraduate TCP/IP networking courses. The effectiveness of these practicals has been evaluated both formally by students and informally in discussion within the teaching team. The implementation of the practicals was judged to be successful because of the positive student feedback and that students improved their test results. Chapter XVI, "Teaching TCP/IP Networking Using Practical Laboratory Exercises," describes the practicals and their impact on student learning and comprehension, based on the author's experiences in undergraduate computer networking courses.

Assessment

In Chapter XVII, entitled "Assessment of ICT Status in Universities in Southern Nigeria," Aduwa-Ogiegbaen and Uwameiye offer readers an insight into factors of faculty affiliation and teaching experience with respect to the use of the Internet. Their results, amplified in this revised study, provide three important findings. First, the faculties of engineering, science and arts (in that order) were the foremost users of the Internet for instructional purposes. Second, the faculties of education and agriculture were the least experienced in using the Internet. And, third, faculty members with less than five years teaching experience use the Internet more than senior faculty members. Readers will most certainly compare their schools with those in this Nigerian study with probably fairly similar results. If such an investigation has not been conducted at your institution, the questionnaire survey and methodology are available in this chapter for your consideration.

In Chapter XVIII, entitled "Using Indices of Student Satisfaction to Assess an MIS Program," the authors demonstrate a methodology by which management information systems (MIS) alumni evaluate the content of courses and their satisfaction with the program. In the initial study first offered in t the *International Journal of Information and Communication Technology Education* (volume 2, no. 2), Chrysler and Van Auken sought to isolate differences in the evaluations of the content of required MIS courses by alumni based upon whether the graduate was using their first year on the job or one's current position as a frame of reference. A factor analysis, a global measure of satisfaction, and a regression analysis were brought into play to measure a student's satisfaction with the entire MIS program. In this updated manuscript, the authors enhance their research by offering implications for evaluating the effectiveness of an MIS curriculum.

Chapter XIX, "How Students Learned in Creating Electronic Portfolios," investigated learning experiences that occurred during development of electronic portfolios for a graduate technology program. Wang and Turner spent time investigating student learning experiences and the problems they encountered in an attempt to understand how they learn in a technology-enriched learning environment. Originally, data were collected through in-depth interviews, participant observations, and documented analyses before, during, and after developing electronic portfolios. Initial findings indicated that creating electronic portfolios support mastery of technology-related knowledge and promote critical thinking and problem-solving skills. This chapter reinforces previous reports that students learn best by doing and even better through collaboration, reflection on artifacts, and synthesis that comes from creating electronic portfolios.

Computer-networked systems create a demand and an opportunity for businesses to approach training and knowledge management from new perspectives. These new training perspectives are driven by the need for businesses to provide the right training quickly and efficiently and to support knowledge systems that are current, accessible, and interactive. In Chapter XX, "Strategic Planning for E-Learning in the Workplace," Berge and Giles discuss strategic planning in terms of the necessary organizational elements and the e-learning requirements to build a framework for sustaining e-learning as a core business process. In this chapter, the process of developing a strategic plan originally posited was augmented with an examination of the internal and external environments that help an organization determine its current situation prospects for business in the future. The chapter examines the two components that guide the future activities of the organization: a mission statement and vision statement. Once this strategic foundation is laid, the organization can go about the business of transforming itself into a learning culture that maximizes the use of technology with an investment in learning that produces outcomes to further business processes and goals.

Summary

Information technology makes it possible for faculty and trainers to improve the manner in which they present materials in both a traditional, face-to-face classroom or via technology-enhanced online teaching. When used properly, technology increases the frequency and quality of instructor-student interaction and, consequently, learner outcomes. *Adapting Information and Communication Technologies for Effective Education* offers its best papers from 2006 categorized as models, educational initiatives, and assessment issues.

The TUI model for faculty development will introduce a three-step staff development program for linking technology training with theory. A blended ICT model adds a developmental framework for use in higher education. The KARPE model will differentiate learning objectives using technology. The ADDIE model will provide the framework for designing both traditional and online instruction. And, the chapter discussing the TRAKS model will synthesize various ideas related to change management, training needs analysis and technology adoption.

A wide range of educational initiatives will be introduced in this text. The development of technology-assisted problem solving packages for engineering will suggest to the reader how interactive multimedia might assist in helping students solve complex engineering dynamics problems. Increasing demands for basic computer skills were examined in light of student perceptions of a required laptop program and will be shared. Benefits for business schools, their students, and faculty from the adoption of geographic information system technology are covered later. The use of a Java-based decision tree algorithm library will report on its successes when integrated into a programming laboratory for junior computer science students. An educational strategy that has the potential to foster the interest and persistence of middle school girls in science, technology, engineering, and mathematics (STEM) careers will be offered. The use of vignettes as a teaching strategy and learning activity in a hybrid online course is exposed. The business plan model is recommended as a viable means for addressing the challenges of diversity and the currency of subjects covered in modern e-commerce courses. Strategies appropriate to meet the needs of students with exceptionalities in today's cyber schools are to be examined along with six dominant themes including communication, student interests, learner focus, the special education label, comparison to other learning methods, and cyber shortcomings. The use of game modifications to deliver new knowledge and fresh information is highlighted. The project management principles explained in one chapter remain applicable after additional investigation; the experience of applying knowledge to real or simulated projects will continue to make important contributions. The shortcomings of the lecture method of instructional delivery are explored and the use of practical exercises found to produce positive student feedback and improve student test results.

Finally, issues of assessment were introduced. The first such chapter will offer an insight into factors of faculty affiliation and teaching experience with respect to the use of the Internet. Another will evaluate the content of required MIS courses using factor analysis, a global measure of satisfaction, and a regression analysis. Electronic portfolios, and how they support mastery of technology-related knowledge and promote critical thinking and problem-solving skills, will be presented to readers. And, the last chapter will discuss strategic planning in terms of the necessary organizational elements and the e-learning requirements to build a framework for sustaining e-learning as a core business process.

As you begin your journey into *Adapting Information and Communication Technologies for Effective Education,* consider how the models, educational initiatives, and assessment issues presented impact your personal understanding of information technology education.

REFERENCES

Grow, G. O. (1996). *Serving the strategic reader: Reader response theory and its implications for the teaching of writing*. An expanded version of a paper presented to the Qualitative Division of the Association for Educators in Journalism and Mass Communication. Atlanta, August, 1994. Retrieved from www.longleaf.net/ggrow.

Cognitive Model (2007).Wikimedia Foundation, Inc. Retrieved April 7, 2007, from en.wikipedia.org/wiki/Cognitive_model.

Section I
Models

Chapter I
Integrating Technology to Transform Pedagogy:
Revisiting the Progress of the Three Phase TUI Model for Faculty Development

John E. Graham
Robert Morris University, USA

George W. Semich
Robert Morris University, USA

ABSTRACT

In a previous article, the authors illustrated a three-step staff development program for linking technology training with theory to transform pedagogy. Essentially, the model identified three key phases: the training phase, application phase, and the integration phase. The focus of this chapter is to update the research on the three-phase model and to highlight the progress Robert Morris University has made to transform the teacher-centered classroom into a technology rich, learner-centered environment. This transformation process will be explained and illustrated for the reader.

INTRODUCTION

Currently, colleges and universities have the obligation and have rightfully assumed the responsibility to provide both their faculties and their students with a knowledge and application of the latest instructional technologies for the enhancement of learning. Wiske (2004) advocated the use of a pedagogical framework that provides criteria for productively using technologies for deepening understanding, while the International Society for Technology Standards, through its National Educational Technology Standards (NETS) project (Thomas, 2004), started to provide educational leaders with guidance in developing national standards for technology. Further, in a study from the U.S. Department of Commerce (2000), the Economics and Statistics Administration, and the National Telecommunications and Information Administration, it was noted that "raising the level of digital inclusion by increasing the number of Americans using technology tools of the digital age is a vitally important national goal" (p. xv). Hence, effective use of technology is being elevated to the highest level of importance on the education continuum.

A common approach to meeting this demand is to provide hardware and software training with the anticipation that faculty and students will see both applied and residual value to this training. Many schools and departments are requiring courses focused on the attainment of specific technology core standards of achievement and mastery of technology tools. In fact, Fox (2005) reports that technology use and access to new and current technologies is on the increase in most schools throughout the United States. However, much of the current research (Oppenhimer, 2005) suggests that placing computers in a classroom has been a waste of time and money in many cases. Specialized training and faculty commitment are extremely important for the effective integration of technology throughout the instructional and learning processes. As noted in a previous article (Graham & Semich, 2006), Robert Morris University has developed some courses to meet specialized technology course requirements with the primary goal to effectively integrate technology across the curriculum. In order for successful technology integration to occur, the authors feel there needs to be a strong link between content and delivery and that the use of technology will be most effective when technology use is based on sound instructional theory. Consequently, the authors' basic contention is to show how educators can bridge training with theory to transform pedagogy. In their article, "Pedagogy and Innovation in Education with Digital Technologies," Johnson, Chapman, and Dyer (2006) connect pedagogy to outcomes as follows: "It is difficult to predict how education will change over the next decade let alone the next century, but there will most certainly be changes in pedagogies which more readily recognize the way young people learn with new technology. Learning outcomes and measurable outputs will need to reflect this" (p. 5).

PROBLEM STATEMENT

Given the increased emphasis and national priority placed on technology, the authors gathered information related to the use of technology for instructional purposes at Robert Morris University. With this data and support from research, they formulated a model faculty can follow to effectively integrate technology across the curriculum and help to transform pedagogy to a more constructivist as opposed to behaviorist model of classroom instruction.

BACKGROUND

Robert Morris University (RMU), in recognizing the importance of technology, has been proactive about providing state-of-the-art presentation classrooms for faculty and students. In these classrooms, faculty has access to an LCD projector, a computer, and a VCR for instructional purposes with the addition of wall-mounted Smartboards, DVD player, and Smartboard software. Digital cameras and specialized software have also been purchased for student use. Likewise, other classrooms on the campus are equipped with appropriate technology resources. For instance, a smart cart, a touch screen, a document camera, cabling for a laptop, and recording equipment are typical devices available for faculty use. By providing these resources, the overall RMU goal remains the same, which is to use technology to transform the learning environment so that it is participative, customized, and student-centered. Recently, the university purchased two laptop carts for classroom use, which include 24 windows operating system laptop computers and a network wireless printer for each cart. In addition, another laptop cart containing the same number of Apple MacIntosh laptop computers with a networked wireless printer will be available for the next academic year. These laptop carts will give faculty greater flexibility to integrate technology outside the traditional lab setting.

An academic technology committee also continues to function and consists of representatives from both the administrative and support sides of the university as well as faculty from each of the six schools. A full-time faculty member, who also serves on the committee, has his schedule reduced half time to serve as the director of instructional technology. While the director works more closely with faculty on an individual basis, the committee, on the other hand, reviews concerns of the faculty and staff, initiates new directives, and oversees the technology planning process, monitors the budget, and sets short and long-range technology goals for the university. This technology committee has been meeting in the new TLC (technology learning center), which is a lab training center that contains the latest technologies including MP3, IPOD and video POD casting, video editing software, and so forth. A recently hired technology staff member is housed in the center and provides training for both faculty and students. Panettieri (2007) recently indicated that "progressive universities are embracing any number of MUVEs (multi-user virtual environments), 3D environments, and immersive virtual reality tools" (p. 27). Ultimately, the technology center will evolve into the focus for all technology training and innovation, including training for all hybrid and totally on-line delivery modalities of instruction as well as other emerging technological initiatives mentioned here.

THE TECHNOLOGY SURVEY

During the 2001-2002 academic year, a survey instrument was created to measure the degree of technology use at RMU for a wide array of pedagogical tools. Particularly, the survey helped to determine the faculty and student use across all programs at the university. Five basic categories of technology usage were surveyed: Internet/Web-based activities, PowerPoint slides, multimedia classroom technology, electronic library searches, and technology integration. These categories were chosen as they were indicative of the variety of technologies and facilities available to faculty and students at RMU. The survey results guided the university in planning for the ongoing integration of technology for instructional purposes. Since this original study, new technologies, new facilities, new staff, and additional new initiatives have helped to cement positive change in technology integration at the university.

The previous study completed at RMU reviewed technology use by both faculty and students in nine academic departments. An important finding from this earlier study revealed that although faculty felt they would continue to use technology in the future, they did not, at the time, have a clear sense for how they would integrate technology through the use of productive pedagogical practices. Based on this evidence, and due to the high rate of inquiry by faculty related to how technology could be integrated into their specialized curriculums and as a result of the increased emphasis placed on technology at all levels, the authors elected to incorporate a three phase technology model that would encourage faculty to work toward not only integrating technology into the classroom but transforming the classroom into a learner-centered, technology supported learning environment.

THE PROCESS

A review of related literature on the subjects of technology training, technology use, and technology integration as they relate to staff development for faculty was significant to the execution of our model. We felt it was important in the development of an applied model to emphasize three separate sequential phases critical to effective technology implementation and enhanced student learning. Combining these three phases into one generic staff program also seemed to be far more chaotic and probably much less residual in terms of successful classroom application since each phase requires a sequential level of competency to develop self confidence and fluidity with faculty. Thus, our suggested model of staff development included these three distinct phases: the training phase, the use or application phase, and the integration phase. It is important to point out that although all phases of the model are distinctive to its success, as with many colleges and universities, the transition to the third phase of technology integration not only requires faculty to learn how to apply technology but also how to transform the classroom into a learning center that promotes student application of technology. This dramatic third phase presented the greatest challenge and subsequently the greatest success since it precipitated both philosophic and operational changes in teaching.

Training Phase

Thorburn (2004) noted that technology integration is taking a long time which can best be described in terms of decades. We know from the research (Kulik, 2002, Waxman, Connell, & Gray, 2002) that when technology is used appropriately, it can improve education. It is also important to realize that without faculty knowledge of how to use the educational technology, instructional time can be wasted in the classroom (Coppola, 2004). Thus, our first phase of the model was the training phase, where we were very basic in our approach to demonstrate to faculty how to use the instructional technologies. Our director of technology implemented small group training sessions in each academic school at the university. These sessions emphasized the step–by-step operations of the technology with numerous opportunities for individual faculty to experience direct hands on training, which highlighted the special features of the technology as a powerful teaching tool. To illustrate, most of our classrooms have a Crestron controller unit with a computer, projection system, recording and playback VCRs, tape recorder/player, document camera, and rear mount recording camera. Shifting through these various modes and demonstrating multiple and single operations of each piece of equipment gave faculty the heightened incentive to move to our second phase of using the new technologies. Since it was also suggested in the research (Shelton & Jones, 1996) that the training should occur outside the school day to remove additional responsibilities of teaching or advising students, sessions were conducted in the evenings and during semester break. In this first phase we felt that the training sessions were the catalyst for other workshops and for progress toward actual classroom use. This phase would eventually contribute toward the integration of technology to transform present pedagogy into a more active learning, constructivist approach in the classroom. In their research on professional development in technology training, Ringstaff and Kelly (2002) strongly supported the need for ongoing professional development if faculty is to move through more applied and integrative technology levels with their students. Further, in the research of Kanaya and Light (2005), they argued that faculty needs sustained assistance in the use of technology throughout all phases including the integration phase.

Technology Use Phase

In the second phase, we wanted faculty to apply the instructional technologies in their classrooms. Since many of the faculty had limited experience with new technologies, we realized this meant that much of their technology use would be extensions or modifications of their existing teaching methods in the classroom. For example, those who were traditionally using the chalkboard or whiteboard would shift to using PowerPoint slides to deliver classroom notes. Essentially, the technology including LCD projectors, computers, the Internet, VCRs, document cameras, a laptop cart with wireless networked printer, digital video editing with iMovie HD, podcasting and video conferencing, MP3 technologies, video podcasting, and new digital cameras would replace older technology such as overhead projectors, flipcharts, and paper copies as handouts. This was the most logical transition since it incorporated using the new technologies introduced in the faculty training sessions in phase one yet did not limit more traditional pedagogical practice. Our primary goal was to have faculty functioning with the new technologies in order that they might see the value of training and applying technologies to update and upgrade their present teaching approaches in the classroom. Judson (2006) stated that teachers who use technology are often the "constructivist-minded teachers" who "maintain dynamic student centered classrooms where technology is a powerful learning tool" (p. 581). Further, in a comprehensive study by Sivin-Kachala and Bialo (2000), it was noted that faculty who receive at least 10 hours in training were more likely to use technology to improve classroom teaching and learning. Consequently, we felt strongly that faculty would see added benefit in working with new technologies as a means of motivating students and perhaps renewing their interest in delivering the course curriculum. In fact, Zhao and Cziko (2001) suggested three conditions necessary for faculty to use technology effectively. They felt technology should meet higher learning goals; technology should support other learning goals; and finally, faculty should have sufficient ability and resources to use the technology. In this second phase we were confident that if the faculty would apply the technology training they received, then we were clearly moving toward the goal of integration of technology and transforming the classroom from a teacher-centered to a student-centered learning environment.

Integration Phase

In our third or final phase of the model, we were able to systematically plan and implement this integration phase of the staff development model in a manner supported by Stager (1995) that places the teacher and learner at the center of the learning experience and provides a meaningful background for learning. This phase is more of a transformation of traditional teaching since it requires a shift in roles placing primary emphasis on the learner. As noted by Honey and Spielvogel (1999), this transformation changes the classroom in that it defines new teacher roles and heightens student interaction. Technology use is not only for faculty but more for student use. Tiene and Luff (2001) described the classroom of the future as a place where teachers could immerse their students and themselves in technology integration. The teacher assumes the role of coach or facilitator while students work in teams collaboratively (Jones, Valdez, Nowakowski, & Rasmussen, 1995; Kupperstein, Gentile, & Zwier, 1999). This new role may involve connecting with other schools perhaps in various locations around the world. It is as Bransford, Brown, and Cocking (1999) suggested that technology can support learning in five ways (p. 195):

- To bring exciting curricula into the classroom that is based on real-world problems and that involves students in finding their own problems, testing ideas, receiving feedback, and working collaboratively with other students or practitioners beyond the school classroom;

- To provide tools and scaffolds that enhance learning, support thinking and problems solving, model activities and guide practice, represent data in different ways, and are part of a coherent and systematic educational approach;
- To give students and teachers more opportunities for feedback, reflection, and revision, including those where students evaluate the quality of their own thinking and products, have opportunities to interact with working scientists, receive feedback from multiple sources which include their peers, and experience cognitive tutors and coaching in areas where improvement is needed;
- To build local and global communities that are inclusive of teachers, administrators, parents, students, practicing scientists, and other interested community people, expanding the learning environment beyond the school walls; and
- To expand opportunities for teacher learning that includes helping teachers to think differently about learners and learning, to reduce the barriers between students and teachers as learners, to create new partnerships among students and parents, and to expand communities of learners that support ongoing communication and professional development of teachers.

Faculty must also begin modeling the technology themselves by implementing technologies like presentation software, online discussions, databases, spreadsheets, and smart boards (Duhaney, 2001; Schrum & Dehoney, 1998). This serves to motivate students to learn and use technologies and provides opportunities for both faculty and students to share new ideas relative to technology application in the learning process. It should be noted that initially faculty use of word processing programs and other software programs were sufficient, but they are no longer considered the only indicator of technology integration (Wetzel, Zambo, & Buss, 1996; Yidrim, 2000). Faculty need to become adept at integrating technology in the classroom, and they need to have students take control of using the new technologies to enhance learning.

THE MODEL

The proposed model was a three phase or stage model that was all inclusive of a comprehensive technology-based staff development program. It is our contention that any attempt to have faculty incorporate all three phases or stages in combined training programs would probably, as previously mentioned, lead to confusion, frustration, and possibly a lack of student applied integration of technology in the classroom. Thus, our model, as noted in Figure 1, is represented in three operational stages that form the foundation for faculty working with students using logical, well-developed pedagogical practices.

These three simple steps provide better opportunities for mastery at the various levels of learning. They are designed to establish competency at each level of the process. It is our feeling that by

Figure 1. Three step technology staff development model (TUI)

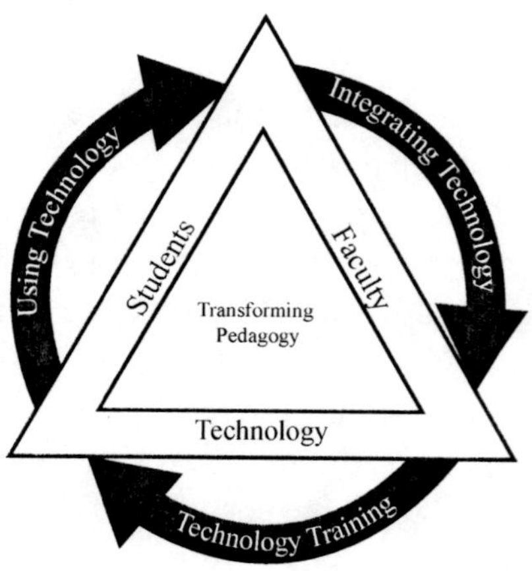

using a sequential staff development program, faculty will have mastered the previous steps before reaching the third or final phase of the model. It is imperative, though, that faculty feel reasonably comfortable with the instructional technologies to progress to the third, more innovative step of integrating technology into their teaching. Each school and each academic discipline will have special faculty resources to help with this major technology shift in the classroom. As part of the model, both individual departmental integration sharing sessions and/or interdepartmental integration sharing sessions would be promoted. Ideally, we would like to see this model evolve into a type of specialized model such as writing across the academic disciplines where various methods, techniques, and approaches are shared within the disciplines. Furthermore, faculty could enter the staff development program at any of three phases, depending on their level of competence. This is probably the most important aspect of our proposed staff training program since faculty have such varying levels of competence and interest in technology application. Granted, integration of technology is the desired goal, but faculty needs to move through the necessary steps to reach this level. As Johnson and Liu (2000) noted, "Everybody is talking about technology integration, but few practicing teachers profess to know exactly how to proceed" (p. 4). They do not know how to proceed because the new technologies present challenges not only in how to use it, but also in how to work with students to truly integrate technology into learning. Our proposed model was designed to lead faculty to a stage that would transform teaching. It is strongly based on constructivist theory which places a premium on active learning and exploring new venues for learning.

Hooper and Rieber (1995) suggested a model for teaching with technology that distinguishes between the constructivist and behaviorist philosophies. Likewise, the focus of their work supports the third step of our TUI model. In the model in Figure 2, the focus is on the technology and the teacher's instruction; whereas, in the contemporary (constructivist) theory section of the model, integration is the first step that leads to reorientation, and eventually evolution.

This model is divided into six distinct phases with integration being the so-called "breakthrough phase" (p. 4). According to Hooper and Rieber, this phase places emphasis on technology to assist in learning, and further, makes the point that without this technology intervention, learning may not occur. In the TUI three phase model,

Figure 2. A model of technology adoption for the classroom

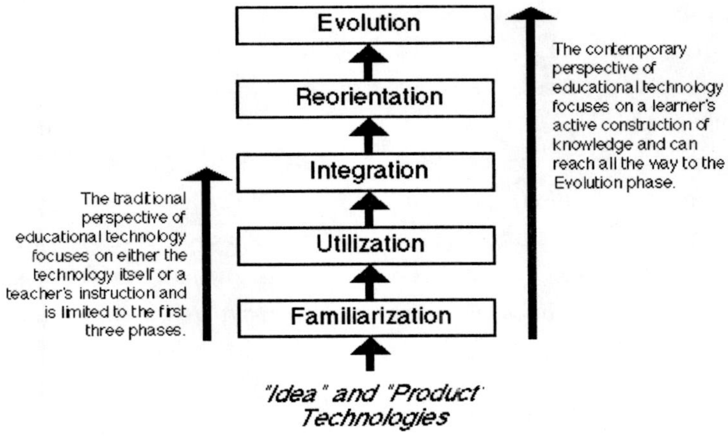

we also view the integration phase as the desired phase that moves toward the construction of new knowledge; however, we take this integration stage one step further since our goal is to truly transform pedagogy. Essentially, faculty at this step should not only integrate technology with their teaching, but also have the students themselves involved in activities with technology in the classroom (Ringstaff & Kelly, 2002). Or, as Marshall (2002) suggests, this phase has technology complement what a good teacher already does naturally in the classroom. In this final phase of integration, faculty can collaborate with their students and with each other to become as Fulton, Yoon, and Lee (2005) describe, a growing network of shared expertise. The following section contains a series of examples that illustrate this collaborative team approach at the university.

Collaborative Team Approach Examples

Once the director of instructional technology demonstrated how to use Web Surveyor, a user-friendly, online survey research tool that facilitates the development of surveys, the collection of data, and the generation of reports, faculty began to utilize the tool for classroom and other purposes. For example, a faculty member teaching in the instructional management and leadership doctoral program utilized the tool for classroom instruction to teach a unit focusing on the development of questionnaires for research purposes. The same faculty member used the software to develop survey instruments to access feedback from cooperating teachers and employers of education graduates for the School of Education and Social Sciences. This example demonstrates how a software application can benefit both students and faculty who are engaged in the research process, which is particularly important at Robert Morris since added emphasis is placed on scholarship.

A faculty member from the School of Business demonstrated an applied statistical software package, SPSS, for use by both faculty and students in conducting research. This presentation provided examples of extended software use and continued individualized support for interested faculty. Subsequent to these sessions, colleagues from two other schools have benefited from the SPSS training and have since designed successful instructional strategies for classroom use.

A guest speaker, Josh Mitchell, from the Pennsylvania Department of Education's Programs for Higher Education, was invited to instruct a group of pre-service teachers and faculty on the various learning opportunities with the Smartboard. This demonstration highlighted a number of different techniques to manipulate various objects for desired effects and focused on special techniques for delivering a lesson using this technology. The expertise of our presenter not only focused on the Smartboard technology but also dealt with how this technology can be used to improve learning in the social studies classroom.

An education faculty member demonstrated a process for enabling students to create an electronic portfolio for employment purposes. Presently, students develop a hard bound portfolio but most have not transitioned to the electronic format. This session has encouraged more education faculty to include course activities that enable students to incorporate various artifacts from student work into an electronic portfolio. Further, as an outcome of this session, faculty learned how to integrate appropriate electronic resources into their teaching, particularly for the increased use of the online delivery method of instruction.

Status of the Model at RMU

Since the model was originally introduced to the faculty and staff at RMU, many positive changes have occurred besides the collaborative efforts of faculty as mentioned in the previous section. The training and use phases of the model are well entrenched into the RMU environment; whereas the integration component of the model continues to evolve. Presently, the associate vice president of academic affairs oversees technology efforts at the university and has encouraged the six schools (business, communication and information systems, education, nursing, engineering and mathematics, and adult/continuing education) to organize separate technology committees to help move the integration process forward. These committees examine ways that pedagogical practices can be improved through the use of technology in the respective disciplines. In turn, faculty and/or administrators from these committees represent the schools at the university-wide technology committee, where all academic disciplines interact. The implementation of these committees is sequential and collaborative in nature, which encourages cross disciplinary planning and team building to implement creative yet practical technological practices in the classroom. In the end, the goal of the integration phase is to encourage the use of modeling and mentoring in place of traditional instruction and demonstration.

The opening of the Educational Technology Center (ETC) on the campus this fall has been instrumental in adding depth and connectivity to instructional technology efforts at the university, and most importantly, the center has helped to highlight the significance of the technology integration phase of the model. Thus, workshops are conducted that emphasize the improved integration of classroom technologies. Additionally, the ETC supports the personal and professional development of faculty, staff, and students. Thus, the ETC has the latest in high-tech equipment available for individual and group use. Examples of ETC equipment are: Smartboards, video conferencing equipment, Ipods, MacBooks, wireless equipment, and Bluetooth equipment. As a service of the center, demonstrations of new technology innovations are held on a regular basis to update faculty on their applicability for instructional purposes. Recently, the center offered sessions on the use of Annenberg media and video on demand, Smart Solutions touch screen technology, and PrismWorks wireless technologies. The availability of these offerings through the ETC will continue to strengthen the faculty member's ability to develop sound, technology-driven pedagogical practices.

Aside from advancing technology efforts at the course level, the university is expanding to integrate technology at the school level and, indeed, at the university level as well. The School of Education and Social Sciences is seeking national professional accreditation through the Teacher Education Accreditation Council (TEAC). As part of these efforts, the school faculty has instituted a plan to integrate technology-based activities into instruction based on the National Education Standards (NETS). Every education course syllabus shows how classroom activities are linked to TEAC as well as to NET standards. Likewise, the School of Business has been strongly committed to making sure technology standards are in place in order to meet those requirements for AACSB accreditation. And further, at the university level, all deans are developing plans to integrate information technology standards at all levels of their curricula in order to address Middle States regional accreditation requirements. Consequently, the technology model presented here, although initiated at the course level, has spread throughout the university and thus has provided positive, collaborative teaching and learning experiences across the disciplines for all students.

SUMMARY

This TUI model has provided faculty with a unique opportunity to transform how they teach in the classroom. By encouraging them to move beyond the training and using phases where they simply changed the tools but preserved the teaching methods, our faculty embraced the idea of integrating the technologies to truly change their teaching. Everyone had a collective goal of creating positive interventions in our students' learning. Earle (2002) noted three stages that faculty move through relative to making change: confidence, competence, and creativity. In our TUI model for technology, our faculty gained the confidence during the training stage with the new technologies. Then, faculty developed a competence by using the technologies. Finally, creativity would become a product of integrating the technologies into the classroom. This final stage would truly transform pedagogy and as Nair (2004) noted, would empower schools to establish new learning environments that reflect a shift from the more traditional teacher-centered classroom to a more student-centered community of learners.

REFERENCES

Bransford, J.D., Brown, A.L., & Cocking, R.R. (Eds.) (1999). *How people learn: Brain, mind, experience, and school.* Washington, DC: National Academy Press.

Coppola, E.M. (2004). *Powering up: Learning to teach well with technology.* New York: Teachers College Press.

Duhaney, D.C. (2001). Teacher education: Preparing teachers to integrate technology. *International Journal of Instructional Media, 28*(1), 23-30.

Earle, R.S. (2002). The integration of instructional technology into public education: Promises and challenges. *Educational Technology, 42*(1), 5-13.

Fox, E. (2005). Tracking U.S. trends. *Education Week, 24*(35), 40-42. Retrieved September 21, 2005, from edweek.org/ew/articles/2005/05//05/35tracking.h24.html.

Fulton, K., Yoon, I., & Lee, C., (2005, August). Induction into learning communities. National Comission on Teaching and America's Future. Retrieved from http://www.nctaf.org/documents/NCTAF_Induction_Paper_2005.pdf

Graham, J., & Semich, G. (2006). A model for effectively integrating technology across the curriculum: A three-step staff development program for transforming practice. *International Journal of Information and Communication Technology Education, 2*(1), 1-10.

Hooper, S., & Rieber, L.P. (1995). Teaching with technology. In A.C. Ornstein (Ed.), *Teaching: Theory into practice* (pp. 134-170). Needham Heights, MA: Allyn and Bacon.

Honey, M., Culp, K.M., & Spielvogel, R. (1999). *Using technology to improve student achievement* {Online}. Retrieved from www.ncrel.org/sdrs/areas/issues/methods/techlgy/te800.htm

Johnson, D.L., & Lui, L. (2000). First steps toward a statistically generated information technology model. *Computers in the Schools, 16*(2), 3-12.

Johnson, J., Chapman, C., & Dyer, J. (2006). Pedagogy and innovation in education with digital technologies. In A Mendez-Vilas, A. Solano Martin, J. A. Mesa Gonzalez, & J. Mesa Gonzalez (Eds.) *Current developments in technology-assisted education* (pp. 135-139). Badajoz, Spain: Formatex.

Jones, B.F., Valdez, G., Nowakowski, J., & Rassmusen, C. (1995). Plugging in: Choosing and using educational technology. *Washington, DC: Council for Educational Development and Research, and North Central Regional Educational Laboratory.* Retrieved from www.ncrel.org/sdrs/edtalk/toc.htm

Judson, E. (2006). How teachers integrate technology and their beliefs about learning: Is there a connection? *Journal of Technology and Teacher Education. Association for the Advancement of Computing in Education (AACE), 14*(3), 581-597.

Kanaya, T., & Light, D. (2005). *Duration and relevance of a professional development program: Using Intel Teach to the Future to illuminate successful programmatic features.* Presented at Society for Information Technology and Teacher Education International Conference 2005. Norfolk, VA: AACE.

Kulik, J. A. (2002). School mathmatics and science programs benefit from instructional technology (InfoBrief, NSF 03-301). Washington DC: National Science Foundation: Retrieved September 21, 2005 from http://dwbrr.unl.edu/iTech/TEAC859/Read/KulikTech.pdf

Kuperstein, J., Gentile, C.K. & Zwier, J. (1999). The connected learning community technology roadmap. Retrieved from www.microsoft.com/Education/vision/roadmap/default.asp

Marshall, J.M. (2002). *Learning with technology: Evidence that technology can and does, support learning.* San Diego, CA: Cable in the Classroom.

Nair, P. (2004). Learning without boundaries — Educational technology planning for the new millennium. retrieved from www.fetc.org/fetcon/1200/nair.htm

Oppenheimer, T. (2003). *The flickering mind: The false promise of technology in the classroom and how learning can be saved.* New York: Random House.

Panettieri, J.C. (2007). Brave new world-advance teaching technologies. *Campus Technology, 209*(4), 27-33.

Ringstaff, C., & Kelley, L. (2002) The learning return on our educational technology Investment. San Francisco: WestEd. Retrieved from www.wested.org/cs/we.view/rs/619

Schrum, L., & Dehoney, J. (1998). Meeting the future: A teacher education program joins the information age. *Journal of Technology and Teacher Education, 6*(1), 23-27.

Shelton, M., & Jones, M. (1996). Staff development that works! A tale of four T's. *NASSP Bulletin, 80*(582), 99-105.

Sivin-Kachala, J., & Bialo, E. (2000). *2000 Research report on the effectiveness of technology in schools* (7th ed.). Washington, DC: Software and Information Industry Association.

Stager, G.S. (1995). Laptop schools lead the way in professional development. *Educational Leadership, 53*(2), 78-81.

Thomas, L.G. "The nets project." *International Society for Technology Standards.* Retrieved November 21, 2004, from enets.iste.org/nets_overview.html

Thorburn, D. (2004). Technology integration and educational change: Is it possible? *Educational Communications and Technology*, February, 2004. Retrieved from www.usask.ca/education/coursework/802papers/thorburn/ind

Tiene, D., & Luff, P. (2001). Teaching in a technology rich classroom. *Educational Technology, 41*(4), 23-31.

U.S. Department of Commerce. (2000). Falling through the net: Toward digital inclusion. A report on American's access to technology tools. Retrieved September 21, 2005 from search.ntia.doc.gov/pdf/fttn00.pdf/

Waxman, H. C., Connell, M. L., & Gray, J. (2002). *A quantitative synthesis of recent research on the effects of teaching and learning with technology on studtent outcomes.* Naperville, IL: North Central Regional Educational Laboratory. Retrieved September 21, 2005 from http://www.ncrel.org/tech/effects/effects.pdf

Wetzel, K., Zambo, R., & Buss, R. (1996). Innovations in integrating technology into student teaching experiences. *Journal of Research on Computing in Education, 29*(Winter), 196-214.

Wiske, S. (2004). Using technology to dig for meaning. *Educational Leadership*, *62*(1), 46-50.

Yidirim, S. (2000). Effects of an educational computing course on preservice and in-service teachers: A discussion and analysis of attitudes and use. *Journal of Research on Computing in Education*, *32*(4), 479-495.

Zhao, Y., & Cizko, G.A. (2001). Teacher adoption of technology: A perceptual control theory perspective. *Journal of Technology and Teacher Education*, *9*(11), 5-30.

Chapter II
Blended ICT Models for Use in Higher Education

L. Drossos
Technical Educational Institute of Messologi, Greece

B. Vassiliadis
Hellenic Open University, Greece

A. Stefani
Hellenic Open University, Greece

M. Xenos
Hellenic Open University, Greece

ABSTRACT

Information transfer is a tradition in higher education. In the information transfer model knowledge is passed from the experts (tutors) to the learners (students) by means of lectures and text books. The hope of increasing the educational impact by using impressive tools based on ICT has the serious disadvantage of increased cost. We argue that new, low-cost educational models based on constructivism can be used in parallel with traditional learning introducing a blended (or enhanced) learning approach. In such a blended environment, organizational, educational, and technological issues need to be considered as a whole. We introduce a light-weight blended educational model based on cooperation and experimentation. We describe the educational background, introduce a development framework and briefly discuss its quality aspects based on the ISO standard.

INTRODUCTION

During the past 10 years the educational community has witnessed a real revolution in the delivery of education. This revolution was mainly technological: high speed networks, powerful hardware available to simple users, multimedia-enhanced material, and free access to informal learning resources are just some of the trends introduced by the amazing advances of technology (Bates, 2000; Bonk & Graham, 2006; Pittinsky, 2002).

Despite the advances in ICT (information and communication technologies), productivity in terms of pedagogy and actual learning gains are not as significant as expected (Groccia & Miller, 2006). Current teaching and learning practices are based on the information transfer paradigm: information is passed from the teacher to the student. Although technology offers impressive possibilities for e-learning other factors, such as the underlying pedagogy, educational models, flexibility, and cost effectiveness are often overlooked. The plethora of advanced tools supporting e-learning and the difficulties in their adoption in real situations has only demonstrated that the primary need is a paradigm shift in the current, information-transfer educational model (Hiltz & Turoff, 2002; Romano et al., 2005; Xenos, Pierrakeas, & Pintelas, 2002).

Many researchers have proposed that this shift should focus on knowledge construction that will enhance, not replace, the traditional information transfer paradigm (Etheris & Tan, 2004; Rodrguez et al., 2007; Warschauer, 2003). Human peers are supported by using different kinds of collaboration technologies and especially, enhanced presence. Human learning is a social process, through sharing and executing tasks. It is a major enabler of the knowledge construction paradigm: active collaboration among learners in order for them to reach a common goal. In this context, learning is not an isolated activity (Hung & Nichani, 2001).

We consider a blended educational paradigm: traditional learning methods are supported by e-services. E-services are designed with the sole purpose of maximising the impact of traditional methods and covering their drawbacks or flaws. A major requirement is both methods should complement each other in the best possible way in administrative, educational, and technological terms. This kind of mixed learning (traditional and Web-based) is not a new concept: major investments in similar learning environments in universities and other higher education institutions across the world have been made in recent years (Bonk & Graham, 2006). Most of these efforts involve small scale, single institute adoption of web based tools which have drawn some useful conclusions (Bender, 2003; Garrison & Kanuka 2004; Haywood et al., 2000; Jefferies et al., 2004; Saunders & Klemming, 2003). Cross–institution (Van Weert & Pilot, 2003) or nation-wide (Demb et al., 2004) efforts were small in number but significant in impact.

Past examples have shown that information technology alone does not generate learning. A community informatics approach where a coordinated effort involving pedagogy and technology planning alike is needed (Jackson, 2004;Warschauer, 2003). Based on our work in Drossos et al. (2006), we theoretically analyze such a single-institute effort which strives to answer more extended questions: how e-learning can enhance the quality of the learning process for higher education students, how such a solution can be cost effective, what are the most appropriate implementation technologies, what are the appropriate pedagogical models and finally how is quality assured. The motivation stems from the vision of creating new, student-centric e-learning models that are both pedagogically and cost effective. We focus on blended experiential learning: experiential learning and cooperation and collaboration. We discuss a lightweight (in terms of costs) educational model, discuss its service functionalities and the technologies that can be used for its implementation. We provide a framework for the development o similar applications and final ways of assuring its quality using the ISO standard.

EDUCATIONAL MODELS, COSTS AND TECHNOLOGY

In order to achieve optimal exploitation of the possibilities provided by modern Web engineering approaches, theories of learning, technology, and management should be incorporated into the planning of a blended learning environment.

Cost and Organizational Considerations of ICT

The enthusiasm of the early adopters of ICT in traditional Higher Education Institutions was soon replaced by scepticism as results where becoming public from impact surveys (Van der Wende & Van de Ven, 2003). Many authors have claimed that the introduction of ICT to traditional higher educational environments may not only boost the quality of teaching but also reduce costs in the middle and long term. However, the second part of this claim is not sufficiently backed up by existing literature since studies contacted have not measured satisfactorily either the cost or the claimed benefits of computer based learning (Boucher, 1998; Groccia & Miller, 2006). Policy makers still seek evidence of mainstream benefits: value and relevance must be demonstrated.

Major cost savings of ICT introduction still remain in theory, while it seems that its greatest pedagogical advantages are the most costly: personalization, real-time communication, and other advanced functionalities lead to significant costs. Other costs may include courseware development costs, incremental capital and recurrent equipment costs, costs associated with provision of appropriate resources, infrastructure costs, maintenance, user support costs, costs of adoption, access costs, security costs, replacement costs, and institutional overheads. This has lead Rumble (1999) to suggest that the cost of utilizing advanced ICT services is nearly the same with face-to-face teaching. This assumption holds for complete distance learning solutions where traditional methods are completely replaced by ICT, but it is our opinion that it also holds for blended learning situations as well. The solution may lay in a consensus between costs and benefits of ICT use. Past efforts have highlighted the fact that the cost to produce and deliver content and services suitable for e-learning is often underestimated (especially update costs) and that costs directly affect the choice of pedagogical methods. Furthermore, academic staff in many countries is often hesitant to use real time tools for delivering content in addition to traditional lectures mainly because this overloads their schedule. Another obstacle is the fact that online presence of tutors requires special training and funding, a burden most institutions are not willing to undertake. For a more in-depth analysis of the cost effectiveness of blended models, the interested reader may refer to (Bonk & Graham, 2006; Cohen & Nachmias, 2006).

A Blended Light-Weight Model

Current teaching and learning practices are based on the information transfer paradigm: information is transferred from the tutor to the student (Figure 1a). In this situation, the student acts only as a consumer of information without being able to easily build knowledge. This static model of learning is supported by most state-of-the-art e-learning tools in the market. Information transfer is popular because it is easily supported by Web technologies but its educational effectiveness is seriously questioned: current e-learning tools offer many impressive functions but they tend to be complex for novice users and are often costly to incorporate, support and expand (Jonassen et al., 2003; Laurillard, 2002; Xenos et al., 2002).

Figure 1. (A) Traditional (information transfer) and (B) knowledge construction learning model

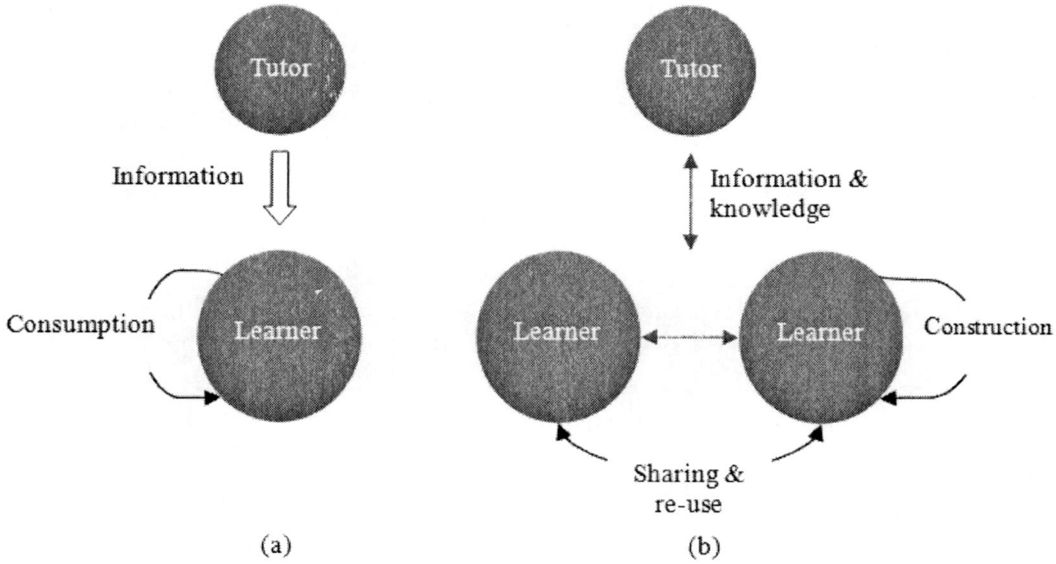

Constructivism is increasingly becoming a very popular enhancement method especially for teaching technological sciences and engineering in higher education (Duffy & Jonassen, 1992). Live experiences and social interaction are the heart of this method: learners construct knowledge by interacting with simulations and cooperate and collaborate with other learners exchanging opinions and facilitating collective activities. The main difference with the information transfer paradigm is that the learner has a more active role being not just the recipient of information but an active participant in the learning process. The same holds for the tutor which becomes a mediator helping learners to construct knowledge, assess learner progress and guide learners to meaningful learning activities (Schwier, 2004). This is somewhat contradictory to the static reproduction of series of didactical sections of the information transfer model.

Knowledge construction is a complicated and not well understood process. It implies dwelling on information, relating it to past experiences and/or building new knowledge, for example, creating and improving ideas (Figure 1b). The best environment for supporting this model is a community where participants share knowledge and debate. The role of interaction and collaboration with other individuals has long been acknowledged as critical for creativity: social networking (Driver, 2002), computer supported cooperative learning (CSCL) (Laat & Lally, 2005) and communities of practice (Wegner et al., 2002) are some of the most popular concepts in this research direction.

We introduce a "lighter" version of this constructivist model, a blended lightweight model which is imposed by organizational and economic factors. We avoid the explicit use of real-time cooperative and collaborative tools which are for many not cost effective to acquire, support, and maintain. This affects the role of the tutor. Since many traditional educational institutions do not provide adequate resources for a complete online experience, the number of tutors that may participate in collaborative sessions is usually small. This means that a major part of the constructivist model cannot be realized to its full potential (since real time assistance from tutors is missing) but this does not mean that it cannot be applied at all. In this model, tutors are present but they

Figure 2. A cost-effective blended learning approach

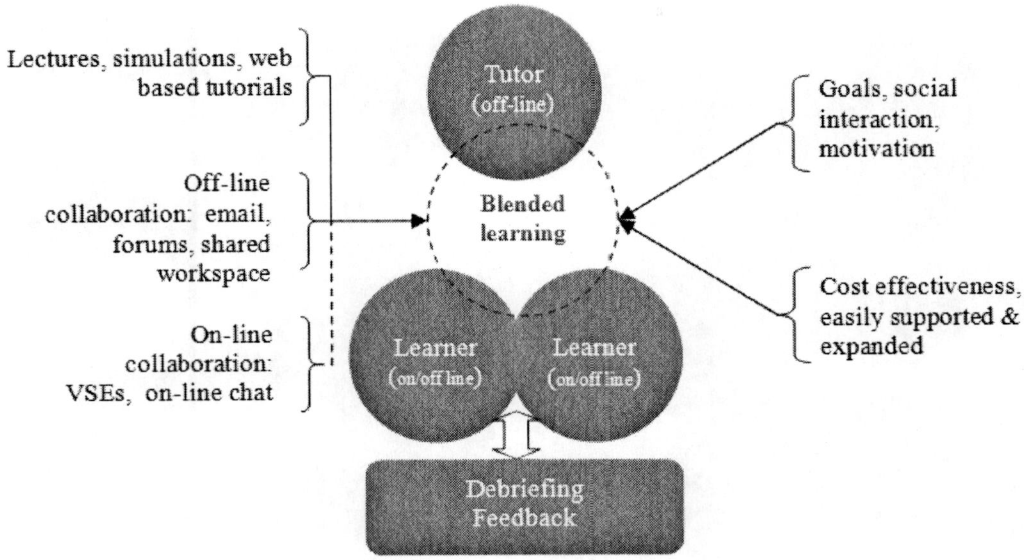

are usually working off-line using tools such as e-mail, forums, and CSCL. Experiential learning, in the form of interactive simulations, is another key factor in our approach and an enabler of the constructivist methodology. As a field of practice, experiential learning has a profound impact on aspects such as theoretical learning models, skill training, life-long learning, and so forth. It is actually a process by which new insights and learning emerge by reflecting on the experience of the learner (Sage, 2000).

Depending on the available resources, automatic or semi-automatic support can be provided in order to compensate for the absence of online guidance by real persons. Apart from these obvious disadvantages, the crucial matter of choosing the right tools and the appropriate educational material, while maintaining cost-effectiveness and maximizing educational impact, needs to be considered. Assistance needs to be closely linked to concrete educational goals and truly support the traditional teaching method of lectures and text books. In the case of science this is, in general, fairly easy to accomplish. Figure 2 presents this approach.

The provision of feedback has also proven to be very important for learners during instructional sessions since even minimal feedback is better than no feedback at all (Collis et al., 2001). Characteristics of feedback include timing (delivery during instruction, after instruction, during evaluation, and after evaluation), purpose (evaluative, instructional) and adaptiveness (based on individualization, difficulty level and test length).

A Design Framework

In this section we present a simple framework as a guideline for the design of the lightweight model introduced in the previous section (depicted in Figure 3).

Collaborative simulations are the most advanced tools for experiential learning. They are also referred to as VSEs (virtual scientific experiments). VSEs may be collaborative or cooperative. First of all, we must specify the context of "collaboration" and "cooperation" which are often used as synonyms. Cooperation is a process in which every member of the group executes a specific task, that is, one portion of the entire as-

figure 3. A simple framework for e-learning

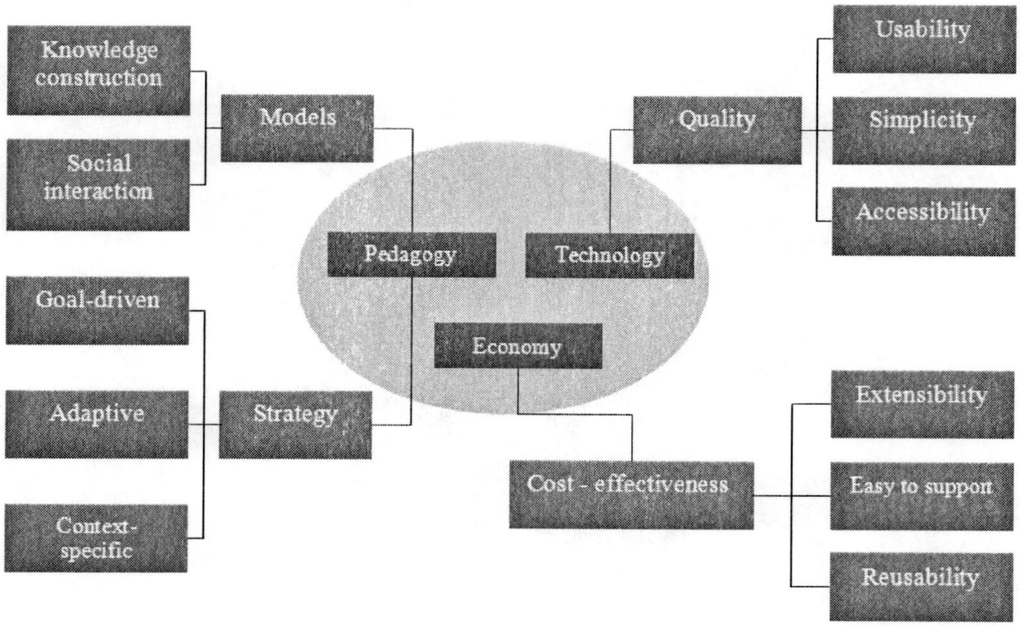

signed VSE; collaboration is a process in which each member of the group works on every part of the total task. Sometimes the boundary between the two types is difficult to distinguish. In either case a VSE should be easily broken down in terms of educational goals and tasks in order to be collaborative or cooperative. Such VSEs are difficult and costly to design and develop, but their educational value is high.

We envisage a service (we call it eCourse) that incorporates experimentation (through VSEs) and collaboration (through Virtual Classrom services). Virtual classroom services (collaborative and social learning) should include functionalities, such as virtual classroom space, private student space, forums, messages, search facilities. Access to educational related material should not be restricted to class members; students from other classes may access resources, if they have the appropriate access rights (knowledge reuse). Since many virtual classes are formed, a virtual pool of information for each course should be constructed. Some information should be restricted and other should be widely available (knowledge sharing). Access to the eCourse services should be made available through a common access point (e.g., a portal). When logged in, the student accesses his/her private integrated and highly personalised space (personalised learning) including:

- **Private shared space (PSS):** Private workspace where learners store learning and other material, search engine, news, forums
- **VSE service:** Participate in an experiment, access experiment history (intermediate results, supporting LOs)
- **Collaborate:** Use online collaboration tools

The eCourse should be operational throughout the duration of the actual course, that is for VSEs to be used both for collaborative and for social learning. VSEs should be modular, comprised of many parts which in turn serve specific learning goals. A student must complete all parts of a VSE. Students can be organized in groups of two to

Figure 4. A three step VSE

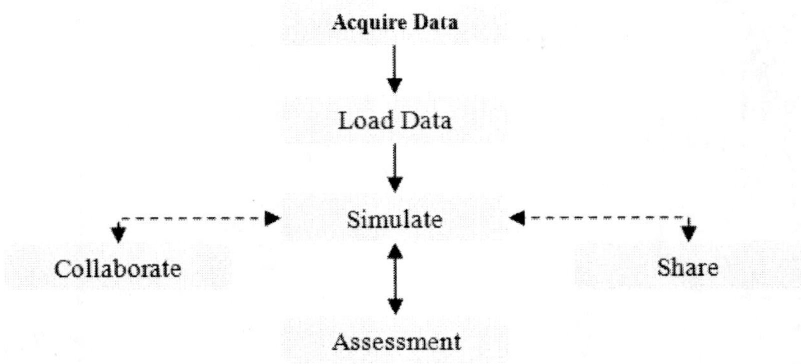

Table 1. Collaborative VSE functions and their characteristics

Functions	Description	Educational Value	Cost
Collaboration			
Forum	Post/Read messages	medium	low
Email	Send/Read messages	medium	low
Chat	Chat with other learners	high	low
Video conference	Video conference with other learners	high	high
Share	Resources (files, results, knowledge)	high	high
Virtual Scientific Experiment			
Load data	Load initial data for simulation (may involve access of remote instruments)	low (medium)	low (medium)
Simulate	Run a VSE	high	high
Save	Save current state	high	high
Configure	Configure VSE parameters	medium	medium
Train	Train for using the VSE's GUI	medium	low
Feedback/ Assessment			
Playback	Playback a VSE	high	high
Test	Take online test	high	low
Ask Tutor	Query the tutor	medium	low
General			
Access LO		medium	low
Search	Search the Internet for learning resources	high	medium
Help	Access the help function	medium	low
Annotate	Attach comments to content, link content to context	high	high

five members, depending on the VSE complexity. Student groups are not static, that is, they may change over time but not during a VSE. In complex VSEs which require the participation of numerous students, roles should be assigned either by the tutor or by the learners themselves. In general, a VSE can be comprised of at least three steps: data acquisition and loading, simulation and final assessment of results (Figure 4).

During the second step, students perform a simulation using the loaded data. Simulation parameters are configurable. The simulation step may include several more steps, depending on the specific experiment. The first step may include live data acquisition from a remote sensor thus requiring management of remote equipment. Online assessment tests should be performed by students between steps. These steps may include multiple choice questions and judgement questions. In the latter case, argumentation can be used to back up student answers including data facts or any kind of evidence. They are used in order to help students assess their own strategies. Feedback should be provided at the end of each test round.

During a VSE learners may communicate with each other using online tools which are provided by eCourse services or external tools. Students may reorganize parts of their repository, create links or construct learning objects (LOs) (self-direct learning). These activities are recorded by special services. An important function is to save a VSE status at any time. Since a VSE is a complex procedure, learners should also have the opportunity to be trained in a test VSE. This collaborative learning phase helps students to understand the online experimentation concepts and introduces them to the concept of collaboration and to the VSE environment. A technologically tedious but educationally valuable option is recording and playback. Playback should be available to learners participating in the experiment and to the tutor. Table 1 summarizes the previously-mentioned functions.

A DEEPER LOOK AT EXPERIENTIAL-LEARNING ASPECTS

Experimentation

Experimentation by way of simulations has been proposed as an effective means for a richer learning experience (Etheris & Tan, 2004; Pohjolainen et al., 2003; Sage, 2000;). Such interactive sessions attract the interest of the user and greatly increase the efficiency of the learning process but, in many cases, they are difficult to support or expand. Nevertheless, their educational value cannot be overlooked. In the words of Albert Einstein, "in natural sciences courses, the first lessons should contain nothing but what is experimental and interesting to see. A pretty experiment is in itself often more valuable than twenty formulae extracted from our minds." This statement underlines the importance of experimentation in many scientific fields. Computer supported experiential learning means use of visual content in order to enhance the learning experience of students and supplement the methods that are already in use (such as text books, online content, synchronous and asynchronous collaboration) (Schwier, 2004).

Experiential learning through cooperation or collaboration is valuable educationally but difficult to realize technologically. Imagine an interactive simulation environment where several students use the same virtual instrument for performing the same experiment. Several problems that would not appear in a real life experiment arise, for example: what happens if one user turns on a button and another turns it off at the same time? The software that supports such an environment should be carefully designed in order to cope with such situations and at the same time retain an adequate level of flexibility and realism.

There are many pedagogical and technological factors that affect simulation use. Pedagogical factors include complexity (e.g., simple, medium, hard), educational context (e.g., mathematics, law), the provision of feedback (e.g., predetermined based on learner's choices or online tests), motivation (how well learners are motivated to use the simulation) and duration (number of sessions required to complete the simulation = reach the educational objectives). The most important factor is how well the simulation is linked to the educational objectives. A weak link will probably reduce significantly the value of the simulation even if its user interface and its collaboration and cooperation capabilities are impressive. Clear feedback is often not considered in many applications although it allows learning to become tangible. Technology can also be misleading. Advanced technological options create over-enthusiasm leading to too complex approaches that are not appropriate for the given educational objectives. Complexity is the main reason for end-user confusion, frustration and disappointment (Xenos et al., 2002). Simulations are not always the most effective means for learning. They may be used as stand-alone e-learning modules or as capstone experienced to classroom lecture, but they excel only in specific contexts (Hung & Nichani, 2001).

Technological factors mainly include the significant difficulty and the accompanying costs to design, develop and support simulations. Depending on the type of simulation (games, virtual laboratory, remote laboratory), its mode (cooperative, collaborative, single user) and adap-

Table 2. Some of the main factors and their effects in using simulations for e-learning

Factor	Description/Effect
Pedagogical	
Complexity	Different levels of complexity serve different pedagogical objectives
Feedback	Feedback is important at all stages in order for the learner to consume/construct knowledge properly
Link to educational objectives	Careful links to concrete educational objectives guarantee success
Context	Simulations maximize their value in some occasions (e.g , mathematics) and perform poorly in others
Motivation	Degree of user engagement, enhancement of user motivation is important for the simulation's success
Duration	A simulation may require one or more sessions to complete. This affects both learner motivation and pedagogical effectiveness
Technological	
User Interface	A simple user interface may attract novice learners
Design and development costs	Simulation are, in general, expensive to design, construct and expand
Group activity	Cooperation/collaboration/single user mode
Training	Amount of training needed to use the simulation environment.
Minimal requirements for use	In many cases, simulations are not only costly to develop but to run to user machines as well (e.g., requirements for h/w, plug-ins, etc.)
Adjustment	Simulation is adjusted to user behaviour providing one-to-one learning. This entails the use of AI techniques but more simulations are not quite flexible
Organisational	
Cost effectiveness	Costs needed to use simulation as a enhanced learning model
Incorporation to existing methodology	Costs related to the inclusion of simulation to existing methods
Support	Human resources needed for supporting simulations

Figure 5. A VSE accessed by a remote user

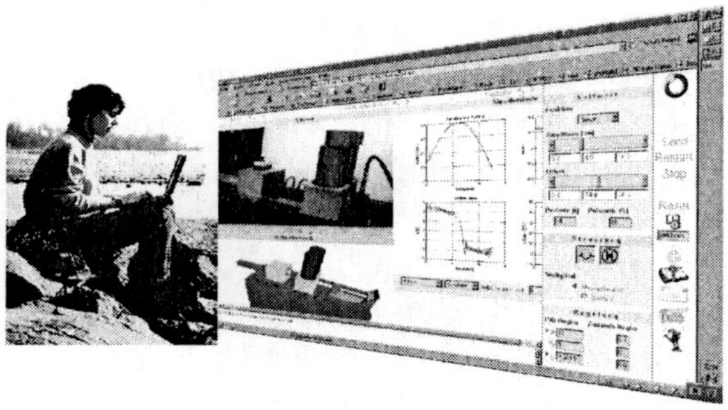

tivity to the learner, costs vary. End user system requirements are sometimes important. Finally, organizational factors should be considered when introducing simulations for an enhanced learning experience: cost-effectiveness, cost for introducing simulations and support. Table 2 summarizes the previously mentioned factors.

VIRTUAL SCIENTIFIC EXPERIMENTS

Simulation and online collaborative experimentation is a difficult educational and technological endeavor. Development, support, and expansion costs are also important when applying these methods in real world cases. Standard Web technology, if properly used, can provide a cost-effective means for enhanced learning even in higher education environments.

A fine paradigm of blended learning are VSEs with incorporated collaboration and cooperation functions (Figure 3). Experimentation takes place using simulations while collaboration and cooperation takes place both between learners and between learner and tutor. The tutor actually becomes a mentor rather than the holder of knowledge. This means that the tutor should be able to employ and encourage social negotiation. Although educational goals for each module that comprises a course are predetermined, the underlying learning model should partially support negotiation rather than imposition of goals and objectives.

Social interaction during VSEs is effectively supported through virtual structures such as virtual classrooms (VCs). The concept of virtual classrooms is difficult to accomplish especially in traditional universities: they are difficult to be formed, maintained, and supported. They also require a significant part of the educational process be focused on the interaction with the instructor and tutor. As mentioned previously, traditional higher education institutes do not have the organization structure to directly support full e-learning solutions by providing specially trained tutors for this purpose. Thus, a consensus should be reached in this case, for example services should not require the online presence of a tutor but rather provide automatic support where possible. Online support by tutors should be provided in rare occasions and only when the institution has anticipated such a role. Furthermore, a lighter version of virtual classrooms (i.e., personalized workspace) should be used for online collaboration and sharing of knowledge. In any case, the administrative and educational burden for the tutors should be as light as possible. Another difficulty in using VSEs is students are used to classrooms, and they need to adjust their learning and teaching styles, respectively. For example, in one class, two students who work at different subjects can both share resources and reuse each others knowledge electronically, a feature not easily supported by traditional learning methods.

In the case where the educational institute decides to support a full VSE option a different method should be used. In our vision, at such a collaboration an eCourse is formed, supported both, by VSEs and VC services. VSEs (experiential learning) should be multi-step experiments closely linked to educational goals and supported by LOs (learning objects). During an experiment conducted by two or more students collaborating together, participants should be able to communicate using synchronous services.

Technologies for VSEs

In open or distance education environments, an efficient and less hardware resource demanding approach is, to replace the real laboratory with a simulated one. This may be realized by the simulation of real world systems and by animation of experiments in a highly interactive environment. Such a virtual laboratory within additional distance education in the form of courses offered across the Internet will fully engage the learners in the learning process through an interactive dynamic environment. This kind of laboratory consists of the simulation of experiments whose output data is indistinguishable from a real experiment data. Moreover, a simulated experiment offers an edge of moving beyond the realm of real hardware. The techniques for implementation of these synthetic learning environments are available.

From the architectural point of view, Internet-based simulation tools fall into three categories: simulation programs that can be accessed remotely through a Web browser, those which are downloaded from servers and run on the client machine, and those which show Internet-based execution. Examples of the second category include Simjava, Simkit, and JSIM, which may be attractive candidates to be used for building specific Web-based virtual laboratory environments due to the code mobility and reusability based on the Java programming language. The third category allows simulation models to be executed over the Internet. Typically, this is performed by a conventional simulator on a server, which is linked to a helper Java applet to the clients. In view of a VSE system in a virtual laboratory, the marriage of this kind of remote simulation with virtual reality technique is essential.

Most existing virtual labs offered across the Web include several fully interactive experiments completely written in Java. The applets embedded into the Web pages comprise the essential physical effects, but cannot claim to be an equivalent substitute of the real experiment, though they are capable enough to demonstrate the underlying principles. This cognitive process promotes the effectiveness of learning. This calls for a close-to-reality environment. Virtual reality (VR) offers a more realistic 3D visual and acoustic environment together with its intuitive forms of interaction. Though VR is typically associated with powerful hardware and deterrent costs, browsers and tools for the virtual reality modelling language (VRML) gives the illusion of immersion in a laboratory environment by creating a closed loop of interaction between the user and the virtual world. This is performed in an intuitive and realistic manner.

Figure 6. A virtual scientific experiment

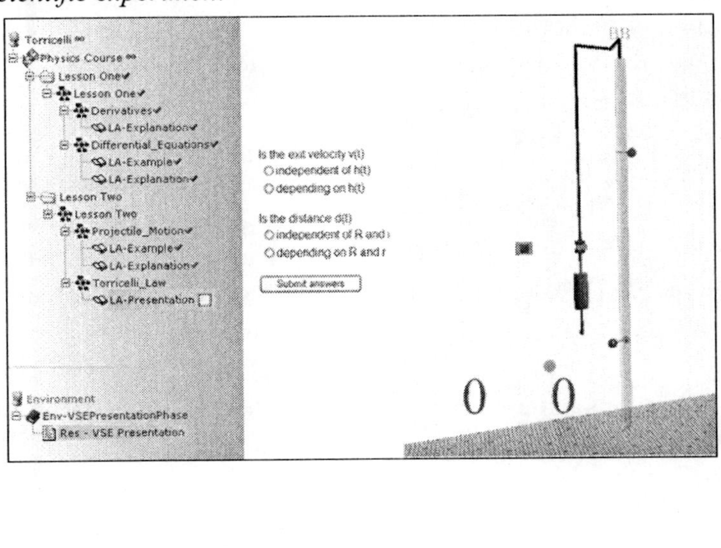

VRML is preferably designed for simulating real world behavior from the visual point of view, but it does not contain any flexible dynamical system simulation elements (E-LeGi, 2007).

Typically laboratory courses are organized for and accomplished by groups. This promotes problem solutions by teamwork, which is a substantial requirement to the abilities, for example, of an engineer. VSE can be accomplished only by a single person. In order to promote teamwork, additional tools are necessary that enable the learners to collaborate in a team. A conventional chat is not the solution, as it does not track and publish the learners operation during experimenting. Figure 6 shows a 3D collaboration environment, where learners and the supervisor can meet them represented by their avatars to have simultaneous access to the experiment. The person empowered by the team to perform the experiment can perform tactile operations, for example, press buttons, turn switches, enter data, etc. and the other person can watch these actions simultaneously.

In such a virtual environment an excellent immersion into an experimental dynamical environment is provided taking multimodal aspects into account. There is the visual information about the 3D scene of the dynamic experimental world, the tactile interaction with virtual plant elements, the real-time information, 3D scene acoustic information about plant noise and eventually haptical information when using a force-feedback device (haptical display). These examples demonstrate that it is possible to overcome the static character of experiment and to make them an attractive place for scientific education. An example of such a tool is VCLab. Figure 6 shows a screenshot of a VSE from E-LeGi project (E-LeGi, 2007).

QUALITY ASSESSMENT OF BLENDED LEARNING

E-learning is a software system and as such, its quality assessment characteristics can be evaluated using the ISO standard. From all ISO standards, only ISO 9126 has a hierarchical structure (defined by quality characteristics and sub-characteristics) that could be used for the assessment of knowledge construction e-learning systems during their operation. ISO9126 has been extensively used as a basis for assessing Web-based systems, so it is well suited as a starting point in our case as well (Nielsen, 2000). However, the versatile nature of the services of an eCource does not fall exactly to the web engineering quality assessment area; so it can be said that e-learning and especially, advanced e-learning services lack adequate quality evaluation metrics.

eCourse services are mostly Web-based and in general follow a "one size fits all" approach. Experience from many surveys and testing of real applications in the general field of Web engineering has demonstrated that a basic success factor is to determine the key factors that determine user acceptance. These factors also define the quality of the services, as they are perceived by the end-user. Past approaches in other disciplines such as e-commerce, took either a technology-centered or a user centered view of quality. The technology–centered view examines the technical specifications of an online system, that is the technological infrastructure needed for successful operation: search engine, adaptation and feedback mechanisms, user interface, security, and so forth.

Formally, software quality is defined as the totality of features and characteristics of a product or service that bear on its ability to meet stated or implied needs. It is worth noting that very few works refer to quality aspects of e-learning systems using formal rules or standards (Louca et al., 2004). In this section we use the eCource services identified in section 2.3 and discuss how to evaluate an e-learning system based on e-learners actions and requirements. In order to assess the quality of e-learning systems the ISO 9126 quality standard is used as a basis to

Table 3. Quality characteristics of ISO 9126

ISO 9126 Quality Model		
Quality characteristics	**Sub-characteristics**	**Explanation**
Functionality	Suitability	Can software perform the tasks required?
	Accuracy	Is the result as expected?
	Interoperability	Can the system interact with another system?
	Security	Does the software prevent unauthorized access?
Reliability	Maturity	Have most of the faults in the software been eliminated over time?
	Fault tolerance	Is the software capable of handling errors?
	Recoverability	Can the software resume working and restore lost data after failure?
Usability	Understandability	Does the user comprehend how to use the system easily?
	Learnability	Can the user learn to use the system easily?
	Attractiveness	Does the interface look good?
	Operability	Can the user use the system without much effort?
Efficiency	Time Behavior	How quickly does the system respond?
	Resource Behavior	Does the system utilize resources efficiently?
Maintanability	Analyzability	Can faults be easily diagnosed?
	Changeability	Can the software be easily modified?
	Stability	Can the software continue functioning if changes are made?
	Testability	Can the software be tested easily?
Portability	Adaptability	Can the software be moved to other environments?
	Installability	Can the software be installed easily?
	Co-existence / conformance	Does the software comply with portability standards?
	Replaceability	Can the software easily replace other software?

produce metrics that are quantifiable parameters for assessing quality.

ISO 9126 is a quality standard for software systems having a hierarchical structure, defined by quality metrics and sub-metrics (ISO, 1999). The ISO9126 structure has six levels of quality namely functionality, usability, reliability, efficiency, maintainability and portability. Although e-learning systems are a sub-category of software systems (actually online systems), they demonstrate some unique characteristics. Thus, although ISO 9126 may be used as basis for e-learning quality evaluation, further analysis and mapping of its characteristics is required. In this work, we use the end-user related characteristics of the ISO 9126 standard to evaluate the services during their operation.

eCourse services are divided in four distinct categories (Stefani et al., 2006): access to resources, specific e-learning services, common services and presentation services. These categories are compared against the first four of the seven sub-characterises of ISO9126, namely functionality, reliability, usability and efficiency. We assume that maintainability and portability are, more or less, common with any software system. Each quality characteristic of ISO9126 is analyzed in several quality sub-characteristics (analysed in Table 3).

The first characteristic, functionality refers to a set of functions and specified properties that satisfy stated or implied needs (Fenton & Pfleeger, 1997). It is decomposed in four quality sub-characteristics: suitability, accuracy, interoperability, and security. The meaning of functionality in an e-learning system can be analyzed as functions and services that the e-learning system provides to the user. As functions in an e-learning system we define:

- The personalization mechanism for different kinds of users (students, teachers, tutors, administrator, quests). Each user should have different levels of permissions and different authorities.
- Search functions: simple search like searching by keyword and logical operators or advances search (searching by category of learning material, metadata-enabled searching, multimedia searching, etc.):
- Multimedia application for digital material
- Collaborative environment
- Knowledge sharing and reuse

All the above factors are affecting the quality of advanced e-learning services measuring technical to pedagogical (although indirectly) parameters. The most important benefit of applying this model is the fact that it provides a formal method for assessing e-learning services according not only to the overall quality, but to each quality characteristic as well. Subjectivity, which is always a significant factor in ISO characteristics, is limited by using strictly quantifiable metrics that can be measured either by man (e.g., evaluators) or machines (special assessment software). The introduction of formal quality metrics during the eCource operation may not only boost the quality of teaching but also reduce management and support costs mainly in the long term.

CONCLUSION

As more powerful, flexible, and affordable technologies become embedded in society, the balance of expectation in higher education shifts to towards their deployment across a range of activities. Advances in the use of ICT in cciences teaching have been reflected in many higher education institutions, albeit with varying degrees of success. The growing importance of ICT in teaching and learning has been fostered by national government investments and a variety of cross-institution support initiatives; however, research indicates that its potential has yet to be fully realized since economic and pedagogical parameters affecting the final technological solutions have not been fully considered.

Web based technology is the technology of choice for e-learning due to its cost-effectiveness, its simplicity and its flexibility. New blended or enhanced models use traditional teaching methods combined with static or dynamic tools based on simple web technologies. Furthermore, new technologies have facilitated collaboration and experimentation enabling the cost-effective introduction of these models in traditional higher education institutions. The ultimate aim of our work was to explore how we can fully integrate tutoring techniques in a computer-mediated collaborative environment. In other words, to use the integration of personal workspace and low-cost off-line collaboration tools as a first step toward developing a fully integrated, low cost environment.

In this chapter we reviewed enhanced educational models and discussed several parameters that affect them. Special attention was given to simulations as an enhanced learning tool. We presented a framework describing the general steps towards a cost-effective blended model. An instance of this model which was used as an example uses collaborative virtual scientific experiments and a set of cost-effective services to realize knowledge building. Although simulations are educationally valuable in several contexts, their introduction poses several educational, technological and organizational questions.

REFERENCES

Bates, A.W. (2000). *Managing technological change.* San Francisco: Jossey-Bass.

Bender, T. (2003). *Discussion-based online teaching to enhance student learning: Theory, practice and assessment.* VA: Stylus Publishing.

Bonk, C.J., & Graham, C.R. (Eds.). (2006). *The handbook of blended learning: Global perspectives, local designs.* San Francisco: Pfeiffer Publishing.

Boucher, A. (1998). Information technology-based teaching and learning in higher education: A view of the economic issues. *Journal of Information Technology for Teacher Education, 7*(1), 87-111.

Cohen, A., & Nachmias, R. (2006). A quantitative cost effectiveness model for web-supported academic instruction. *Internet and Higher Education, 9,* 81-90.

Collis, B., Boer, W.D., & Slotman, K. (2001). Feedback for web-based assignments. *Journal of Computer Assisted Learning, 17,* 306-313.

Demb, A., Erickson, D., & Hawkins-Wilding, S. (2004). The laptop alternative: student reactions and strategic implications. *Computers & Education, 43*(4), 383-401.

Driver, M. (2002). Exploring student perceptions of group interaction and class satisfaction in the web-enhanced classroom. *The Internet and Higher Education, 5,* 35-45.

Drossos, L., Bassisliadis, B., Stefani, A., Xenos, M., Sakkopoulos, E., & Tsakalidis A., (2006). Introducing ICT in a traditional higher education environment: Background, design, and evaluation of a blended approach. *International Journal of Information and Communication Technology Education, 2*(1), 65-78.

Duffy, T.M., & Jonassen, D.H. (1992). *Constructivism and the technology of instruction: A conversation.* Lawrence Erlbaum Associates.

E-LeGI. (2007). *European Learning GRID Infrastructure project.* Retrieved from http://www.elegi.org

Etheris, A.I., & Tan, S.C. (2004). Computer-supported collaborative problem solving and anchored instruction in a mathematics classroom: an exploratory study. *Int J Learning Technology, 1*(1), 16-39.

Fenton, N., & Pfleeger, S. (1997). *Software metrics a rigorous and practical approach.* Thomson Computer Press.

Garrison, D.R. & Kanuka, H. (2004). Blended learning: Uncovering its transformative potential in higher education. *The Internet and Higher Education, 7*(2), 95-105.

Groccia, J.E., & Miller, J.E. (2005). *On becoming a productive university: Strategies for reducing cost.* Bolton, MA: Anker Publishing Company.

Cohen, A., & Nachmias, R. (2006). A quantitative cost effectiveness model for web-supported academic instruction. *The Internet and Higher Education, 9*(2), 81-90.

Haywood, J., Anderson, C., Coyle, H., Day, K., Haywood, D., & Macleod, H. (2000). Learning technology in Scottish higher education — a survey of the views of senior managers, academic staff and experts. *ALT-J,* 8(2), 5-17.

Hiltz, S.R., & Turoff, M. (2002). What makes learning networks effective? *Communication of the ACM, 45*(4), 56-58.

Hung, D., & Nichani, M. (2001). Constructivism and e-learning: Balancing between the individual and social levels of cognition. *Educational Technology, 41*(2), 40-44.

ISO (1999). Information technology—evaluation of software—quality characteristics and guides for their use. *International Standard, ISO/IEC 9126.*

Jackson, S. (2004). Ahead of the curve: Future shifts in higher education. *Educause Review, 39*(1), 10-18.

Jefferies, A., Thornton, M., Alltree, J., & Jones, I. (2004). Introducing web-based learning: An investigation into its impact on university lecturers and their pedagogy. *Journal of Information Technology Impact, 4*(2), 91-98.

Jonassen, D.H., Howland, J., Moore, J., & Marra, R.M. (2003). *Learning to solve problems with technology.* Pearson Education.

Laat, M., & Lally, V. (2005). Investigating group structure in CSCL: Some new approaches. *Information Systems Frontiers, 7*(1), 13-25.

Laurillard, D. (2002). *Rethinking university teaching: A conversational framework for the effective use of learning technologies* (2nd ed.), London: Routledge Falmer.

Louca S., Constantinides, C., & Ioannou, A. (2004). Quality assurance and control model for e-learning. *In Proceedings of Computers and Advanced Technology in Education* (pp. 468-472).

Nielsen, J. (2000). *Designing web usability: The practice of simplicity.* Indianapolis, IN: New Riders Publishing.

Pittinsky, M.S. (2002). The wired tower: Perspectives on the impact of the internet on higher education. Financial Times/Prentice Hall.

Pohjolainen, S., Hautakangas, S., Ranta, P., Levasma J., & Pesonen, K. (2003). A learning experiment in mathematics using A&O-learning environment. *Int. J. Cont. Engineering Education and Lifelong Learning, 13*(1&2), 57-74.

Rodrguez, D., Sicilia, M. A., Cuadrado-Gallego, J.J., & Pfahl, D. (2007). e-learning in project management using simulation models: A case study based on the replication of an experiment. *IEEE Transactions on Education 49*(4), 451-463.

Romano, J., Wallace, T.L., Helmick, I.J, Carey L.M. & Adkins L. (2005). Study procrastination, achievement, and academic motivation in web-based and blended distance learning. *The Internet and Higher Education, 8*(4), 299-305.

Rumble, G. (1999). Costs of networked learning: What have we learned? *In Proceedings of the FLISH'99. Conference on Flexible Learning on the Information Superhighway.* Sheffield, England. Retrieved from http://www.shu.ac.uk/flish/rumblep.htm

Sage, S.M. (2000). A natural fit: Problem-based learning and technology standards. *Learning & Leading with Technology, 28*(1), 6-12.

Saunders, G., & Klemming, F. (2003). Integrating technology into a traditional learning environment. *Active Learning in Higher Education, 4*(1), 74-86.

Schwier, R.A. (2004). Virtual learning communities. In G. Anglin (Ed.), *Critical issues in instructional technology.* Portsmouth, NH: Teacher Ideas Press.

Stefani, A., Vassiliadis, B., & Xenos, M., (2006). On the quality assessment of advanced e-learning services, *Interactive Technology & Smart Education, 3*(3), 237-250.

Van der Wende, M., & Van de Ven, M. (2003). *The use of ICT in higher education: A mirror of Europe.* Utrecht, Lemma Publishers.

Van Weert, T.J., & Pilot, A. (2003). Task-based team learning with ICT, design and development of new learning. *Education and Information Technologies, 8*(2), 195-214.

VClab. (2007).Retrieved from http://www.esr.uni-bochum.de/VCLab/

Warschauer, M. (2003). *Technology and social inclusion: Rethinking the digital divide*. The MIT Press.

Xenos M., Pierrakeas C., & Pintelas, P. (2002). Survey on student dropout rates and dropout causes concerning the students in the course of informatics of the Hellenic Open University. *Computers & Education, 39*(4), 361-377.

Chapter III
The KAR–P–E Model Revisited:
An Updated Investigation for Differentiating Teaching and Learning with Technology in Higher Education

Lawrence A. Tomei
Robert Morris University, USA

ABSTRACT

Since 1996, the KAR-P-E model has served to differentiate teaching and learning of technology. It is offered here as an archetype for other institutions seeking to develop their own comprehensive technology program. Knowledge, application, research, practice, and evaluation (KAR-P-E) offer the necessary dichotomy among instructional technology programs for undergraduates, graduates, and doctoral candidates. Similar to other more well-known taxonomies, the KAR-P-E model is progressive and assumes mastery and competency at previous levels. Readers are exposed to the ISTE technology standards for teachers as well as how particular institutions implement the set of competencies in their individual programs of study. By establishing how technology skills are addressed in higher education, readers will be able to transfer the KAR-P-E model to new initiates at all levels of instructional technology education, business, and corporate as well as traditional education.

INTRODUCTION

The phenomenon of technology-based learning has dramatically changed the direction and delivery of education in the past decade. Pastore (2001) estimated that, by 1999, 1,500 colleges and universities were offering Web courses, with that number expected to double by 2005. The U. S. Department of Education found some 26,000 online courses with an estimated 100 new college courses going online every month (James & Voigt, 2001).

To meet the increasing demands for technology at all three levels, technology-based education programs have been implemented for pre-service (undergraduate) learners, in-service professionals (graduate) students, and post-graduate (doctoral) candidates. Technology courses across three levels beg questions in the minds of faculty and students alike as they move through their formal education agendas. Specifically:

- When it comes to technology skills and competencies, what can I expect to learn differently as a graduate or doctoral candidate than I did as a freshman?
- Is there a different set of skills and competencies appropriate for each of these levels?
- If I take undergraduate technology courses am I sufficiently prepared (i.e., competent) to use technology throughout an entire career?

REVIEW OF THE LITERATURE

Standards and Instructional Technology Education

The International Society for Technology in Education recognizes three distinct levels of personal technology development in higher education (ISTE, 2004). At the outset, technology *foundations* are suitable for all learners as they advance their own strategies for acquiring knowledge. At mid-level, skilled *practitioners* acquire the tools they need to exercise their chosen avocation. At the highest level, *professionals* seek the competencies necessary to share a lifetime of service and experience with peers and colleagues and thereby advance their profession.

Table 1 illustrates the emphasis placed by the ISTE on the various skills and competencies expected of educators. At the outset of the educator's career, stress is laid on grasping technology operations and concepts and the importance of using technology for enhancing professional growth. Currently, 42 of the 50 states have adopted the ISTE technology standards for their professional staff. As more and more teachers prepare to take the reins of a classroom, the impetus on technology shifts to the learning environment and the surrounding social issues. First-year teachers are expected to have mastered the basics of technology, turning their attention to the curriculum and assessment.

The Technology for All Americans Project (TfAAP) was created by the International Technology Education Association (ITEA) through funding from the National Science Foundation (NSF) and the National Aeronautics and Space Administration (NASA). The project began in 1994 with the first of three phases (Table 2). Phase I established the philosophical foundations for the study of technology in classrooms and articulated the essential role of schools in developing technologically literate citizens. Phase II emphasized content standards for the study of technology built around a cognitive-activities base and included knowledge, abilities, and the capacity to apply both knowledge and abilities to the real world. Phase III addressed such important topics as student assessment, professional development, and program enhancement.

SUCCESS is a comprehensive, integrated, education-focused program for the infusion of technology into the curriculum of a school. The program was developed courtesy of a five-year grant by the Vira I. Heinz Endowments and focused initially on private, then public schools in western Pennsylvania. SUCCESS guarantees increased student achievement if participating schools adhere to the basic principles of the program, including: full participation of every teacher in the school, key leadership by the school principal, completion of a five-day summer workshop followed by observed and evaluated lesson presentations during the school year, cooperation with a technology advisor during the entire first year of the program, and commitment to the program for all three years.

SUCCESS schools integrate technology-based teaching strategies by selecting two teachers to attend a formal education program in Instructional Technology, then assume a technology leadership role in their respective schools. These teachers are enrolled in a graduate-level certification program and, upon completion, receive the Pennsylvania Department of Education Instructional Technology Specialists certificate. Most choose to complete their master's degree in instructional technology with an additional five courses. Together, workshop teachers and those selected for further study receive instruction in (K)nowledge, (A)pplication, and (E)valuation of technology-based lessons to meet the objectives shown in Table 3.

Taxonomies and Teaching with Technology

In 1956, Benjamin Bloom offered a rubric for developing cognitive-oriented student learning objectives. The taxonomy of educational objectives (Bloom, Englehart, Furst et al., 1956) proposed six progressively complex levels of higher order thinking: knowledge, comprehension, application, analysis, synthesis, and evaluation. After years of application and dozens of research efforts to validate the classification scheme, Bloom's Taxonomy is now among the most widely used strategies for developing classroom goals for teaching and learning. Krathwohl and Kibler would expand on the initial premise with further development of the affective and psychomotor domains, respectively.

In 2005, Tomei offered the latest rendering of taxonomies, this time in the technology domain. The newest classification system addresses how educators can prepare their charges for the classrooms of tomorrow. The taxonomy for the technology domain includes literacy, collaboration, decision-making, instruction, integration, and tech-ology as progressive levels of complexity for teaching technology and constructing increasingly multifaceted student learning outcomes.

Figure 1. The 3-tiered KAR-P-E model

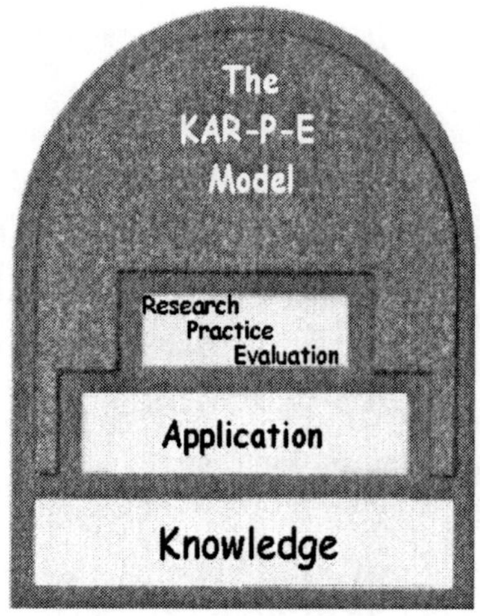

Combining instructional technology standards with the six levels of the taxonomy produced a schemata that has become the basis for the KAR-P-E model, an overarching schemata for differentiating levels of teaching and learning with technology. It serves as an archetype for those developing their own comprehensive technology education programs at any level of post-secondary education.

The KAR-P-E Model

Knowledge, application, and research, practice and evaluation (KAR-P-E) offer the necessary dichotomy among technology education programs for undergraduates, graduates, and doctoral candidates. As with other perhaps more well-known taxonomies, the KAR-P-E model: (1) applies to all learners in all disciplines; (2) develops the learner in progressive, sequential steps; and, (3) assumes mastery and competency at previous levels before advancing up the hierarchy.

The *knowledge* level of the model introduces candidates to technologies as personal learning tools and, therefore, is the domain of the undergraduate program. At the knowledge level, learners acquire the technology skills that serve to enhance both their immediate and life-long learning needs.

At the *application* level, candidates master technology-based skills for immediate inclusion into the everyday roles and responsibilities of the adult learner. At the application level, practicing classroom teachers, for example, acquire technology skills that benefit their own learners. Success is measured as observable increases in learner achievement and classroom outcomes.

At the third and highest level of the KAR-P-E model lies *research, practice* and *evaluation.* The professional scholar explores the rich knowledge base and comprehensive review of the literature (i.e., research) to support the implementation of technology as teaching and learning tools. In addition, they are also charged with changing the way technology is experienced (the practice of the discipline) in the classroom and work environment. Finally, scholars must keep an ever-watchful eye on verifiable learner achievement (evaluation) as they assess educational opportunities supported by technology.

THE STUDY

Statement of the Problem

The study examined whether greater incidents of student learning objectives in information technology courses taught at the undergraduate, graduate, and doctoral levels of higher education could be found at the three respective levels of the KAR-P-E model. In other words, it was expected that knowledge objectives would occur more at the undergraduate level, application objectives at the graduate level; and, research-practice-evaluation objectives at the doctoral level.

Specifically, this chapter presents the results of an initial study validating the use of the KAR-P-E model and its contributions to the art and science of teaching and learning with technology. By establishing how technology skills are addressed in higher education, readers are able to transfer the KAR-P-E model to new initiatives at all levels of information technology education, business and corporate as well as traditional education. The study was conducted in two phases; the findings, conclusions, and recommendations included in the paper summarize both segments of the research.

Participants

Seventeen schools participated in the initial phase of this research project. Thirteen of the initial 17 schools (76.4 percent) conducted IT programs on two or three of the academic levels and two of the institutions examined were outside the United States. Since the initial study was completed (and its findings presented at the 2005 IRMA International Conference), an additional 52 schools were randomly selected for review and added to the research. Most (42/52 or 80.7 percent) sponsored IT programs on at least two of the three academic levels reviewed in the study and eleven of the 52 were located outside the United States.

To recap, a total of 69 school programs are included in the chapter's final discovery. Some 55 of the 69 (79.7 percent) schools offered IT programs on at least two of the three academic levels reviewed: undergraduate, graduate, and doctoral.

Conduct of the Study

The research included an investigation of first selected, then randomly identified, programs in instructional technology. The investigation considered only core technology courses and only technology-oriented learning objectives that comprised the curricula of undergraduate, graduate, and post-graduate technology programs of study were categorized for further analysis.

An e-mail requesting participants to submit learning objectives in their respective programs was sent to the initial cadre of schools. Additional programs were then selected on the basis of information presented on their Web sites, online programs of study, and downloadable course syllabi. Data from these sites were reviewed and specific learning objectives found in course syllabi were categorized as having a knowledge, application, or research-practice-evaluation focus. Two separate examiners reviewed the programs to enhance inter-rater reliability.

Phase one of data analysis began November 1, 2004. The initial conclusions and recommendations phase of the study commenced December 1, 2004 and a preliminary analysis was completed in time to meet the required 2005 IRMA International Conference submission deadline of January 2005. The second phase of the data collection, analysis, conclusions, and recommendations began immediately and concluded in July 2005. The results of this initial phase of the study were presented as a session within the Information Technology Track of the Conference in May 2005. Feedback has been rewarding, encouraging, and stimulating. Positive comments were received both during and immediately following the conference presentation as well as email reactions from readers of the article published in the *International Journal of Information Communication and Technology Education* (IJICTE), (vol. 2(2), April-June 2006).

By far, the majority of the feedback focused on the (relatively) limited review of undergraduate programs and a longitudinal view of learning objectives over time. This revision to the IJICTE article addresses the first area; that is, an increased examination of undergraduate programs. Since the original article was first published, a third phase of the investigation was conducted and another 18 (mostly undergraduate) programs were examined. Additions to the data originally reported as well as its impact on the findings and recommendations originally posited will be noted.

Overall, 1286 courses and nearly 12,000 objectives were identified, categorized, and analyzed during the initial phases of this research (Table 4). A third phase of the investigation, as already noted, added another 18 programs increasing the number of programs reviewed from 69 to 87; the number of courses examined from 1286 to 1542; and, the number of learning objectives classified from 11,925 to 14,344 (Table 5).

Findings

The investigation supports the assertion that knowledge is the essential building block of technology at the undergraduate (bachelor's) level of higher education. Likewise, research, practice, and evaluation were paramount to post-graduate scholars. However, it could not be confirmed that application was the most critical category of objective for graduate (masters) candidates. For the most part, further investigation during the third phase of the study confirmed these assertions. Knowledge continues to be associated most closely with undergraduate education (10 of 18 programs added to the research). In a similar vein, research-practice-evaluation objectives continued to align with post-graduate doctoral scholars (another 2 of 18 programs reviewed). Notably however, it was found that Application was more apparent in the additional 6 of 18 graduate (i.e., master's) programs examined. It appears that programs are evolving (especially at the graduate level) with respect to the KAR-P-E model taxonomy. The findings that follow reflect updated results incorporating the new data.

Undergraduate Programs

At the bachelor's level, it was discovered that knowledge indeed played the most pronounced role. Of the 1,300 objectives reviewed at this level, nearly half (.49) of the learning objectives were categorized as knowledge-based outcomes. Another 30 percent moved the undergraduate to explore the application of technology while 21 percent of the objectives considered research, practice, or evaluation. During the third phase, percentages at this level changed little. Still, half of the objectives were found to be knowledge-based. Similarly, undergraduate percentages at the application and research-practice-evaluation levels remained consistent at .32 and .20, respectively. Examples of typical knowledge outcomes at this level include:

1. Undergraduate students will correctly define terminology related to computers and technology in their written and oral communication.
2. Undergraduate students will operate a multimedia computer system with related peripheral devices to successfully install and use office productivity software.

Graduate Programs

For many graduate programs, knowledge objectives were the most prevalent aspect of their learning experience. Some 43 percent of the objectives examined at this level remained at the bottom rung of the KAR-P-E model. Initially, it was hypothesized that the majority of the objectives would be found on the application level; however, application only came in a distant second at 33 percent. With additional 18 programs reviewed, the revised statistics are at least notable. More learning objectives were noted on the level of application (.43) inverted from the previous study when knowledge objectives were at .43. Typical application outcomes follow:

1. Using the principles of instructional system design, master's candidates will develop and implement an eight-page, text-based, student workbook containing essential elements appropriate for their selected classroom lesson.
2. Given a Portfolio Exercise and diskette file, master's candidates will prepare an intelligent portfolio for use throughout the program in instructional technology. This portfolio will be exhibited and assessed during the course of the student's program of study to evidence the understanding of the concepts and principles presented in this course.

Post-Graduate Programs

As expected, most (.44) of the doctoral learning objectives examined were found the R-P-E level: nine percent research, 18 percent practice, and 17 percent evaluation. Even during the most recent phase of the investigation, the relationship of KAR-P-E for this category of programs shifted little; percentages remained statistically unchanged. Example objectives at this level included:

1. Using Internet-based data, candidates will be able to correlate learner achievement scores and the ratio of learners-to-computers (Research).
2. Candidates will develop a visual presentation suitable for directors and technology coordinators that provides an overview of instructional technology and its potential impact on decision-making (Practice).
3. Doctoral candidates will assess selected educational software packages in core academic areas and appraise their content coverage, effective use of technology, and impact on learner outcomes (Evaluation).

The analysis presented in this chapter (with the inclusion of the third-phase data) continues to establish the importance of knowledge-based skills and competencies across the spectrum of higher education, ranking first or second at all three levels. Individually, all programs examined infused knowledge, application, and research, practice, and evaluation objectives to some degree into their respective curriculum.

Summary of Findings

Additional undergraduate, master's, and doctoral programs in instructional technology remain unaccounted for in the population of post-secondary IT programs; therefore, further data collection remains important. The findings now include some 87 programs, 1542 courses, and over 14,000 learning objectives. Added to the initial investigation, the sample size continues to be sufficient to generalize the KAR-P-E model to all higher education programs in technology education.

RECOMMENDATIONS

Inferential statistical analysis of the data collected is recommended to determine if the differences were statistically significant. First, a study of correlation is recommended to measure the linear relationship between the learning objectives collected and the three academic levels of IT programs. A correlation coefficient should be calculated to determine if a relationship exists between these variables. Second, a t-test is necessary to compare mean values of two sets of numbers. For this study, a t-test will isolate any statistically significant difference with respect to knowledge, application, and research, practice, and evaluation in undergraduate, graduate, and doctoral IT programs. Statistical analysis of this magnitude will establish the relevance of the model to all higher education IT programs and will dictate whether the final results of this investigation deserve broader dissemination.

The final step in this investigation of learning objectives would be a review of programs over time. Specifically, at the master's level, it is hypothesized that as teachers become more versed in the applications of technology for teaching and learning, IT programs will necessary alter their focus from knowledge to more application objectives. A longevity study to validate that assumption is suggested for follow-on research.

CONCLUSION

The KAR-P-E model for differentiating teaching and learning with technology appears worthy of consideration in the practice of post-secondary teaching and learning with technology. These findings identified an apparent connection between knowledge-based learning objectives and undergraduate IT programs; the follow-on phases confirmed the original findings. As well, there seems to be a relationship between research, practice, and evaluation for doctorate-level learners.

Not confirmed at first was the anticipated link between application and graduate candidates, However, the third phase of the investigation did support a shift as in-service (i.e., active classroom) teachers, comprising the majority of learners at this level, establish their technology competencies and propel higher education programs and migrate their own learning objectives from knowledge to more applications-based outcomes.

The acceptance or rejection of the KAR-P-E model as a schema for higher education is supported by the investigation. Its potential to provide differentiated instruction for professional programs is important. However, even with this comprehensive review, further investigation is encouraged. And, of course, further longitudinal research in the quality of student learning outcomes seems appropriate.

REFERENCES

Dargan, C. P. (2003) *Designing online courses: Reflections from the cyber trenches* (pp. 1-2). (Graduate project, University Of Northern Iowa).

International Society for Technology in Education. (2000). *Educational Technology Standards and Performance Indicators for All Teachers.* Retrieved January 2005 from: http://cnet.iste.org/teachers/t_stands.html

International Technology Education Association. (2002). (ITEA) is the professional organization of technology teachers. Retrieved January 2005 from: http://www.iteaconnect.org/TAA/TAA.html

James, M., & Voigt, M. (2001). Tips from the trenches: Delivering online courses effectively. *Business Education Forum, February*, 56-60.

National Education Association. (2000). *A Survey Of Traditional and Distance Learning Higher Education Members.* NEA, Washington, DC. Retrieved November 22, 2002 from: http://www.nea.org/he

Pastore, M. (2001). Companies, universities moving toward e-learning. *Cyberatlas*.com. Retrieved October 12, 2002 from http://cyberatlas.internet.com/markets/education/article

The SUCCESS Program/ (2004). An unpublished manuscript by Melissa B. Tomei, Duquesne University.

APPENDIX: TABLES

Table 1. Profiles for technology-literate teachers (ISTE, 2005)

	I. Technology operations and concepts	II. Planning & Designing Learning Environments & Experiences	III. Teaching, Learning, & Curriculum	IV Assessment & Evaluation	V. Productivity & Professional Practice	VI. Social, Ethical, Legal, & Human Issues
General program preparation	14	0	8	4	13	4
Student teaching or internship	4	8	8	7	6	9
Point of initial licensure	1	8	9	6	3	2
First year of teaching	4	9	8	8	5	6
	23	25	33	25	27	21

Table 2. Standards for technology literacy (ITEA, 2003)

Standards	Student Assessment Standards	Professional Development Standards	Program Standards
Assessment of student learning will be explicitly matched to the intended purpose.	X		
Assessment of student learning will be systematic and derived from research-based assessment principles.	X		
Assessment of student learning will reflect practical contexts consistent with the nature of technology.	X		
Assessment of student learning will incorporate data collection for accountability, professional development, and program enhancement.	X		
Professional development will provide teachers with educational perspectives on students as learners of technology.		X	
Professional development will prepare teachers to design and evaluate technology curricula and programs.		X	
Professional development will prepare teachers to use instructional strategies that enhance technology teaching, student learning, and student assessment.		X	
Professional development will prepare teachers to design and manage learning environments that promote technological literacy.		X	
Professional development will prepare teachers to be responsible for their own continued growth.		X	
Professional development providers will plan, implement, and evaluate the pre-service and in-service education of teachers.		X	
Technology program implementation will facilitate technological literacy for all students.			X
Technology program evaluation will ensure and facilitate technological literacy for all students.			X
Technology program learning environments will facilitate technological literacy for all students.			X
Technology program management will be provided by designated personnel at the school, school district, and state/provincial/regional levels.			X

Table 3. Standards for SUCCESS (SUCCESS Publication, 2004)

		K	A	E
1. Basic Operations and Concepts	Nature and operation of technology systems including the role of hardware, software, and connectivity in learning and problem solving	X		
	Correct computer terminology	X		
	Identify hardware components; explain their purpose/use	X		
	Basic software applications; identify the appropriate program for task at hand	X		
	Proper use and care of hardware/ software resources	X		
	Basic troubleshooting strategies for hardware/ software	X		
	Role of technology in society and its implication for their future personal and professional lives	X		
2. Productivity Tools	Effective keyboarding skills and techniques	X		
	Word processing applications: create and edit documents, inserting graphics and tables, use of basic desktop publishing techniques		X	
	Database application to collect, organize, and sort information		X	
	Spreadsheet application to create tables or analyze and present statistical data in graphic form		X	
	Drawing or paint application to create graphics, enhance presentation of information or ideas		X	
	Presentation application to design multimedia presentations employing audio, video, and graphics		X	
3. Research/ Information Searching	Basic parts of the Internet and WWW	X		
	Internet terminology in Web-based discussions	X		
	Harvesting information and graphics from sites selected by the teacher for classroom assignments		X	
	Search tools effectively to locate information	X		
	Simple Webpages for classroom projects		X	
	Accuracy, relevance, appropriateness, comprehensiveness, and bias of Websites	X		
	Legal and ethical use of the Internet, explain the consequences of misuse			X
4. Communi- cation Tools	Technology to illustrate thoughts, ideas, and stories		X	
	Compose and send e-mail to collaborate with others and to request and send information for research	X		
	Cooperation and collaboration with peers using technology		X	
5. Additional forms of Technology	Video presentations	X		
	Digital cameras and scanners to enhance projects	X		
	Peripheral devices such as graphing calculators, data collection probes, environmental probes, plotters, projectors, and exploratory environments to support learning and research	X		
	Basic operations for audio, audiovisual and robotic technology	X		
	Computer assisted instruction		X	

Table 4. Initial data results presentation

Program of Study	Nr of Learning Objectives K	Nr of Learning Objectives A	Nr of Learning Objectives R-P-E	% of Learning Objectives K	% of Learning Objectives A	% of Learning Objectives R-P-E
Bachelor (Undergraduate) Programs Reviewed 1308 objectives and 271 courses	636	398	274	49	30	21
Masters (Graduate) Programs Reviewed 5155 objectives and 492 courses	1944	1763	1448	43	33	24
Doctoral (Post-graduate) Programs Reviewed 5462 objectives and 523 courses	1749	1297	2416	32	24	44
Totals Reviewed 11,925 objectives and 1,286 courses	4329	3458	4138	36	29	35
Shaded area represents actual findings from this study						
Boxed area represents hypothesized majority of objectives at each level						

Table 5. Revised data results presentation (Phase 3 Data included)

Program of Study	Nr of Learning Objectives K	Nr of Learning Objectives A	Nr of Learning Objectives R-P-E	% of Learning Objectives K	% of Learning Objectives A	% of Learning Objectives R-P-E
Bachelor (Undergraduate) Programs Reviewed 2008 objectives and 370 courses	964	642	402	48	32	20
Masters (Graduate) Programs Reviewed 6598 objectives and 632 courses	2045	2837	1716	31	43	26
Doctoral (Post-graduate) Programs Reviewed 5738 objectives and 540 courses	1607	1491	2640	28	26	46
Totals Reviewed 14,344 objectives and 1,542 courses	4329	3458	4138	36	29	35
Shaded area represents actual findings from this study						
Boxed area represents hypothesized majority of objectives at each level						

Chapter IV
Applying the ADDIE Model to Online Instruction

Kaye Shelton
Dallas Baptist University, USA

George Saltsman
Abilene Christian University, USA

ABSTRACT

This chapter assembles best ideas and practices from successful online instructors and recent literature. Suggestions include strategies for online class design, syllabus development, and online class facilitation, which provide successful tips for both new and experienced online instructors. This chapter also incorporates additional ideas, tips, and tricks gathered since the paper was originally published in the October 2004 issues of the International Journal of Instructional Technology and Distance Learning *as "Tips and Tricks for Teaching Online: How to Teach Like a Pro!"*

INTRODUCTION

Online education has quickly become a widespread and accepted mode of instruction among higher education institutions throughout the world. Although many faculty who teach traditional courses now embrace some teaching methods popularized by online education such as incorporating online quizzes and discussion boards, some instructors may still feel intimidated when asked to develop a course offered entirely online. Even the best lecturers may find that teaching online leads to feelings of inadequacy and being ill-prepared. While providing training, offering tools for ePedagogy, and sharing success stories are good ways to build faculty confidence, solid instructional course design is still a necessary process for quality online instruction.

The ADDIE model, described by Molenda (2003) as "a colloquial term used to describe a systematic approach to instructional development, virtually synonymous with instructional systems development" (p. 34), is a generic instructional design model that provides an organized process for developing instructional materials. This systemic model is a five-step process that can be used for both traditional and online instruction. The five steps, analysis, design, develop, implement, and evaluate, provide an ideal framework to discuss solid instructional design techniques for online education. In addition to this discussion, this manuscript offers tips and tricks for designing and teaching an online course, gathered from conversations and interviews with online instructors, current literature, conference presentations, and the authors' personal experiences as distance educators.

ANALYSIS

The analysis phase, though one of the most essential in the ADDIE model, is often overlooked. Like any significant project, excitement to get started often overtakes methodical planning, and the eagerness to see the finished results can put relevancy and quality at risk. Undertaking something as involved as developing an online course demands careful analysis. For the purpose of this book, we divided the phase into three segments: analysis of the learners, analysis of the course (including its goals and learning objectives), and analysis of the online delivery medium.

Figure 1. The ADDIE model

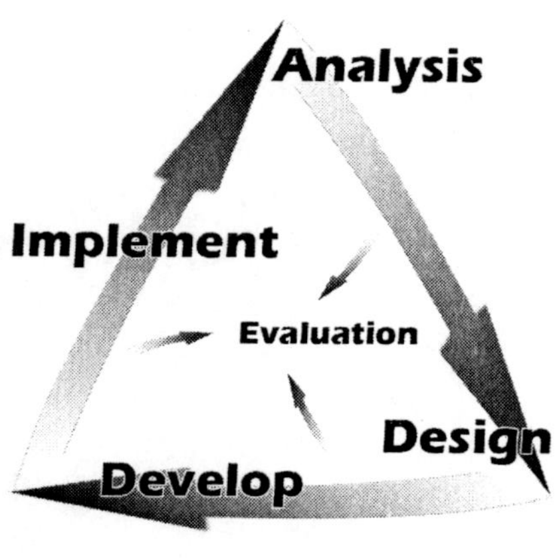

Analysis of the Learners

In this part of the analysis phase, the course designer or design team should perform an audience analysis to provide focus on the learners, their needs, and their learning preferences. In fact, Olgren (1998) reminds us that "if learning is the goal of education, then knowledge about how people learn should be a central ingredient in course design" (p. 77). The course developer should examine ways in which online learners are similar to learners in traditionally offered courses and how they are different as this also leads to an understanding of audience needs within the course. As far as demographics, Gilbert (2001) describes a typical online student as being over 25, employed, a caregiver, and already with some amount of higher education experience (p. 74). However, the demographics are changing at many institutions as more online courses are being offered and traditional full-time students are electing to take online courses as part of their regular course load. Therefore, both andragogical (adult learning theory) and pedagogical methods of course design as well as some mix of experiential, problem-based, and constructivist approaches to learning should be considered.

Students enrolled in online courses often have different expectations than when enrolled in traditional courses. These expectations, described by Lansdell (2001), include increased levels of feedback, increased attention, and additional resources to help them learn (as cited by VanSickle, 2003). In response to meeting these expectations, alternative methods of instruction and class facilitation have evolved to support student cohesiveness and encourage learning. To successfully challenge the online student, increased communication is required between instructor and student (White, 2000). While much of that communication is created in the later phases of the ADDIE model, a careful analysis of the required communication elements will ensure that the intended communication is on target and appropriate for the audience at hand.

Analysis of the Course

In most cases, online courses are not new to the institutional curriculum but existing courses which are being created for a new medium. Therefore, course goals and learning objectives already exist and may not need modification. However, since the course developer will be, in essence, recreating the course from the ground up, the course developer should review the learning objectives for the course and how that relates to other courses and the overall program curriculum. A working knowledge of the goals and objectives is a must as these will be the guiding principles for all content creation.

The course developer should seek answers to the following questions: Why does this course exist? What does it seek to accomplish? Who is the course for? What are the learning objectives? In what ways does this course fulfill degree requirements? The answers to these questions provide the proper perspective in the following instructional design phases as well as provide a working set of priorities to be used in course design and development.

Analysis of the Online Delivery Medium

Online courses, being a relativity new medium of instruction, have yet to achieve a universally understood definition. It is helpful for the course developer and others involved in formulating a working definition of online delivery. For example, for the authors, online delivery is assumed to have the following characteristics:

- The course is held online during a regularly defined class semester or quarter or an established amount of weeks.
- The course is broken up into separate learning modules.
- Student participation is required within a set time period—each content module is presented with a given start and end time.
- Learning takes place as students synthesize the prepared material and interact in class discussions with peers and the instructor(s) within the required time period described above.

Online course delivery offers exciting possibilities, as well as frustrating limitations. Without an analysis of the delivery medium, the online course can result in what Fraser (1998) calls "shovelware"—content that is simply moved from one medium to another without regard for the capabilities of that medium. To fully understand the concept, consider Fraser's (1999) analogy: "When the motion picture was invented, early practitioners saw it primarily as a means of distributing existing material, such as stage performances. It was some time before movies were recognized as a new medium with expressive possibilities, which while overlapping existing media, went far beyond anything previously attainable" (p. B8). Could you imagine Star Wars as a stage performance? Just as film transformed storytelling, online education is reshaping education.

DESIGN

The design phase begins to organize strategies and goals that were formulated in the analysis phase. It also provides details which enhance the course delivery process. Brewer, DeJonge, and Stout (2001) found that course planning and preparation directly influences course effectiveness and really hinder student learning.

Before designing an online course, it is helpful for instructors to view existing courses already offered online. Not only does this familiarize the course developer with the basic components of an online course, it usually inspires ideas that generate excitement about the design process. A Web search can find open examples, but may be limited since most courses are located within password protected courseware management systems. However, there are two open initiatives which can be readily accessed: the MIT's Open Courseware (ocw.mit.edu) and Carnegie Mellon's Open Learning Initiative (www.cmu.edu/oli); both Websites offer many courses in various disciplines that can help instructors with their own course design. A third, The University of Calafornia Berkeley, provides online material in the form of Webcasts and enhanced podcasts (http://Webcast.berkeley.edu/courses).

The design phase is most analogous to that of the creation of a blueprint, a plan for construction that helps guide all involved toward the intended outcome. For online instruction, that blueprint is the course syllabus. The syllabus is the heart of the design phase; careful preparation of the syllabus prepares the learning environment and discourages confusion and miscommunication. For this phase, the major components are examined within the framework of a typical online course syllabus.

Ko and Rossen (2004) relate the syllabus to a course contract and observe that new online instructors do not usually include enough information. McIsaac and Craft (2003) term the syllabus as the roadmap for the course and remind us that students will be frustrated if they try to work ahead only to find out the syllabus has changed within the course. They suggest having a structured syllabus available before the course starts so students can be prepared for course expectations.

Within the syllabus, student expectations should be clearly defined along with well-written directions relating to course activities. These expectations should be stated in the opening orientation material as well as in the course syllabus. Preparation includes clear definitions of the following within the syllabus: contact information, course objectives, attendance requirements, a late work policy, the course schedule, orientation aids, grading scales and rubrics, communication practices, technology policies and overall course design.

Contact Information

The syllabus should include administrative information such as available office times, phone number and e-mail address, and preferred mode(s) of contacts. However, unlike a traditional course, instructors should be very clear about "online office hours," or hours of unavailability. Boettcher and Conrad (1999) suggest an online instructor not be available 24 hours a day to students, but establish a framework for turnaround response. This framework should offer recommendations for how long a student should expect to wait before repeating an e-mail request that has gone unanswered and as Jarmon (1999) suggests, how quickly students should expect a response.

If there is a specific time when the instructor will be online, he or she should include a "fastback" time, or online office hours. A fastback time is a time period when students can expect a quicker than normal e-mail response, usually within the hour or soon after the message is received. Many instructors offer online office hours where they enter the class chatroom and wait for questions. It is often reported by instructors that students under-utilize this time of interaction, choosing to send e-mail as their questions arise, rather than waiting until a prescribed time in the future.

An alternative to using the virtual office hour for questions is to use the chatroom for social conversation. A virtual social experience helps create a closer bond with instructor and classmates, and strengthens the learning community. This is a form of a "cyber sandbox" as described by Palloff and Pratt (1999). The cyber sandbox is defined as a generic discussion area for students to just hang out and talk about movies, jobs or other interests. The creation of a social outlet not only helps keep regular class discussions on topic, but Palloff and Pratt (1999) found that the social connection promotes group cohesion.

Course Objectives

Well-defined course objectives, derived from the analysis phase, are an important element to be published in any course syllabus. However, clearly stated objectives are even more imperative as students do not have the opportunity to participate in "first day of class syllabus discussions" so common in many traditional courses (Jarmon, 1999). The communication of course objectives is also important because in an online course, much of the responsibility for learning is placed upon the student. Failure to properly inform the student of the objectives leaves them feeling confused and puzzled about assignments, and moreover, where the entire course is headed.

Attendance Requirements

Attendance requirements should be clearly stated, as attendance is necessary for successful online learning communities. Palloff and Pratt (2001) advise, "If clear guidelines are not presented, students can become confused and disorganized and the learning process will suffer" (p. 28). The online learning community requires students to take active roles in helping each other learn (Boettcher & Conrad, 1999). Students who do not participate not only cheat themselves, but also those in the learning community.

If instructors expect good participation, then the requirements must be clearly defined. Ko and Rossen (2004) observed that when students were not graded, their participation was less than adequate. In fact, some students may think that if they take an online course, they can take a vacation and still catch up with their coursework upon their return or do a few modules ahead of time before they leave. While online courses do allow for flexibility, students must participate regularly with their instructor and classmates. Students may ask if they can post ahead of the other students or take the course on a self-paced schedule. Because of the prevalence of this question, online instructors should have a policy regarding early posting and state it clearly in the syllabus.

Participation in online courses is inherently different from traditional courses. Students do not automatically understand how to participate in online courses. Course participation requirements should be defined in the syllabus and with each assignment. Where possible, assignments should be grouped into familiar categories such as class discussion, Web searches, quizzes, reading assignments, and so forth. Creating a sample discussion, or model, may increase students' understanding of the participation requirement and how credit is assigned.

Late Work Policy

A policy for late assignment submissions and missed exams should be created. Students who are not actively participating in the learning community are not supporting other students. Because of this interdependence, some instructors have a "no late work accepted policy," while others assign reduced credit. Another option is to create alternative assignments or exams for past due work. To facilitate course management, these alternative assignments could be offered at the end of the course for those who missed assignments during the normal time period.

Course Schedule

One of the most important elements of an online syllabus is the course schedule. The course schedule defines each learning module with beginning dates and due dates, assigned reading, assessment, and other activities. The course schedule becomes the course map for the student and should be included with the course syllabus and placed redundantly throughout the course. In fact, Ko and Rossen (2004) assert that "in an online environment, redundancy is often better than elegant succinctness" (p.76). If the Website or course management system allows linking from the syllabus, then link each course content module to the schedule making it readily available to the student. Students should be encouraged to print out and carefully follow their course schedules. Similarly, Johnson (2003) suggests that instructors should also "keep a schedule of activities for themselves: when to interact with students, when to respond to questions, when to grade assignments, and when to give feedback on performance" (p. 112).

The instructor should allow for flexibility and revisions of the schedule based on the progress and needs of the class but should avoid adding additional assignments not covered in the course syllabus. Careful consideration of course assignments should be given before the course starts to be sure that students meet the required learning objectives.

Orientation Aids

Orientation notes for success in the class should be available for the student (Jarmon, 1999). This may include hints for time management and good study practice. Frequently asked questions (FAQ) support self-help in answering questions (Jarmon, 1999) as it allows students to look for information before e-mailing the instructor. In fact, McCormack and Jones (1998) suggest an FAQ can significantly reduce questions. One doesn't need all the questions or answers up front, as over time as questions arise and answers are provided, a comprehensive FAQ will emerge that can be utilized in future semesters.

Grading Scales/Rubrics

Grading scales and rubrics should be defined for each assignment. If the courseware management system allows, each assignment could be linked to the rubric for clarity. When group assignments are utilized, instructors should use a grading rubric

Table 1. Sample online course schedule

Sample Online Course Schedule				
Session	**Date Begins**	**Content**	**Assignments**	**Due Date**
1	January 15	Chapter 1 of text	Post Introductions to Class Discussion	January 21
2	January 22	Chapters 2-3 of text	Class Discussion	January 28
3	January 29	Chapter 4 of text	Class Discussion	February 4
			Outside Reading Summary	
			Review for Exam	
4	February 5	Exam I over Chapters 1-4	Exam I open 3 days only Feb 11-13	February 13

for the students to grade each other as well as the entire group. This motivates students to participate and provides for equity in group work grading. It is also helpful if the instructor assigns groups or teams the first time as the class should get to know each other before self-selection is allowed.

Communication Practices

An inbox consistently full of e-mail will be overwhelming to most instructors. Therefore, it is important to include in the syllabus, guidelines for class behavior and posting to the discussion boards, e-mail protocols, and assignment submission procedures. Establishing e-mail protocols and communication guidelines will assist the instructor in classroom management. Many instructors require the course session number or identifier in the subject line so that the e-mail related to the course can be filtered to a separate mailbox. If students need immediate attention, the word "Help" should be placed into the subject line so the instructor knows to open that e-mail first, assuring prompt instructor response.

Many instructors create individual e-mail sub-folders for each online student. E-mail that has been answered or graded can be filed away, providing for a record of all course correspondence. Another tip for instructors is to read their mail backwards from newest received to oldest. In many cases, students have solved their problems so that earlier questions become irrelevant. Students may also be asked to use their institutional e-mail address so that instructors are not confused by address changes mid-term or are forced to deal with bounced mail from full inboxes.

Technology Policies

Technology policies should be stated in the syllabus directing students to a helpdesk or resource other than the instructor for technology difficulties. Additionally, instructors should encourage students to create draft postings of assignments in a word processor and save them before posting to the class. This will minimize spelling and grammar mistakes and provide a backup copy for the student in case of technical problems. Students should also be reminded to save all work on a computer hard drive and to a removable device, such as a floppy disk or USB flash drive. Saving work to a USB drive allows the student portability between home, office, and campus systems, and a chance of recovery if systems go down. Finally, students should be instructed to monitor spam filters that may prohibit them from receiving their online course e-mail.

Course Design

The online course design should provide an intuitive navigation path for the student. Students should be able to locate the syllabus, calendar, assignments, and other required activities quickly. Individual content items can be easily identified for the student by adding a consistent icon each time it is used. For example, each time a reading assignment is presented, an icon of a book could be used. Using the same colors and design for similar items will aid the student as well. As a final suggestion, each module of content should have an overview page for organizing the unit of material (Hirumi, 2003).

DEVELOPMENT

Development is a rewarding phase in that the results are concrete and visible. The development stage will include a review of the course objectives, an analysis of the textbook, content module development and content chunking, the creation of content, the development of learning objects, student assessment and additional resources. As a side note, development is also a stage where faculty members may be the most dependant upon outside assistance due to the skilled creation of graphical and multimedia elements commonly found in online courses. In every other stage, even though coaching and mentoring are highly recommended, faculty are usually capable of completing the requirements alone and with skills that are already within their repertory.

Course Objectives

The online course objectives should be clearly identified within the analysis phase and built into the syllabus in the design phase, and now robustly used to guide the course developer during the development stage. Each lesson unit should be designed with the overall course objectives in mind and the objectives should be stated at the beginning of each lesson unit informing the student of the content to be covered. The learning outcomes of the lesson unit should also match the course objectives and appropriate degree objectives, where applicable. Methodologies for assessing these objectives can be altered for the online classroom. If any activities such as the use of online group collaboration or asynchronous class discussion will be needed to meet course objectives, they should be identified in this phase.

Textbook

The textbook is an important asset for an online course. The instructor should examine the text from the perspective of online delivery and understand that in most cases, the text will be a primary source for content delivery. The text should be a strong, stand-alone resource for the course and ideally offer ancillary support for the student such as Website links and review quizzes. In many cases, textbooks will provide additional resources for both faculty and students. Textbooks that offer the instructor assistance in the form of a CD-ROM, test bank, lecture outlines, PowerPoint slides, or Website material give added support in creating an online course. Some textbooks published by Prentice Hall, Irwin-McGraw Hill, and others, offer these licensed resources free-of-charge should the instructor adopt the text. Other textbooks offer course cartridges of content that import directly into courseware management systems like Blackboard or WebCT.

Instructors are sometimes reluctant, when transitioning a course from traditional to online, to adopt a new textbook, but if the result is easier course conversion, they usually concur. The course text book should be chosen early enough in the process for the instructor to become familiar with the contents of the textbook, and, of course, should support the core objectives of the course. Changes in the text may require extensive changes in the supporting course content. Additionally, if the instructor should decide to change textbooks, all of the publishers' licensed or copyrighted material must be removed from the course and replaced with content from the new text or from other sources. It advisable to clearly document each resource with its original source so that it can be easily found should it need to be removed down the line.

Lesson or Module Unit

When designing the course schedule, the course should be broken into lesson units. These are often one-week periods, but can be shorter or longer, depending upon the course. Ideally, a good lesson unit has many parts such as introduction, session objectives, reading assignments, instructional content, handouts, class discussion, written assignments, quizzes and exams, and a unit summary.

The flow of the course should be intuitive, transitioning from week-to-week, or session-to-session without the student feeling lost or isolated in the process. The total number of sessions in the course has a great impact on the course design. Just adding or eliminating as few as two sessions can lead to total course redesign. If the number of course sessions changes often, consider using smaller content chunks (see "Content Chunking") that can be combined into a single unit. Redundancy of key course information is important.

Each learning module should contain a checklist to facilitate student completion. This should be "print ready" so that students can print and read them offline. Course content that presents an easy-to-find and understandable checklist will save numerous e-mails later from students inquiring about due dates and pleading for deadline extensions.

Content Chunking

Content chunking is more of an instructional design process, rather than a theory. It uses modular design in the delivery of online content. Each "chunk" of material is broken into small, understandable lessons or vignettes for the students to absorb. An example of chunking would be to break apart a lecture (that would amount to five written pages for example) covering several topics into smaller pieces (perhaps one or two pages each). The entire lecture, if left un-chunked, would be a tedious Webpage to scroll through, and more importantly, too much information to absorb in one session. Instead, the concept of content chunking would break the lecture into perhaps five or six smaller concepts. When a lecture is broken into topics or ideas and put on separate pages, research shows the student is more likely to understand the content. In the online format, students can navigate through the session exercising personal preferences; for example, to skip the lecture and take the quiz first. It is to their best benefit if the content is organized and easy to move through logically.

Quality course content should be a constant concern for the institution. Course content contributes highly to the success of students and the online education program. Course content can be obtained from several methods such as purchasing from peer institutions or for-profit entities. However, most of the pioneering institutions in online education use internal sources for content creation.

Content Creation

Using rich media such as online graphical models and video can be impressive but is time consuming and expensive. Text-based content is easy to create, but cumbersome for the student to read, especially if it cannot be printed. Often, online students will print out the lectures and highlight or mark the text as they read; therefore, text-based lectures should be designed with this in mind.

Some institutions have created a style guide for the development of online courses. A style guide recommends colors, font styles, icon usage, and the placement of certain institutional information in each of the courses. This consistency throughout the program conveys institutional ownership and endorsement of the courses and the materials in these courses. More importantly, it allows students to find the material they are looking for quickly and without unnecessary inquiries to the course instructor.

Along with the guidelines for a consistent look and feel, the style guide may also suggest the format in which the course material is presented. The institution often recommends an instructional design theory for the creation of course materials and publishes this in the style guide along with examples. When students are presented with a familiar learning unit layout, they are more able to focus on the content and learning objectives, which should increase student learning.

Learning Objects

In regard to online learning objects and interactive learning elements, there are three options: buy, borrow, or build with the latter consuming the most of this section. Should a faculty member elect to buy or borrow an element, module or course, there are many choices now readily available. While they may not be *exactly* what the faculty member had in mind from the analysis and Design stages, textbook publishers and online content brokers offer many choices, although some disciplines may be better represented than others. In the borrow category, learning object repositories such as MERLOT and Wisc-Online provide the course designer with peer-reviewed modules and most are free.

So what exactly are learning objects? According to the IEEE Learning Standards Committee (2001), a learning object is "any entity, digital or non-digital, which can be used, re-used or referenced during technology-supported learning." Many free resources for learning objects are available online, or learning objects can be developed specifically for each course. The following is a list of repositories:

- Apple Learning Interchange: http://ali.apple.com/ali/resources.shtml
- Campus Alberta Repository of Educational Objects: http://www.careo.ca/
- The Connexions Project at Rice University: http://cnx.rice.edu
- Multimedia Educational Resource for Learning and Online Teaching (MERLOT): http://www.merlot.org
- Wisc-Online Learning Object Project: http://www.wisconline.org

For many faculty, the choice to build from scratch is the option they elect to exercise frequently. The creation is often team-based, where one or more instructors partner with one or more instructional designers and/or graphical designers. Team-based approaches help alleviate the need for support by spreading the burden across multiple individuals with multiple talents. Team-based courses can also allow for improved course content and more complete materials due to the broader range of expertise and experiences from multiple individuals.

Assessment

The distance element of online education adds a unique twist to assessment of student learning. The online platform and ubiquity of technology among students affords the course developer a host of electronic tools. Online assessment tools are usually provided with a courseware management system as well as commercial vendors such as:

- **Questionmark:** Questionmark Perception (http://www.questionmark.com)
- **Respondus:** Respondus (http://www.respondus.com/)
- **Software Secure:** Securexam (http://www.softwaresecure.com/)

These vendors support high stakes testing with products that do not allow students to print exams or open additional browser windows; however, there are many ways around these safeguards such as secondary computers, digital cameras, and countless other ways to beat the system.

Obviously, the proctored testing environment provided by having all the students in a single location under the direct supervision of the course instructor is difficult to duplicate online. Some institutions, especially those with local audiences, still require on-campus proctoring of exams or work with institutions within testing consortia to provide such services. While this is an option, it does not really fit within the ideal of a completely online and time-flexible course. Therefore, many course developers have looked for alternative assessments and opportunities to examine student learning with alternatives to traditional testing methodologies. One suggested method is called authentic assessment which is defined as "a form of assessment in which students are asked to perform real-world tasks that demonstrate meaningful application of essential knowledge and skills" (Mueller, 2006). This method works exceptionally well in online environments and should be considered whenever possible. A good resource can be found at jonathan.mueller.faculty.noctrl.edu/toolbox/Index.htm.

Provision of the grading rubrics used for scoring assessments within the course material is also highly recommended. Students should be aware of grading criteria and allowed to self-evaluate wherever possible. One commendable practice is to have student pre-score their work and submit their assessment along with their work at the time of submission. This allows the faculty member to focus discussion on points of disagreement, helping guide students to better critical evaluation and awareness of their own work.

Additional Resources

In the connected world of the Internet, outside resources are easily built-in to the course. Linking to Websites and online resources is obvious. Other resources, such as library resources, online reserve materials, and institutional support resources such as a writing center and tutoring centers provide students with just-in-time resources and referrals. Even referrals and links to technical support and helpdesk resources may be provided to the students where course developers anticipate certain technical tasks might prove challenging to students.

IMPLEMENTATION

The next phase, implementation, includes opening the course and initiating instruction. An enthusiastic and engaging opening week of class is a great way to start the course. This time period is fragile; disruptions or unnecessary interferences may set a tone that stifles learning for the remainder of the course. It is important to create an initial impression that will stimulate the development of the learning community and nurture the students to maturity. Hirumi (2003) suggests the following goals for students in the first week of the course:

- Have a good understanding of course requirements and expectations,
- Can locate and interpret relevant policies and procedures,
- Are confident in their ability to use various tools and course features, and
- Can identify challenges associated with and discuss strategies for facilitating virtual teamwork" (p. 87).

The course should begin with a welcoming e-mail and announcement, instructions for class-wide introductions, emphasis on the syllabus, a tone of excellence established, and nurturing the learning community.

Welcome E-Mail and Announcement

Moore, Winograd and Lange (2001) offer several tips for the first session of class: send a welcome e-mail that invites the students to join the class, telephone students who do not appear the first week, and duplicate your welcome e-mail in a class announcement if the course management system allows. The announcement should also encourage students to regularly check their e-mail. The first week should have fewer assignments to allow students to post introductions and get to know each other. Technical issues should be resolved immediately.

Introductions

The instructor should spend time getting to know the students individually during the first week of class and encourage students to do the same. An introductory discussion inviting the participants to share something in particular with the group is a successful strategy for building learning community. The instructor should participate heavily in this discussion (being careful not to dominate) and should respond to one or two comments in each student's introductory posting. Ko and Rossen (2004) suggest the "initial postings in the discussion forum, your first messages sent to all by e-mail or listserv, or the greeting you post on your course home page will do much to set the tone and expectations for your course. These 'first words' can also provide models of online communication for your students" (p. 189). To assist with personal connection, the instructor should print out the introductory discussion and keep it near the computer for reference. When responding to a student's question, the instructor should occasionally refer to the discussion and reference a personal note to the student such as "How is your son who plays college baseball doing this season?" The discussion following the question, leads the student to feel as though he or she is talking one-on-one with the instructor.

Offering an icebreaker in the first session, such as "share your silliest moment in college" or "name the animal you most identify with," helps to alleviate nervousness and provides insights to the fellow students' personalities. Several good icebreakers that also provide an instructor with student information include the VARK learning styles (http://www.vark-learn.com/english/index.asp) and the Keirsey temperament sorter (http://www.keirsey.com). The Kingdomality profiler (http://www.kingdomality.com) provides not only a medieval vocational assessment, but is fun and generates discussion. Each of these Websites offers instant feedback, and the students can post their results and a short paragraph whether they agree or disagree. Many other Websites allow students to discover their commonalities and similarities and can be found with a simple Internet search.

Emphasize the Syllabus

A great tip for the first class session is to create a syllabus quiz or scavenger hunt that "teaches students how to navigate your course" (Schweizer, 1999, p. 11). Next, offering bonus points to assess syllabus comprehension is a successful way of engaging the student in the first class session. Encouraging students to review the syllabus more thoroughly can alleviate confusion later in the course as they familiarize themselves with the course requirements. For example, an art appreciation course requires outside visits to an art museum. This requirement is clearly listed in the syllabus; however, students sometimes want to visit a Web museum instead. This is type of question should be clarified with a syllabus quiz to alleviate any disappointment, confusion, or scheduling conflicts.

Establish a Tone of Excellence

The first several weeks also set the tone for academic participation. Instructors should grade discussions/assignments stringently in the first few assignment cycles. Establish a tone of excellence early and encourage students to do their best. "Students want to receive timely and personal feedback" (Boettcher & Conrad, 1999, p. 97) early in the course because they may not be able to assess their progress as easily online (Boaz, 1999).

Nurturing the Learning Community.

As the course progresses, the learning community will still require nurturing from the instructor. A learning community becomes self-sufficient when an instructor provides ample communication, facilitates the discussion board, treats each student as an individual, adds emotion and belonging, responds quickly to questions, models required behavior, creates appropriately sized groups, and clearly outlines expectations for group activities.

Provide Ample Communication

Online students are eager for communication as it is "the foundation of successful distance learning courses" (Johnson, 2003, p. 113). In fact, Johnson (2003) also suggests that communication throughout the course "must be ongoing, regular, continuous, and easy" (Johnson, 2003, p. 113). Lack of instructor-student communication early on will create a negative learning community, thus debilitating the learning process. Instructors should use class-wide announcements, group e-mails, and chat archives to facilitate accessible, public communication in the online course. As the course continues, students should be encouraged to facilitate the discussion and assume some of the roles previously controlled by the instructor.

Communication must be both reflective and proactive. Many courses use class-wide journals or summaries to bring closure to modules. Sending out class-wide summation,introduction, and transitional e-mails at the end of each module, wrapping up the previous content, and introducing the next module provides for a sense of transition. Reminding the students of requirements for the current module, such as projects or exam dates, is helpful to the students and it only takes about 10 minutes a week for either of these tasks. Proactive communication yields fewer questions, saving dozens of hours answering the questions individually. Johnson (2003) recommends that students should be taught to communicate early any questions or confusion they may have due to the lack of body language available in the online environment. Instructors cannot see looks of confusion or frustration.

Instructors should keep their interaction with the class as accessible as possible. Using the "Course Announcement" area frequently for reminders and duplicating important information in e-mails will increase open communication and provide the entire class access to the information. It is also important to communicate to students each time grades are posted. This creates a "don't call me, I'll call you" communication pattern for grade information and alleviates individual e-mails from students requesting grades on their assignments. Students will quickly realize the instructor will post a notification when grades are posted, so requests are unnecessary. Within that communication, students should be reminded to contact the instructor if they notice a missing grade. This places the responsibility back with the student for finding and submitting any missing work.

Facilitate the Discussion Board

Bischoff (2000) reminds us that "the key to online education's effectiveness lies in large part with the facilitator" (p. 58). Likewise, for class discussion to be successful, the instructor should become a facilitator and review discussions without controlling them. Many online instructors have found that too much activity can be as harmful as none at all. This particular role of the facilitator in the online classroom can be difficult for a traditional instructor. A traditional instructor may be accustomed to dominating or controlling the discussion through lecture; however, in an online class, all students have equal opportunity to participate in the discussion and may outside of the instructor's influence. It takes a good deal of time for some instructors to feel truly comfortable in allowing the discussion to take place without their intervention; experience will eventually guide them.

For good discussion board facilitation, the instructor should randomly reply to students and provide prompt explanations or further comments regarding the topic of discussion. Johnson (2003) found that "when a professor shows interest in discussions by commenting on students' ideas and insights, students feel valued and encouraged to participate more" (p. 113). The instructor should provide feedback in the discussion even if it is merely a "cheerleading" comment, redirection, or guideline submission. The instructor should intervene when the discussion seems to be struggling or headed the wrong way (Palloff & Pratt, 2001), but should not over-participate in the discussion, as this will be considered stifling and restrictive. Some instructors prompt absentee or "lurker" students with a gentle reminder e-mail or telephone call. According to Bischoff, (2000), "A phone call may prove more timely and effective" (p. 70) in helping a student engage in the discussion.

Many instructors assign assistant facilitators and summarizers for each discussion session, providing opportunities for different kinds of student involvement. Other instructors use "coaching teams" made up of students or tutors as the first line of support, then invite the students to ask the instructor for clarification or further assistance. Under favorable circumstances, the "discussion will end in acceptance of different opinions, respect for well-supported beliefs, and improved problem-solving skills" (Brewer et al., 2001, p. 109). McIsaac and Craft (2003) remind us that class discussions should take place after the reading assignments; students may also need to be reminded of this before they participate.

Treat Each Student as an Individual

Instructors should value individual contributions and "treat their students as unique" (White, 2000, p. 11). A simple technique is to use the students' preferred names or nicknames in all correspondence. It is also important to add positive emotion and visual cues. The online environment can be limiting when the communication is mostly text-based. Typing the cues in an e-mail can serve the same purpose as nodding a head in agreement or offering a welcoming smile as would occur in a traditional course.

Add Emotion and Belonging

When online learning is facilitated incorrectly, students can feel isolated and cheated. This could lead to feelings of separation and disappointment that negatively impact learning. White (2000) advises that "a positive emotional climate can serve as a frame of reference for online students activities and will therefore shape individual expectancies, attitudes, feelings, and behaviors throughout a program" (p. 7). Since there are no visual clues in the online classroom, one suggestion for communication is to type out the emotion expressed in parentheses (*smile*) or to include emoticons, such as :-) for happiness or :-0 for surprise or dismay. It is also possible to describe body language in e-mail. Salmon (2002) offers this example: "When I read your message, I jumped for joy" (p. 150). This descriptive effort shows the students the instructor's personality and positively stimulates the online community. It is also beneficial, as Hiss (2000) suggests, for online instructors to remember to keep their sense of humor.

Respond Quickly

Time delays in a threaded discussion can be frustrating for students. This is especially true if a response was misunderstood and students have attempted to clarify. Instructors should try to post daily or on a regular schedule that has been communicated to the students. Some instructors create homework discussion threads for content support, which provides a forum for students to help each other.

Model Behavior

Instructors who engage students in collaborative groups should facilitate development of social skills. This begins at the onset of the course when the learning community is formed and students recognize the online classroom as a safe place to interact. Group skills should be modeled by the instructor and outlined in the course syllabus. For example, if a two paragraph introduction is expected, the instructor should model that in their own introduction to the class.

Create Appropriately Sized Groups

Most students enjoy the online social interaction and find that it encourages their learning experience. Independently minded students discover that the asynchronous nature of the course enables them to participate more readily than in the face-to-face classroom. In creating groups, Ko and Rossen (2004) recommend that instructors divide students into groups instead of allowing students to pick their own. Students may find it difficult to meet online and form groups quickly. Many instructors search the introductory material to find common elements among students to hasten group cohesion.

Groups should not be too large or too small. The most effective group size appears to be four students per group. Utilizing these suggestions, groupwork should begin early to promote a positive learning experience in the classroom. The actual process for completing the project should be outlined by the instructor, but the final outcome should be the group's responsibility.

EVALUATION

The final stage of online instruction is for evaluation and assessment. Evaluation is a rewarding experience where one can observe learning occurring in the minds of students and reminds many instructors why they choose this as their career. Evaluation is a time of reflection and satisfaction for a job well done. At this stage, instructors should assess each student's performance against course objectives, including what worked well and what should be improved. This is often accomplished by evaluating the course with a "best practices" online course rubric, keeping a journal and by soliciting feedback on instruction and course content.

Online Course Rubrics

With faculty teaching online for over a decade, online course rubrics have been developed to help evaluate quality in online courses. These rubrics examine best practices for design, requirements for interaction, and attempt to measure the overall quality of the course. Currently, there are several excellent rubrics but we can thoroughly recommend the following: California State University Chico's Rubric for Online Instruction (www.csuchico.edu/celt/roi/index.html), Blackboard's Exemplary Course Rubric (www.connections.blackboard.com/), and Quality Matters (www.qualitymatters.org).

Keep a Journal

Self-examination with contemplative thought is a successful approach for course improvement. A recommended practice is to keep a journal that records items that should be redesigned or altered the next time the course is taught. The instructor should make notes of assignments that worked well and those that were difficult, and critically evaluate the effectiveness of content and instruction.

Solicit Student Feedback on Instruction

Student feedback improves instruction. A good place to gather the feedback is inside the course management system. It is helpful to survey for student feedback during the course, not just at the end with course evaluations. The instructor can develop a discussion thread for students to post feedback about the course anonymously, including possible suggestions for improvement. If a student does offer feedback, the instructor should acknowledge the feedback and be appreciative for the remarks.

Feedback instruments should provide the students with a way to communicate what they like the best or least about the course instruction. Schwartz and White (2000) suggest a mid-course feedback process by enlisting a student volunteer to send an e-mail message to the class soliciting feedback. They also suggest the following questions be used, encouraging honesty and participation:

- List three areas that are working well in this course
- List three ways to improve the class. (p. 175)

The student volunteer would gather the messages, remove names, and send them to the instructor. If possible, course changes in response to students' comments will allow students to feel empowered through taking an active role in their education. The feedback should also be used to change subsequent courses taught.

Solicit Student Feedback on Course Content

All online instructors should look for possible course revisions. Course content should never remain static. Moore et al. (2001) propose that "because online course design and teaching are so new, evaluating the effectiveness of your course and then refining it based on the results of that evaluation become imperative" (p. 12.3). If using end-of-course summary feedback, the instructor must receive this feedback in time to reevaluate the course for the next semester and modify, if necessary. Another possibility is an end-of-session discussion regarding the focus of the next session, thus allowing for minor course revisions even as the course continues to be taught.

CONCLUSION

Online teaching has brought a new modality to education. It has also brought frustration and anxiety to instructors attempting this new method of instructing students. Moore et al. (2001) shared that "one faculty member who had only just finished her course online said it was like diving into a great chasm, blindfolded" (p. 11.3). Instructors who are comfortable with the traditional methods for teaching in the classroom may still struggle to engage students over the Internet. While many of the same techniques apply, teaching online requires additional techniques for success. The ADDIE instructional model provides a basic path for developing and teaching an online course: analyze the course objectives and audience; design and develop the materials and activities; implement the course materials and encourage learning, and finally, evaluate the effectiveness. In the online classroom, the environment is prepared with a carefully designed syllabus and policies and the learning community is nurtured to grow and become self-sufficient. By utilizing these strategies for teaching online effectively, an instructor will engage the online learner, nurture a successful learning community, and alleviate the frustration and fear that goes along with teaching online.

REFERENCES

Bischoff, B. (2000). The elements of effective online teaching. In K. W. White & B. H. Weight (Eds.), *The online teaching guide* (pp. 57-72). Needham Heights, MA: Allyn and Bacon.

Boaz, M. (1999). Effective methods of communication and student collaboration. In *Teaching at a distance: A handbook for instructors* (pp. 41-48). Mission Viejo, CA: League for Innovation in the Community College.

Boettcher, J. V., & Conrad, R. M. (1999). *Faculty guide for moving teaching and learning to the Web.* Mission Viejo, CA: League for Innovation in the Community College.

Brewer, E., DeJonge, J., & Stout, V. (2001). *Moving to online: Making the transition from traditional instruction and communication strategies.* Thousand Oaks, CA: Corwin Press, Inc.

Fraser, A.B. (1999). Colleges should tap the pedagogical potential of the world-wide web. *Chronicle of Higher Education, 45*(48), B8.

Gilbert, S. D. (2001). *How to be a successful online student.* New York: McGraw-Hill.

Hirumi, A. (2003). Get a life: Six tactics for optimizing time spent online. In M. Corry & C. Tu (Eds.), *Distance education: What works well* (pp. 73-101). New York: The Haworth Press.

Hiss, A. (2000). Talking the talk: Humor and other forms of online communication. In K. W. White & B. H. Weight (Eds.), *The online teaching guide* (pp. 24-36). Needham Heights, MA: Allyn and Bacon.

IEEE Learning Technology Standards Committee. (2002). The learning object metadata standard (WG12). *IEEE Learning Technology Standards Committee (ILTSC),* Piscataway, NJ: IEEE. Retrieved December 29, 2006, from ieeeltsc.org/wg12LOM/lomDescription

Jarmon, C. (1999). Strategies for developing effective distance learning experience. In *Teaching at a distance: A handbook for instructors* (pp. 1-14). Mission Viejo, CA: League for Innovation in the Community College.

Johnson, J. L. (2003). *Distance education: The complete guide to design, delivery, and improvement.* New York: Teachers College Press.

Ko, S., & Rossen, S. (2004). *Teaching online: A practical guide* (2nd ed.). Boston: Houghton Mifflin.

McCormack, C., & Jones, D. (1998). *Building a Web-based education system.* New York: John Wiley & Sons, Inc.

McIsaac, M., & Craft, E. (2003). Faculty development: Using distance education effectively in the classroom. In M. Corry & C. Tu (Eds.), *Distance education: What works well* (pp. 73-101). New York: The Haworth Press.

Molenda, M. (2003). In search of the elusive ADDIE model. *Performance Improvement, 42*(5), 34-36.

Moore, G., Winograd, K., & Lange, D. (2001). *You can teach online.* New York: McGraw-Hill Higher Education.

Mueller, J. (2006). Authentic assessment toolbox. *Authentic Assessment.* Retrieved December 30, 2006, from http:/jonathan.mueller.faculty.noctrl.edu/toolbox/index.htm

Olgren, C. H. (1998). Improving learning outcomes: The effects of learning strategies and motivation. In C. C. Gibson (Ed.), *Distance learners in higher education* (pp. 77-96). Madison, WI: Atwood.

Palloff, R. M., & Pratt, K. (1999). *Building learning communities in cyberspace: Effective strategies for the classroom.* San Francisco: Jossey-Bass.

Palloff, R. M., & Pratt, K. (2001). *Lessons from the cyberspace classroom: The realities of online teaching.* San Francisco: Jossey-Bass.

Salmon, G. (2002). *Developing e-tivities: The key to active online learning.* London: Kogan Page, Ltd.

Schwartz, F., & White, K. (2000). Making sense of it all: Giving and getting online course feedback. In K. W. White & B. H. Weight (Eds.), *The online teaching guide* (pp. 167-182). Needham Heights, MA: Allyn and Bacon.

Schweizer, H. (1999). *Designing and teaching an on-line course: Spinning your Web classroom.* Upper Saddle River, NJ: Prentice Hall.

VanSickle, J. (2003). *Making the transition to teaching online: Strategies and methods for the first-time, online instructor.* Morehead, KY: Morehead State University. (ERIC Document Reproduction Service No. ED479882)

White, K. (2000). Face to face in the online classroom. In K. W. White & B. H. Weight (Eds.), *The online teaching guide* (pp. 1-12). Needham Heights, MA: Allyn and Bacon.

Chapter V
TRAKS Model:
A Strategic Framework for IT Training in Hierarchical Organizations[1]

Shirish C. Srivastava
National University of Singapore, Singapore

Thompson S. H. Teo
National University of Singapore, Singapore

ABSTRACT

Introduction of new information technology (IT) in organizations is a necessary but not a sufficient condition for organizational success. The effective adoption and use of IT by organizations is dependent to a large measure on the strategic planning for using the technology, including long-term planning for training the organizational members. Despite the strategic nature of technology training in organizations, most existing studies on technology training address only the operational issues, for example, training needs assessment, learning, delivery methods, and so forth. The strategic concerns of IT training for enhancing business productivity are largely not addressed by the current literature. To address this gap, we explore the strategic role of IT training in hierarchical organizations. We synthesize various ideas in the literature on change management, training needs analysis and IT adoption to evolve a 'strategic IT training framework' for hierarchical organizations, namely the TRAKS model. The proposed framework recognizes the differences in IT training requirements for different levels of employees. Further, the model suggests tracking training requirements based on attitudes, knowledge, and skills for different segments of employees and planning training accordingly. The study provides an actionable and comprehensive tool, which can be used for systematically planning IT training for enhancing productivity of organizations.

INTRODUCTION

New information technology (IT) introduction in an organization is a necessary but not sufficient condition for its effective adoption, usage, and implementation. For the new technology to make an impact on the organizational performance, it has to be used effectively in a planned manner (Devaraj & Kohli, 2003). The capability to effectively absorb and use the new technology depends to a large measure on the learnability and absorptive capacity of the organization (Cohen & Levinthal, 1990). For developing such an organizational capability, it is imperative to realize the importance of training and treat it as a strategic objective for achieving long term organizational success (Gallivan et al., 2005; Kang & Santhanam, 2003; Swartz, 2006). Despite the need for conceptualizing technology training as a strategic concern, most existing studies on technology training address the operational issues of training process, for example, training needs assessment (Nelson et al., 1995), learning styles (Bostrom et al., 1990), and delivery methods (Compeau & Higgins, 1995; Sein & Bostrom, 1989). The strategic issues related to IT training (e.g., what kind of training is required for employees? Should the training given to all employees be similar in content and delivery?) remain relatively unexplored in past research. In this chapter, we explore these strategic concerns of IT training for hierarchical organizations[2]. We reiterate the strategic objectives of IT training which are usually lost sight of in the mundane and routine training activities in organizations.

Need for Systematic IT Training

In most organizations, IT training is a matter of chance rather than a planned initiative. In contrast to this practice, the definition of training refers to a planned effort by a company to facilitate the learning of specific knowledge, skills or behaviors that employees need to be successful in their current job (Goldstein, 1992). The pressure for better training is expanding due to the increasingly popular view that people, rather than technology, represent the primary source of enduring competitive advantage (Ford, 1997). Although the need for training is being realized by many organizations, in many cases, the training for "new technology" is not in tandem with organizational requirements. Some employees do receive IT training but it is mostly a result of the personal initiative of that particular employee, in the field of his or her interest. This field may or may not be of direct consequence to his or her job. In some cases, it is the mere persuasion of the "training provider" which initiates the training nominations from these firms. Consequently, the content and context of IT training is often decided by the "training provider" and not by the firm. This results in incongruence between training outcomes and organizational requirements. Effective training has to be in consonance with existing organizational structures and practices. There is a need to consider the interface between the organizational system and training for the outcome to be fruitful and effective (Goldstein, 1992; London, 1989; Vonk, 2006).

In many cases, IT training is thought of as a "necessary evil" and not as a strategic tool for enhancing productivity. For example, Indian Railways, which is one of the biggest employers in the world with over 1.6 million employees, does not have a systematic IT training program for its employees, though it is one of the biggest users of IT resources. Employees are imparted with IT training on the basis of their emergent 'skill needs,' rather than as a part of a well thought strategic plan. Some firms are proactive in realizing the importance of IT training, but are still not able to plan their training modules systematically for want of "critical knowledge" about the "who and what" of IT training, that is, which employees should be trained in what aspects for better leveraging IT resources. An example where the firm's success can be attributed to its well thought out and planned IT training is the Housing Development Board (HDB) in Singapore. HDB realized the importance of systematic IT training for its employees and was able to leverage training for its success. One of the major contributing factors was the top management's proactive attitude towards IT adoption and training (Teo, 1999; Teo & Ranganathan, 2003).

There is no doubt about the fact that everyone in an enterprise does not require the same kind of training in IT for effective adoption and performance, especially in the context of hierarchical organizations, which have a well defined chain of command and the position of employees in the organizational hierarchy determines their responsibilities (Srivastava & Teo, 2004). The proposed framework seeks to identify the training requirements for different segments of employees so that customized IT training programs can be designed to facilitate speedy and fruitful IT adoption by these enterprises. Effective training requires a systematic approach to training needs assessment which determines not only who to train but also what to train (McGhee & Thayer 1961). McGhee and Thayer (1961) also cite a lack of theoretical models for providing systematic training. Surprisingly, this gap in IT training literature has still not been addressed in a systematic and convincing way. This study seeks to present a comprehensive, conceptual, actionable strategic IT training framework for business enterprises, which will help in efficient and effective IT proliferation and usage.

TRAKS MODEL: STRATEGIC IT TRAINING FRAMEWORK

Training has long been recognized as a necessity for effective adoption and usage of IT by organizations (Davis & Bostrom, 1983; Gallivan et al., 2005; Lucas, 1975). Most current research on IT training discusses the modes and content of training from an operational perspective (e.g., Gist et al., 1987; Keil, 1998; Olfman & Mandiwalla, 1994). Though studies analyzing the impact of IT training have found mixed results, many of them have shown that training *does* influence users' skills and acceptance of IT (Davis & Davis, 1990; Gallivan et al., 2005; Nelson & Cheney, 1987).

Goulding and Alshawi (2004) highlighted the importance of treating IT training as an important organizational strategy for gaining a competitive advantage. In their paper, they introduced the generic processes involved in developing an IT training framework in order to support and deliver the business strategy. Noe and Ford (1992) also stated the need for training practice to be used as a part of the strategic planning process of the firm. In contrast to this philosophy, most firms view IT training as an operational or a functional necessity rather than as a strategic tool to gain competitive advantage. In line with the changing market conditions, the training systems in organizations also have to continuously evolve. Using training as a strategic tool is valid not only for IT, but for other functions as well. Relative to other functions, the scenario of IT training presents a yet more challenging endeavor because it calls for a complete transformation of many of the existing organizational systems. For example, the proliferation of enterprise resource planning (ERP) and customer relationship management (CRM) techniques are often based on the concept of business process re-engineering which require a major revamping of the existing systems. The rate of evolution for all new technology tools and methods, including IT has to be definitely at a much faster pace. Tannenbaum and Yukl (1992) have stressed on the need for training to be viewed as a system embedded in the organizational context. Training should be conceptualized as integral to the strategic goals of the organization (Schuler & Walker, 1990) and a component of the human resource planning process (Jackson & Schuler, 1990).

The orientation of training has typically been micro in its orientation, with a focus on individual learning development and change. This is true despite the fact that at the conceptual level, training needs assessment (McGhee & Thayer, 1961), evaluation (Kirkpatrick, 1967) and instructional design models (Goldstein, 1992) state that training should be aligned with the organizational goals. Lepore et al. (1989) in their qualitative study found training to be one of the factors shaping users' acceptance and usage of IT. Further, they also established that the importance of training in creating an impact was moderated by different factors such as users' occupational status (i.e., professional versus clerical employees) and specifics of the implementation strategy (i.e., top-down versus bottom-up implementation). In a similar vein, Harrison and Rainer (1992) and Compeau and Higgins (1995) also established the importance of IT training, but found that in addition to training, many individual attributes play an important role in determining IT acceptance and usage. Thus, planning IT training as per the organizational strategy, taking into account the individual requirements, will provide maximum benefits to the organization. This brings forth the importance of segregating employees for IT training based on their level in organization. Hence, in the context of IT training, for maximum effectiveness, it is not only important to focus on "what is to be learned" but also "by whom" (Campbell, 1988; Lepore et al., 1989).

McGhee and Thayer (1961) and Goldstein (1992) argue that a thorough need analysis for training should include: (a) organizational analysis; (b) task analysis; and (c) person analysis. In this study, at the organizational level, we are concentrating on organizations that are hierarchical in structure. At the task analysis level, we are considering the job requirements of various levels of management and at the person analysis level, we are generalizing the personnel at different levels. Ostroff and Ford (1989) applied a multilevel perspective to needs analysis, and noted that the previous three facets may reside at different or even multiple levels of analysis. The training program of the organization needs to be linked to the organizational business strategy (Brown & Read, 1984), the changes in the strategic plan should be reflected in the revised training objectives (Hussey, 1985) and the needs assessment must incorporate a future orientation (Scheinder & Konz, 1989).

Levels in an Organization

All personnel in an organization can be classified into three distinct levels based on the kind of work that level performs. Anthony (1965) made the distinction between the three levels of management based on their decision-making functions (strategic, tactical and operational decisions). The three levels into which all the employees of an organization can be classified are top, middle and frontline. The top level includes the CEO and different unit heads. They are the people who are responsible for spelling out the roadmap of the company. Their decisions have long-term implications not only for the company but also all its employees. The role of this level in smaller organizations like small to medium enterprises (SMEs) is even more important because here they are not only aware of the key strategic problems of the company, but the smaller size of the company brings them closer to the actual workplace; hence they are able to monitor the effects of their decisions also. The middle level includes the functional managers. They are largely responsible for the smooth functioning of the areas under them within the broad framework of policies and guidelines spelt out by the top management. They are required to plan and source the various resources for production and marketing. This group of personnel requires having a thorough knowledge of working procedures for the industry. The frontline personnel include all the employees excluded from the upper two categories. They include the supervisors, inspectors and workers. They are the employees who are actually involved in the day-to-day business operations and are required to have well developed skills in handling the various devices and systems, which they operate.

Since different levels of employees have different kinds of functions to perform, it implies that these three levels have different "informational needs" in relation to their function. Hence their training needs are also quite different from each other (Daft, Lengel, & Trevino, 1987; Srivastava & Teo, 2004; Swartz, 2006). Further, the different levels require different kinds of attitudes, knowledge, and skills (AKS). The different types of knowledge acquisition require different types of training methodologies. Anderson (1982) made a distinction between declarative knowledge, which is fact knowledge (knowing what), and procedural knowledge, which is knowledge of procedures (knowing how). The frontline level may require more of the procedural knowledge whereas as we go higher, the personnel may require more of declarative knowledge related to IT.

Figure 1 presents a strategic IT training framework, namely the TRAKS[3] model, for organizations which takes into consideration the hierarchical nature of many organizations. The framework recognizes that different levels of employees in an organization have different functionalities; hence, the IT training requirements for the different levels are quite divergent in terms of content. The three broad contents of IT *training requirements* are *attitude* towards IT, *knowledge* of IT, and on job IT operational *skill*. The model proposes to 'track' the functional requirements for different levels and impart training accordingly. The change in the breadth of the triangle and quadrilateral in Figure 1 indicates the change in requirement of the training content for different levels of hierarchy.

The proposed framework seeks to offer answers to questions regarding training component for different levels of the organization and serves as a practical tool for hierarchical organizations in planning their IT training initiatives. The profound problem with IT training has been that in most of the cases, the training is not directed to the informational needs of that level and often there is a mismatch. This mismatch of the IT training content with the informational needs of the employee results in a twofold wastage. First, the money spent on training that employee is wasted since it will not help him in his job. Second, the time spent on the training is also a wasted resource.

Figure 1. TRAKS model: Strategic IT training framework

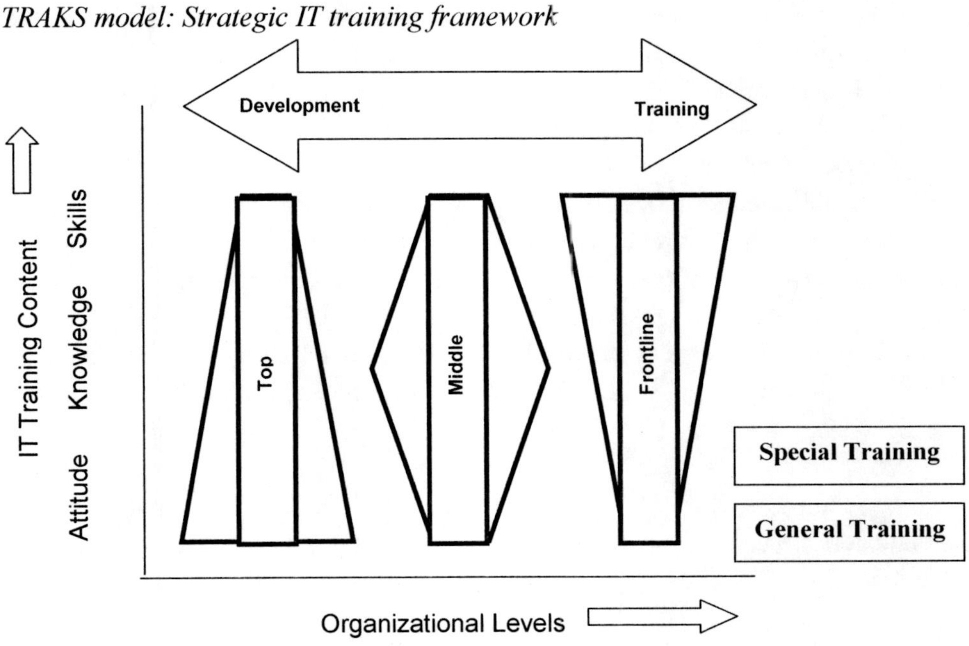

Table 1. Summary of training content

Training content	Fundamental question	Explanation with example
Attitudes	*Why?* The answers to such questions determine the "strategic direction" of the firm and are usually provided by the top management	It seeks to explain the importance of IT and why it should be adopted by the organization, the kind of benefits (long term as well as short term) that can be derived from the use of IT. The emphasis is more on molding the views towards leveraging IT to improve business productivity and competitiveness. In the case of ERP, such training will inform the participants about the significant benefits that IT is capable of giving to the firms. It seeks to develop the enthusiasm and remove inhibitions by informing about the "real business value" of IT. The trainees are also taught about the different technologies available as well as their potential impact, so that they can better decide on the choice of technology for the company.
Knowledge	*What?* These decisions determine the "tactical course" of action of the firm and are mostly in the domain of the middle management of the firm	This seeks to inform about the details for a particular technology. It aims at empowering the trainees with the requisite background to distinguish and decide which among the options available for a particular technology may be beneficial and suited for their business. Going further with the ERP example, the knowledge component of the training provides the ability to decide among various choices of ERP systems available to suit their needs.
Skills	*How?* The frontline workers require this expertise to "operate" the various systems in an enterprise	This aspect of training provides the necessary "ground tools" to the workers to actually work on the chosen systems. It provides the workers with the necessary expertise to operate the specific software and hardware chosen by the company. An example of skills may include the techniques for operating the different modules in SAP ERP system. This "skills training" logically comes after the two vital preceding decisions have been taken (1) to use ERP system in the company and (2) among available ERP systems to use SAP

The proposed framework (Figure 1) explores the IT training for different levels of organizational personnel with regard to the training content. Training content expounds the broad parameters (in terms of attitudes, knowledge, and skills) on which the "planners" should organize the training for its different levels of employees. A summary of the training content is illustrated with illustrative examples in Table 1 in the context of ERP implementation.

This differential hierarchical IT training of employees has been successfully implemented by the Housing and Development Board (HDB), Singapore which encompasses formal and structured IT training programs for different levels of staff from junior officers to the CEO (Teo, 1999). The IT training programs are designed as per the job and informational requirements of the level of personnel.

Top Level

The top level managers are usually the perpetrators of "underlying currents" and "culture" in an organization. In most traditional organizations, the top managers are often viewed as "trendsetters" whom all employees in the organization try to emulate. Hence it is very important for top management to have positive and favorable attitude towards IT and new technology adoption. This has implications on the training content for these top level managers. This group of people requires more of attitudinal training towards IT (Table 2). They should be able to realize the importance of IT and the impact that it can have in transforming their enterprise. They require relatively little IT-specific knowledge or skills. These leaders should be trained in a way so that they understand the potential benefits of IT adoption as well as the potential costs of not adopting IT. Such understanding by top management would enable them to be better able to enthuse and motivate their employees for IT adoption. The tapering pyramid in the proposed framework expounds training mostly in understanding "whys" (i.e., attitudes). The requirement of training about "what" (i.e., knowledge) and "how" (i.e., skills) is comparatively lesser.

Table 2. Training requirements for different levels of employees

Requirement/ Level	Top	Middle	Frontline
Attitudes	• Positive belief towards IT relevance • Enthusiasm for IT proliferation • Creative, innovative and risk taking attitude • Ability to enthuse and motivate others for IT adoption	• Positive, proactive and enthusiastic towards IT adoption • Attitude to learn and teach new things for better efficiency and productivity	• Positive and enthusiastic towards new learning and IT adoption
Knowledge	• The latest developments and trends in IT • In depth business knowledge and emerging IT standards for their industry	• In depth knowledge about the 'capabilities' of the available hardware and software systems • Knowledge about the 'implementation impediments' for different IT systems • Latest developments and trends about IT usage in similar industries	• Generic knowledge about the capabilities of IT for their industry and specific knowledge about the IT systems which they have to work on
Skills	• General office and communication software e.g., e-mail, word processing, spreadsheet etc.	• General office and communication software e.g., e-mail, word processing, spreadsheet etc. • Understanding of the operational requirements for the software and hardware systems being used by their department • Specific skills for the critical IT systems in their department	• General office and communication software e.g., e-mail, word processing, spreadsheet etc. • Understanding of the software and hardware systems being used by their department • Specific specialist operational skills for the software and hardware systems being used by them • General skills and ability to handle related IT systems in their organization

Their preparation should be aimed more on the "developmental" dimension rather than on the "training dimension." They require having a broad understanding of the various ways in which IT can help in their business. They should be aware of the various kinds of IT available in the market and the latest trends in the industry worldwide. They should have enough knowledge so as to decide about the kind of systems relevant for the business.

The objectives of employee development are not necessarily tied to a specific job or task. London (1989) defined development as courses, workshops, seminars and assignments that influence personal and professional growth. Development is less focused on specific skills instead it focuses on the comprehensive knowledge and attitudes required for improving the long term personal effectiveness of the employee which results in an overall benefit to the firm. The top management in a firm is responsible for deciding the course of action for the enterprise; hence their overall development in IT will result in empowering them with the right attitudes for executing this function effectively and efficiently.

Middle Level

The middle level is mostly concerned with the tactical decisions in an enterprise. Middle level managers are required to make decisions on how to best utilize the existing systems in an enterprise as per the directions of the top management. Suppose the top management has been imparted an "attitudinal training" in IT and they decide that ERP system is suitable for their enterprise. They give necessary directions to the middle level management to implement ERP in their organization. Now the middle level management should have the "critical knowledge" to appreciate the functionality of ERP system. They should be able to spell out the relative benefits of using ERP system and consequently help the top level management in choosing the required system, consultants, and so forth. Thus their training sequence is next in importance to the top level and their training content is more focused on the "knowledge" aspect of training. They require having a thorough knowledge and understanding of the various systems of the firm and the IT capabilities and more importantly how they can be integrated.

The training program planning procedures need to identify and consider the technical as well as managerial skills needed for advanced technologies well in advance of its implementation (Kozlowski, 1987; Majchrzak, 1988). This requires knowledge of planning techniques that are not well represented by the conventional needs assessment models (Kraut, Pedigo, McKenna et al., 1989). The hexagon shown in the framework emphasizes the need for having a greater emphasis on "knowledge related" aspects of IT in their training rather than attitudes or skills. Once they are able to spell out what is to be adopted by the enterprise then the frontline workers can be imparted the specialized training of "skills set." Thus the training programs for the middle level managers should be more knowledge related so that they are able to comprehend the IT options available in the industry and are able to make "informed decisions" (Srivastava, 2001). Many German midsize firms are adopting Linux as their cost effective platform (Blau, 2003). Such a decision can come only from a well-informed middle management which has a thorough knowledge of the various options and has the capability to make a comparison.

Frontline Level

The frontline workers are the actual executors of the various tasks in an organization. The top level management brings in the "idea" (concept) in the enterprise, the middle level management gives "form" (methodology for operationalizing the concept) to that idea and the frontline workers actually "execute" (operationalize) this idea. The frontline workers should have rigorous training in the actual systems and IT modules related to their job. If we consider the ERP implementation example again, then the frontline workers require requisite "skills" for operating the selected modules of the IT systems chosen. Their training may be very specialized depending on the skills set required for operating the particular systems. As shown in the proposed framework, they require training in the specialized skills the maximum and comparatively less of knowledge and attitudes related training. There is no doubt about the fact that they do require having a positive attitude towards IT, and this attitude can be instilled in them through "socialization" and "proliferation" from the top management. Their training need not be directed towards IT related "attitudes" and "knowledge" but should be focused towards the specific skills required by them for execution of the particular job. Since their skill acquiring activity can begin only after the "top management is prepared to embark upon the IT odyssey and the middle management has chosen the ship for this journey," hence logically the sequence of their IT training in an enterprise is after the top and middle management.

General and Special Training

The IT training requirements of personnel in any organization can also be classified as general and special. General training (composed of attitudes, knowledge and skills) is the common training component that has to be imparted to all employees for efficient functioning in the organization, whereas special training is given as per the specific job requirements of the employee. Specialist training can be person, group or level specific. Goulding and Alshawi (2002) highlighted the need for differentiating between 'generic' and 'specific' IT training in an organization. The strategic training framework in Figure 1 captures this in terms of general and special training for all levels of employees. From the framework, it is clear that all categories of employees require some basic grounding in IT related attitudes, knowledge and skills for efficient functioning. The only difference in the content in the specialist and general training is that top management may require negligible specialist training in IT skills whereas the frontline management may require very less specialist training in attitudes. At this point, we would also like to emphasize that it is not possible to achieve IT success in an organization without imparting some general IT training in all the three aspects (attitudes, knowledge and skills) to all categories of employees. For example, in an organization, the top management may require general IT skills like checking e-mails, working on a word processor and a spreadsheet, but may require specialist training in attitudes, for example, being more creative and proactive towards new technologies, risk taking ability, perseverance and persistence, and so forth. On the other end of the spectrum, the frontline staff may require specialist knowledge about the various IT related systems that it is using for different operational requirements, for example specialized software packages, and functional knowledge of ERP modules. Their requirement of operational knowledge does not discount their basic attitudinal requirements of their enthusiasm for learning and using new technologies. The point about general and special training for different levels of organizational personnel is also highlighted in Table 2, which charts the requirements for different levels of employees.

Huang (2002) has also highlighted the importance of training employees in certain fundamentals of information technology which will remain nearly constant even in a dynamic technological environment. The general training in the proposed framework (Figure 1) is similar to general technology education and the special training has been captured in the business application training and just in time training (Huang, 2002)

Again considering the case of IT training in HDB, Singapore; the training categories are divided into four levels (basic, advanced, extended and continuing) depending on the job requirement and computer literacy of the individual staff member (Teo, 1999). This is done with a view to provide a better fit between the actual training imparted and the job requirements (Brown & Read 1984; Kirkpatrick 1967). The attitudinal training in HDB is also brought about through seminars, conferences and discussions and also through the "promotion of professionalism" among IT users through formal certification (accredited by the Singapore Computer Society) of its staff (Teo, 1999; Teo & Ranganathan, 2003).

CONTRIBUTIONS AND CONCLUSION

In the present day context, most organizations view IT training as a functional requirement rather than a strategic tool for gaining competitive advantage. The motivation for this chapter is to provide a theoretical basis for providing a strategic IT training framework applicable for hierarchical organizations. Through this study, we make following contributions to theory and practice. *First,* building from the literature on training, change management, and IT adoption, we provide a theoretically driven actionable, conceptual strategic IT training framework, namely the TRAKS model. This framework is one of the first IT training frameworks that views IT training as a strategy based on the functional requirements for different levels of employees. TRAKS model suggests that *training requirements* be based on *attitudes, knowledge,* and *skill* needs for different levels of employees. Using this model, organizations can keep their IT training on the right 'track.' Such training provides maximum benefit to the organization. The proposed framework provides a direction to future researchers to further explore the strategic impacts of IT training. In addition to the researchers, the proposed framework has implications for practitioners as well. As systematic training is an important input for IT adoption in enterprises, we hypothesize that the presented IT training framework will help in transforming "technological shyness" to "technological savviness" leading to enhanced business productivity and competitiveness. We have highlighted the applicability of the proposed framework by drawing some examples from HDB, Singapore, which is an organization recognized for its efficient and effective IT training programs (Teo, 1999; Teo & Ranganathan, 2003).

Second, we reiterate that training should be viewed not only as a means for serving operational needs but it should also be used as a strategic tool (Noe & Ford 1992; Schuler & Walker 1990). The proposed training framework segments organizations in the traditional hierarchical structure and identifies the broad content of IT training in context of these levels of employees to facilitate IT adoption in a systematic way. The top management personnel of an enterprise who are supposed to provide a strategic direction to the enterprise are the ones who should have a "positive attitude" towards IT adoption and should understand the tangible and intangible benefits that IT offers to them in the short as well as long term. They should not only be the first ones in an enterprise to be trained in IT but their training should also be focused towards empowering them with "attributes" that result in fruitful IT adoption by these enterprises. Once the top management sets the ball rolling with their right attitudes, the middle management should be in a position to execute the IT plans in an enterprise. Hence they must have the "right knowledge" to make the right decisions about the choice of platforms, software, and so forth. Their training should therefore infuse in them the knowledge to understand and make decisions best suited for the firm. The role of the frontline workers is at the delivery stage of the IT plan, conceptualized by the top management and operationalized by the middle management. These frontline workers should be "skilled" in operating the chosen software and hardware systems, so that right results are delivered to the firm by IT adoption. Hence their training requirement is more on the skills aspect and actual performance at the delivery stage.

Third, enterprises are faced with the problem of dwindling resources and increasing competition. The proposed framework provides guidelines to practitioners and managers to efficiently deploy their resources on fruitful IT training. It gives a direction to the managers for planning IT training of its personnel so that there are no wastages and the various levels of personnel get the IT knowledge which is "functionally and strategically relevant" for them.

Fourth, the proposed framework reiterates that not all employees in hierarchical organizations require similar kinds of IT training and is especially applicable in the context of developing countries. The "informational needs" of top, middle, and frontline level personnel are very different from each other. Hence the IT training programs for different levels must be designed according to their "roles and requirements" to avoid wastage of scarce resources. Systematic IT training as per the proposed framework will make these enterprises competitive in the global economy. Overall, the framework provides researchers and practitioners with a useful tool to better understand the different training requirements for different levels of the organization. Such understanding would pave the way for more effective usage of scare resources to ensure that personnel at various levels are adequately trained to leverage IT effectively to improve business productivity and enhance competitiveness.

There are three main limitations of this framework. First, in the present day world, organizational structure is itself undergoing a major transformation. We are gradually moving towards flatter organizations, where the classification as per the traditional structure may not hold good for many organizations. However, still many organizations tend to have a hierarchical structure. Second, some organizations are relatively small and the top management at times may also be performing the operational and tactical role, apart from the strategic role. Hence the framework has to be suitably modified for such enterprises. Third, we have assumed that IT adoption should be driven from the top. Sometimes, the middle level and frontline personnel are the ones who bring to management attention what the competitors are doing with regard to the deployment of IT. Nevertheless, top management support for IT is an essential element for successful IT deployment. Such support would be difficult if top management does not have favorable attitudes toward IT adoption.

Future research can identify the detailed elements of AKS required for the various levels of personnel for particular IT system implementation for example ERP and CRM. Extensions of this chapter can also be done by studying some of the successful organizations and analyzing their IT training strategy for its employees in comparison to the proposed strategic IT training framework.

REFERENCES

Anderson, J.R. (1982). Acquisition of Cognitive Skill. *Psychological Review, 89,* 369-406.

Anthony, R.N. (1965). *Planning and control systems: A framework for analysis.* Boston, MA: Harvard University Graduate School of Business Administration.

Blau, J. (2003). German Midsize Firms go for Linux. IDG News Service. Retrieved from http://www.infoworld.com.article/03/10/29/HNgermangoforlinux_1.html

Bostrom, R.P., Olfman, L., & Sein, M. K. (1990). The importance of learning style(S)? In end-user training. *MIS Quarterly, 14*(1), 101-119.

Brown, G.F., & Read, A.R. (1984). Personnel and training policies — some lessons from western companies. *Long Range Planning, 17*(2), 48-57.

Campbell, J.P. (1988). Training design for performance improvement. in J.P. Campbell, R.J. Campbell, & Associates (Eds.), *Productivity in organizations* (pp. 177-216). San Francisco: Jossey-Bass.

Cohen W., & D. Levinthal. (1990) Absorptive capacity: A new perspective on learning and innovation. *Administrative Science Quarterly, 35*(1), 128-152.

Compeau, D.R., & Higgins, D.A. (1995). Computer self-efficacy: Development of a measure and initial test. *MIS Quarterly, 19*(2), 189-211.

Daft, R.L., Lengel, R.H., & Trevino, L.K. (1987). Message equivocality, media selection, and manager performance: Implications for information systems. *MIS Quarterly, 11*(3), 355-366.

Davis, S. A., & Bostrom, R. P. (1983). Training end users: An experimental investigation of the roles of the computer interface and training methods. *MIS Quarterly, 17,* 61-85.

Davis, D. L., & Davis, D. F. (1990). The effect of training techniques and personal characteristics on training end users of information systems. *Journal of Management Information Systems, 7*(2), 93-110.

Devaraj, S., & Kohli, R. (2003). Performance impacts of information technology: Is actual usage the missing link. *Management Science, 49*(3), 273-289.

Ford, J.K. (1997). Advances in training research and practice: A historical perspective. In J. K. Ford & Associates (Eds.), *Improving training effectiveness in work organizations* (pp. 1-16). Hillsdale, NJ: Lawrence Erlbaum Associates.

Gallivan, M. J., Spitler, V. K., & Koufaris, M. (2005). Does information technology training really matter? A social information processing analysis of coworkers' influence on IT usage in the workplace. *Journal of Management Information Systems, 22*(1), 153-192.

Gist, M. E., Schwoerer, C., & Rosen, B. (1987). Effects of alternative training methods of self efficacy and performance in computer software training. *Journal of Applied Psychology, 72*(2), 307-313.

Goldstein, I.L. (1992). *Training in organizations: Needs assessment, development, and evaluation* (3rd ed.). Pacific Grove, CA: Brooks/Cole.

Goulding, J. S., & Alshawi, M. (2002). Generic and specific IT training: A process protocol model for construction. *Construction Management and Economics, 20,* 493-505.

Goulding, J. S., & Alshawi, M. (2004). A process driven IT training model for construction: Core development issues. *Construction Innovation, 4*, 243-254.

Harrison, A. W., & Rainer, R. K. (1992). The influence of individual differences on skill in end user computing. *Journal of Management Information Systems, 9*(1), 93-111.

Huang, A.H. (2002). A three-tier technology training strategy in a dynamic business environment. *Journal of End User Computing, 14*(2), 30-38.

Hussey, D.E. (1985). Implementing corporate strategy: Using management education and training. *Long Range Planning, 18*(5), 28-37.

Jackson, S.E., & Schuler, R.S. (1990). Human resource planning: Challenges for industrial/organizational psychologists. *American Psychologist, 45*, 223-239.

Kang, D., & Santhanam, R. (2003). A longitudinal of training practices in a collaborative application environment. *Journal of Management Information Systems, 20*(3), 257-281.

Keil, M., Cule, R. E., Lyytinen, K., & Schmidt, R. C. (1998). A framework for identifying software project risks. *Communications of the ACM, 41*(11), 76-83.

Kirkpatrick, D.L. (1967). Evaluation of training. in R.L. Craig & L.R. Bittel (Eds.), *Training and development handbook* (pp. 87-112). New York: McGraw-Hill,.

Kozlowski, S.W.J. (1987). Technological Innovation and Strategic HRM: Facing the challenge of change. *Human Resource Planning, 10*, 69-79.

Kraut, A.I., Pedigo, P.R., McKenna, D.D., & Dunnette, M.D. (1989). The role of manager: What's really important in different managerial jobs. *Academy of Management Executive, 3*, 286-293.

LePore, S. J., Kling, R., Iacono, S., & George, J. R. (1989). Desktop computerization and quality of work life: The role of implementation process and infrastructure. *In J.I. DeGross, J.C. Henderson, & B.R. Konsynski (Eds.) Proceedings of the Tenth International Conference on Information Systems* (pp. 223-235). Atlanta: Association for Information Systems.

London, M. (1989). *Managing the training enterprise.* San Francisco: Jossey-Bass.

Lucas, H. C., Jr. (1975). Performance and the use of an information system. *Management Science, 21*(8), 908-919.

Majchrzak, A. (1988). *The human side of factory automation.* San Francisco: Jossey-Bass.

McGhee, W., & Thayer, P.W. (1961). *Training in business and industry.* New York: Wiley.

Nelson, R. R., & Cheney, P. H. (1987) Training end users: An exploratory study. *MIS Quarterly, 11*(4), 547-559.

Nelson, R.R., Whitener, E.M., & Philcox, H.H. (1995). The assessment of end user training needs. *Communications of the ACM, 38*(7), 27-39.

Noe, R.A., & Ford, J.K. (1992). Emerging issues and new directions for training research. *Research in Personnel and Human Resource Management, 10*, 345-384.

Olfman, L., & Mandviwalla, M. (1994). Conceptual versus procedural software training for graphical user interfaces: A longitudinal field experiment. *MIS Quarterly, 18*(1), 405-426.

Ostroff, C., & Ford, J.K. (1989). Assessing training needs: Critical levels of analysis, In I. L. Goldstein (Ed.), *Training and development in organizations* (pp. 25-62). San Francisco: Jossey-Bass.

Scheinder, B., & Konz, A. (1989). Strategic job analysis. *Human Resource Management, 28*, 51-63.

Schuler, R.S., & Walker, J.W. (1990). Human resources strategy: Focusing on issues and actions. *Organizational Dynamics, 19*(1), 5-19.

Sein, M.K., & Bostrom, R.P. (1989). Individual differences and conceptual models in training novice users. *Human Computer Interaction, 4*, 197-229

Srivastava, S.C. (2001). IT training needs for mechanical department over Indian railways. *In proceedings of Seminar on Training Needs for Indian Railway Officials, Indian Railways Institute of Mechanical & Electrical Engineering (IRIMEE)* (pp. 8-16). Jamalpur.

Srivastava, S.C., & Teo, T.S.H. (2004). IT training as a strategy for business productivity: A framework for small and medium-sized enterprises in Asia. *In proceedings of the 8th Pacific Asia Conference on Information Systems, (PACIS 2004)*, Shanghai, China.

Swartz, N. (2006). Enterprise-wide records training: Key to compliance, success. *Information Management Journal, 40*(5), 35-44.

Tannenbaum, S., & Yukl, G. (1992). Training and development in work organizations. *Annual Review of Psychology, 43*(1), 399-441.

Teo, T.S.H. (1999). Managing information technology at the housing and development board in Singapore: A holistic approach for the public sector. *Journal of Information Technology Cases and Applications (JITCA), 1*(3), 26-44.

Teo, T.S.H., & Ranganathan, C. (2003). Leveraging IT resources and capabilities at the housing and development board. *Journal of Strategic Information Systems, 12*, 229-249.

Vonk, R. (2006). Six ways to link *training* with business performance. *Strategic HR Review, 6*(1), 20-23.

ENDNOTES

1. Adapted and revised version of "Srivastava, S. C., & Teo, T. S. H. (2006). IT training as a strategy for business productivity in developing countries. *International Journal of Information & Communication Technology Education, 2*(4), 39-51."
2. Organizations which have a clear chain of command and levels of execution
3. Acronym for **training requirements** based on **attitudes, knowledge,** and **skills** needs. The TRAKS model tracks the functional needs for different levels of employees and proposes that training should be based on it.

Section II
Educational Initiatives

Chapter VI
Technology Assisted Problem Solving Packages for Engineering

S. Manjit Sidhu
University Tenaga Nasional, Malaysia

S. Ramesh
University Tenaga Nasional, Malaysia

ABSTRACT

This chapter presents the development of technology-assisted problem solving (TAPS) packages at University Tenaga Nasional (UNITEN). This project is the further work of the development of interactive multimedia-based packages targeted for students having problems in understanding the subject of engineering mechanics dynamics. One facet of the project is the development of engineering mechanics dynamics problems for core undergraduate engineering courses. This chapter discusses the development of an interactive multimedia environment for solving relative motion of a rigid body using rotating axes. More specifically this chapter outlines the framework used to develop the multimedia package, highlighting our multimedia design process and philosophy.

INTRODUCTION

The influence of the computer is best seen in its multimedia configuration which includes an integration of multiple media elements that is, text, graphics, images, audio, video and animation into a coherent learning environment which in turn transform student learning and problem solving approach (Janson, 1992). Previous studies have shown that traditional learning (classroom teaching) could not engage the learners in visualization tasks and perform virtual experiments

(Cairncross, 2000). In contrast, multimedia learning aids have the potential to promote interactivity through its wide range of graphical environments. Additionally, the learner can control the rate of delivery and sequencing of the material being presented, that is the learner can learn at his or her own pace without loosing interest in the subject matter.

The present study discussed pertinent issues of a technology-assisted problem solving (TAPS) engineering environment project at University Tenaga Nasional (UNITEN). Our past research has led to the implementation of structured three-dimensional (3-D) environment that enhanced visualization coupled with real-time motion by integrating 3D animations with multimedia technology. This problem-solving environment has been extended to 3D virtual worlds where the user could freely explore and learn-by-discovery.

Newer emerging technology, such as virtual reality (VR), is also being researched for its effectives in education. VR systems were first introduced in the learning environment in mid 90s (Macpherson, 1998). The term 'virtual reality' is currently used to describe a range of computer-based systems in which a user can explore hardware or software generated 'micro world' (artificial environments) that allow close resemblance to reality. VR extends the interaction-oriented features of multimedia by the concept of cyberspace, that is, modeling objects and their behavior in virtual environments, integrating position-tracked human-computer interaction devices and performing numerically intensive computations for real-time navigation.

The prime feature of VR is 'interactivity.' Special VR hardware and software are thus required to allow human-computer interaction to permit input of the user's actions and movement to the computer and to provide corresponding simulated feedback to the user. An early application of such system was the flight simulator used to train pilots. However, it is in the area of hi-tech computer games that many of the application developments in this field have taken place. Although VR has been used for educational purposes (Bell & Scott, 1995, Dede et al., 1996; Kim et al., 2001), the potential of VR is just beginning to be exploited by a few science and engineering educators (Manseur, 2005).

The long-term objective of this work is to develop realistic 2D and 3D virtual TAPS packages where a user could learn-by-discovery and gain better knowledge by doing meaningful tasks. Our present research aimed to improve and define new patterns of interactions by adding interactivity to realistic 2D and 3D environment. It is believed that interactivity could enhance user learning by giving the virtual environment the capability to coach and provide feedback.

CURRENT STATE OF TEACHING AND LEARNING ENGINEERING COURSES

In general, education, in higher learning institutions in Malaysia still focuses on older educational models of linear progression or surface learning, whereas counterparts from other nations provide predominantly high-impact audio-visual perception.

The western countries, particularly the UK and USA, have used computers and CAL packages to motivate students of higher learning institutions since the 1960s (Ismail, 2001). Although encouraged by the government's policy towards the use of new technology in teaching, several academicians in Malaysia commented that they do not have the experience in developing multimedia-learning materials (Julia et al., 2002).

However, since the emergence of newer hardware and software technologies for multimedia and VR, educational practitioners began to study on the pedagogical effectiveness of these technologies. In a developing country such as Malaysia, multimedia technology was first briefly introduced in the late 1990s and became popular with the launch of Multimedia Super Corridor

(MSC) (Norhayati et al., 2001). Subsequently VR hardware and software are being used in various research fields' such as medicine, manufacturing, and for scientific visualization.

Malaysia is devoting this massive MSC to create the perfect environment for companies and education sector wanting to develop, distribute, and employ multimedia products and services. One of MSCs primary areas of multimedia applications includes "*smart schools,*" where educational software packages are being customized to facilitate teaching and student learning purposes in primary and secondary schools. In general, although the educational sector is aware of the presence of MSC, these new technologies (multimedia and VR) are not exploited in the teaching of engineering subjects. Additionally, multimedia and VR systems are not available on a large scale to support learning environments. Since these technologies are in its infancy, especially in the higher learning institutions, further research is needed to address its usefulness and benefits.

For engineering and technology education, multimedia and VR applications can include computer simulation, numerical analysis, computer aided design (CAD), computer-aided manufacture (CAM) and electronic communications (Palmer, 2000). At present the teaching methods at UNITEN are dominated by the conventional overhead projectors with transparencies and the use of white boards. However in order for developing TAPS environments to become an effective learning aid at UNITEN, certain criteria must be met. For example the instructors should be prepared to accept and encourage students to use the technology as an additional learning aid in an effective manner. The full potential of interactive multimedia learning packages cannot be realized if computers are merely used by students for preparing lab reports and assignments. Although many computer-based learning environments have emerged in general, in this chapter, the term technology assisted problem solving (TAPS) environment will be used to refer to the packages that were developed in this research and used by students to assist them in their learning. The TAPS packages aimed at coaching students in learning particularly slow learners (students who have difficulties in understanding the concepts in engineering mechanics dynamics subject), on the best approach to solve a particular engineering problem in step-by-step or logical approach.

Problems Encountered by Engineering Students in UNITEN

UNITEN is a higher-learning institution that provides both academic courses and engineering technology skills training within the same campus. The university's courses are focused on engineering, information technology, business management and related areas.

Though relatively new as a university, UNITEN is moving rapidly towards establishing itself as the regional center of educational excellence. The support and commitment it receives from its holding company, Tenaga Nasional Berhad (TNB), has enabled it to accelerate the infrastructure development of the campus.

The engineering courses offered at UNITEN includes, electrical engineering, mechanical engineering, civil engineering, and electrical power engineering. This research focuses the problems faced by students in the field of mechanical engineering, particularly in the engineering mechanics dynamics course. This subject is chosen because the instructors noted that many students had difficulties in visualizing dynamic motion of particles or rigid bodies.

The problems that many undergraduate students face while studying the engineering mechanics dynamics course is the difference in understanding with regard to what is being taught in the classroom. Generally, undergraduate students often expect a variety of teaching methods to be used in their learning. The lecture method is a common way of disseminating knowledge to students but it treats all students at the same level of basic acquired knowledge. However, in general most of these students do not bring to the course the same academic preparation (do not have the same motivation, interest, ability to learn and grasp) and come from different disciplines, cultures, regions with limited exposure to modern technology, have widely varying learning styles and different level of proficiency in material learned at the foundation level. This results in different starting points, progress rates and ultimately different levels of satisfaction and academic progress and performance.

Additionally, some entry-level undergraduate students do not have very strong grades in science and mathematics that makes certain engineering subjects difficult for them to understand and this discourages learning from taking place. As a result of this problem, if the lectures are taught too fast, this group of students may not be able to keep pace with the rest of the class. In this situation some students are left discouraged, and often the instructors are forced to find alternative methods (for example conducting extra classes) to help these students in understanding the subject matter. Since slow learners may take more time to understand the problem solving techniques and may require the topic to be repeated several times, it was proposed to employ and use multimedia and VR technologies to help them visualize and understand the engineering problems.

A major challenge facing instructors in teaching the engineering mechanics dynamics course is helping students relate the theory to the physical world. Past experiences in teaching first and second year students indicated that there were students who found it difficult to visualize some of the more difficult concepts in engineering and to apprehend theory and practical. In general engineering textbooks cannot represent mechanical actions in the form of dynamic illustrations (animated forms) such as movement of linkages, pistons, and crankshaft.

Yet another challenge that UNITEN currently faces is the training of instructors to integrate these new technologies for the teaching of engineering. In general, instructors must be computer literates to develop and utilize multimedia and VR technologies. Therefore, trained instructors are required if multimedia and VR technologies are to be integrated and implemented within the classroom.

Proposed System for Visualizing Engineering Concepts

Experience through practice by doing is extremely important in the development of basic and advanced skills, particularly in engineering, where the student needs to practice solving a wide range of problems and handle different equations and theories. A highly interactive virtual environment encourages the students to explore complex relationships and increases the development of advanced skills through self-motivated discovery activities (Manjit et al., 2003). For example, some useful learning-by-doing approach for mechanical engineering problem solving with a TAPS environment is presented in Table 1.

The engineering TAPS environment implemented and presented in this chapter is used to illustrate a 2D and 3D virtual environment to allow students to learn as well as to visualize and ultimately to solve problems pertaining to relative motion analysis using rotating axes in mechanics dynamics course.

THE NEED FOR TAPS PACKAGES

The mechanical engineering course is largely based on practical skills and requires the acquisition of basic skills and domain knowledge before applying them on real problems. In order to design and develop a technology assisted problem solving (TAPS) package particularly to guide students in learning and solving engineering problems, it is necessary to be acquainted with its development and its process of realization in practical terms

Table 1. Learning-by-doing approach with a TAPS environment

Learning-by-Doing Approach	User Activities	Coach Virtual Learning Environment
Interaction	The student interacts and observes meaningful tasks, e.g., the motion of a rider, jumping of a platform.	• Animated video files are integrated with audio files and graphics. • Student is narrated to explain the question during the motion.
Steps & Solutions	A sequence of steps and solutions of the problem is presented to the user. The user moves forward to the next step or back to the previous step or solution.	• Animated page showing steps and solutions are created and integrated with the tool. • The tool guides the user to manage the sequence of steps the user should perform to solve the problem and control the 2D animated mechanisms i.e., play, stop, reset and pause.
Simulations	The student experiences a problem-solving environment in a virtual manner through the accumulation of his actions and the behavior of the animated mechanisms in a 2D environment.	• The simulations are integrated with 2D graphics that are embedded with audio files. • The tool manages the state of the 2D animated mechanisms and the user's interactions. • The tool further provides graph for users to view data and interpret in a pictorial form.

in computer software. Therefore, it is imperative to examine some of these approaches in order to discover the extent to which they help engineering students in their learning. This examination includes an overview of good practice in the positioning and operation of navigational features, visual screen presentation, the nature of presentation, help and feedback and views on the role of the learner in using the TAPS packages.

Research has shown that in general students studying physics and engineering subjects encounter many difficulties in understanding the concepts of engineering mechanics. For example in kinematics topic, in a study of student understanding of two-dimensional motion, diagrams of trajectories of moving objects were shown to five students in an introductory university course and to five physics faculty (Reif & Sue, 1992). The participants were told whether the objects were speeding up, slowing down, or moving with constant speed and were asked to draw the acceleration vectors at specified points. The novices did very poorly at these tasks; even the experts had some difficulties. A detailed analysis of how the two groups approached these tasks enabled the investigators to identify the underlying knowledge and skills required for successful performance.

Some investigations have focused on student understanding of the graphical representations of motion. A descriptive study that extended over several years and involved several hundred-university students helped identify a number of common difficulties encountered by students in making connections between the kinematical concepts, their graphical representations, and the motions of real objects (McDermott et al., 1987). Another study identified that students have difficulties with the graphical representation of a negative velocity (Goldberg & Anderson, 1989).

On the other hand, the topic of dynamics and misconceptions about the relationship between force and motion has been extensively studied. Difficulties students have in interpreting the relationships between force and more complex concepts, such as work, energy and momentum are less documented. Some samples of investigations reported in the literature on student understanding of mechanics course can be summarized as follows:

- Prior to instruction, more than 100 students in an introductory university mechanics course were given a short-answer test on concepts of force and motion (Champagne et al., 1980). The test used a technique abbreviated as DOE (demonstration, observation, explanation). The results revealed that the students, who had previously studied physics, had mixed ideas such as a force will produce motion; a constant force produces constant velocity and the magnitude of the velocity is proportional to the magnitude of the force; acceleration is due to an increasing force; and in the absence of forces, objects are either at rest or slowing down. The results of another study also indicated that both before and after an introductory course in mechanics, many students seemed to believe that motion implies a force (Clement, 1982):
- In a study involving curvilinear motion and trajectories of moving objects, about 50 undergraduates were asked to trace the path that a pendulum bob would follow if the string were cut at each of four different positions along its path (Caramazza et al., 1981). Only one-fourth of the students gave a correct response.
- Other studies have examined student difficulties with situations involving gravity. A study of several hundred first-year university students in Australia involved in the use of simple lecture demonstrations related to gravity (Gunstone & White, 1981). For example, students were asked to compare the time it would take for an equal-sized steel and plastic balls to fall from the same height. On this task, 75 percent of the students gave different answers.

Since engineering subject involves a simultaneous mix of mathematics and physics, Vallim (2006) pointed out that some instructors are beginning to explore and develop multimedia computer aided learning packages for teaching. The following are some difficulties experienced by the instructors in using conventional teaching methods in engineering:

- In the area of engineering, the traditional communication model follows a one-dimensional, linear path that focus on the instructor/lecturer as the most important element of a communication transaction. This model does not take into account the level of the learners. In addition, traditional learning methods could not engage the learners in visualization tasks and work on virtual experiments (Janson, 1992; Kahn, 1992).
- Engineering mechanics dynamics, like many other engineering subjects, is fundamentally about problems solving through the application of scientific principles. The engineering problems are often complex, and relationships among the variables of an experiment can be difficult to visualize (Scott, 1996).
- Traditionally, problems in engineering dynamics are presented to the student as a combination of schematic diagrams and text descriptions. The shapes and lines that make up the schematic diagram have very specific engineering meanings, and the words accompanying the diagram also give rise to student error because critical information about the solution of the problem is often concealed in the text in unexpected ways (Scott, 1996).
- Theory oriented approach results in some disparity between text coverage and student comprehension (Ratan & Mitty, 1997).
- One of the difficult issues to deal with engineering within the curriculum at the introductory level is the process of abstraction of real and practical situations into mathematical models (Gramoll, 2001).
- Although many forms of learning aids have been used by educationalist to support them in their teaching (Fogler et al., 1992; Squires

et al., 1992), there is a need to provide better-enhanced learning aids. For example multiple tools such as calculator, glossaries of words, and electronic notepad can be integrated in a single learning package that can perform multiple tasks simultaneously, is user-friendly, and caters learner's requirements and could guide the learner when reaching an impasse (wrong answers).

- In general, the feedback that students receive on their homework is relatively ineffective. Feedback usually comes too late; solutions are often made available to students after the week's homework was completed (Steif, 2003).
- The engineering dynamics subject is difficult to understand from the textbook alone because there are many cognitive steps that lead from a problem through a series of steps to solution. Subsequently, this scenario creates additional educational difficulties, such as some learners lack the ability to translate mathematical word problems into the form necessary for effective computation and poor visualization of the problem that ultimately leads to lack of interest in the subject matter.

Although there are many conventional computer aided learning (CAL) packages available in the field of mechanical engineering, much of the efforts in the engineering CAL packages have attempted to replace the lecture and not focus on problem solving skills. Multimedia based technologies have the potential of providing a mean for dealing with the aforementioned issue in a dynamic (animated), provocative, and cost-effective manner that not only will increase the effectiveness of the educational program but will also increase the quality of the resulting students.

Development of TAPS Packages

This section describes the approach taken to integrate computer-based technologies in problem solving learning environment, subsequently termed as technology assisted problem-solving (TAPS) packages. While many software packages have been developed and used for the purpose of student learning in engineering, these packages do not provide the user adaptability, in particular to students experiencing difficulties in studying mechanical engineering, that is, students who normally need more time to understand a particular concept in engineering. As a result, these packages fail to provide adequate feedback, as they do not guide students to solve the engineering problem in a step-by-step approach. Additionally students who use such packages do not know if they have applied the appropriate formulas to solve the problem (some may use wrong formulas or working approach) even though the answer given by them could be correct.

TAPS packages are developed to include multimedia features and simple intelligent functions such as alerting a student by displaying messages (hints) on screen if a wrong formula is applied or a wrong answer given in solving the selected engineering problem. However, if the user still cannot solve the problem, the student could approach the TAPS package by clicking on "solve" button to aid the student in solving the problem. The solution is given in a step-by-step manner showing how the answer is obtained. Additionally, desktop virtual reality features were incorporated to encourage students to interact and engage with the TAPS package. These efforts have focused on conveying technical knowledge to the student solving the engineering problem in such a way so as to support the acquisition of theoretical knowledge.

To help students experiencing difficulties in the subject matter, further improvements were carried out in the previous developed TAPS packages and newer ones were implemented. These TAPS packages can be classified as cognitive tools for learning, problem solving, testing, and simulation. The reasons for employing TAPS packages can be summarized as follow:

- To use and store the knowledge of experienced instructors (human) and make the same easily accessible to the students;
- To develop a suitable user interface for simplifying the difficult engineering concepts;
- To help slow learners acquire problem solving skills;
- To provide encouragement to students in independent learning by incorporating simple intelligence (expert system like rules) in the TAPS packages;
- As an attempt to improvise the limitations of the already existing computer based learning packages thereby making them more acceptable as effective learning aids in UNITEN.

The TAPS packages developed for this research employs selected engineering problems that are difficult to understand by first year mechanical engineering students at UNITEN. Since the information, diagrams and sketches are presented in a static way in engineering textbooks, multimedia, and desktop VR technologies were found to be a suitable alternative in delivering technical information to students in the subject matter. For example, each problem-solving step in the TAPS package can be narrated and shown in an animated form to help students understand the problem being presented.

Desktop Virtual Reality Environment Taps Package (Design Approach 1)

Since the nature of the engineering mechanics dynamics subject requires the theory to be applied

Figure 1. The main interface (virtual environment) of the mechanics dynamics problem of the DVR TAPS package

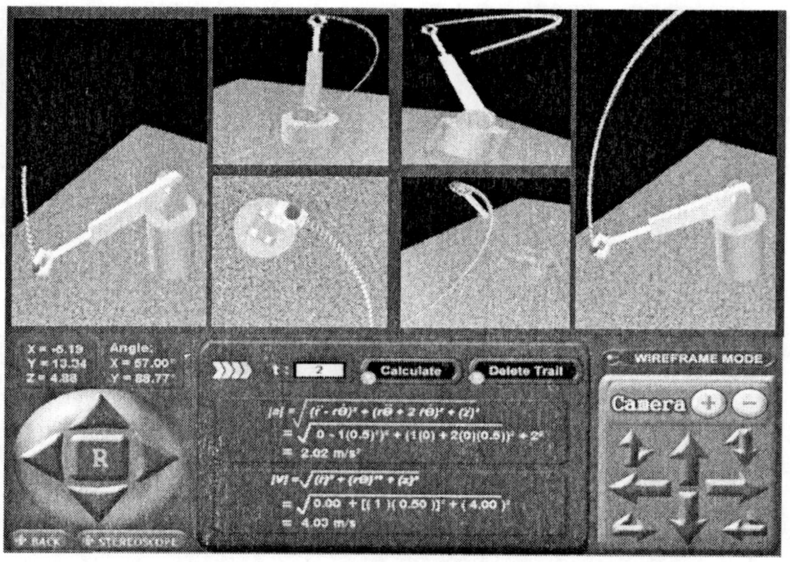

to physical problems before it is better understood, similar sample problems were covered in tutorials before the selected engineering problems were constructed using authoring tools such as Macromedia Director® and 3D modeling tool such as Alias Maya®. In the desktop virtual reality environment TAPS package, design approach 1 (problem based on curvilinear motion), progress was made to implement a 3D problem-solving model that was tested in a desktop virtual reality (DVR) environment for greater interaction and visualization. Every effort was made to give clear explanations on linear and curvilinear motion in this package. In the brief tutorial of this TAPS package, 2D animated examples illustrating the motions are displayed and narrated to the students. Additionally to make the tutorial more interesting, 3D animated example models are used to explain the concepts of the motions. The student is then explained about the components of a particle that experiences curvilinear motion.

The main interface (virtual environment) as illustrated in Figure 1 is a 3D model of a robotic arm that can be viewed and interacted with in the 3D environment. The 3D interface provides an interactive environment in which the students visualize the mechanics dynamics problem. It allows the student to move, resize, rotate and interact with the robotic arm on the display screen. In addition the user can adjust viewpoints, that is, solid or wire-frame mode (without texture) of the 3D robotic arm and change the display options such as changing the background colors from a color palette list.

The DVR TAPS package has a user-friendly environment that is built on six major modules, namely the action interpreter, the assessor, the interface, motion path generator, problem-solving engine and a randomized multiple-choice questions quiz. Outputs from the problem-solving engine consist of all the equations necessary to solve the problem. These equations are then used by the action interpreter and assessor to provide appropriate hints. The action interpreter module interprets the student's problem solving action in the context of the current problem and determines the type of feedback to provide. For example, if the student has input the wrong formula, the student will be prompted if a hint is needed. If the student attempted to answer without approaching the hint button and still gives the wrong answer, a *solve button* will be visible on the screen. The student can then click on the *solve button* to allow the TAPS package to guide the student in solving the problem. If the answer given by student is correct, the student may then proceed to the next step. If a complete solution has been accomplished, except for numerical substitution, the student could choose the *solve button* for the TAPS package to do the appropriate substitution.

However, one of the significant contributions shown in this TAPS package is the motion trail (path) algorithm to show curvilinear motion. The

Table 2. The robotic arm path algorithm

1	*Accept time (valid 1 ~ 10 seconds)*
2	*Temporary transform and assign "object" to temporary variable and initialize to world position*
3	*Temporary variable name = model*
4	*Assign properties i.e. height, width and length to model*
5	*Set model height to 0.1*
6	*Set model width to 0.3*
7	*Set model length to 0.3*
8	*Temporary variable = (model)*
9	*Assign (Temporary transform) to world postion*
10	*Model = model + 1*

11	*IF (glob_lift_limit = 180) THEN*
	normal_mode = FALSE
	reverse_mode = TRUE
	larger_angle = -1
	ELSE

12	*IF (glob_lift_limit = 0) THEN*
	normal_mode = TRUE
	reverse_mode = FALSE
	larger_angle = -1
	ENDIF

Figure 2. Original position and path generated to depict the curvilinear motion exhibited by the virtual robotic arm

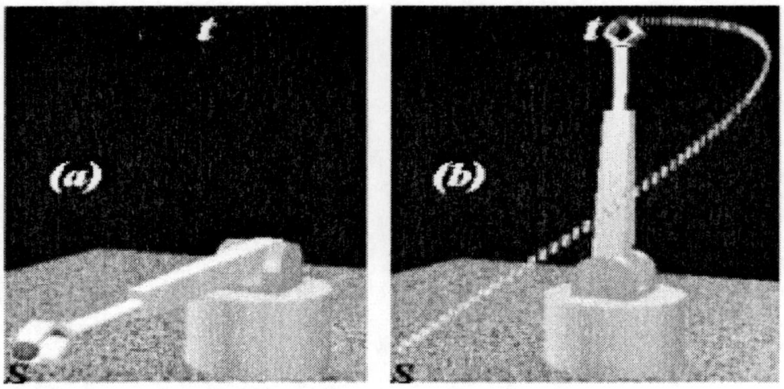

problem-solving engine contains the object motion path algorithm to show curvilinear motion taken by the robotic arm, say for example from the start to the end point of a path. An example of the algorithm is shown in Table 2. This algorithm was tested and found to be suitable to construct a short sequence of intermediate motions to transform and rotate the robotic arm from point ***s*** to ***t*** as shown on Figure 2. These motions can serve to fill in the intermediate scenes between ***s*** and ***t***, thus such scene can greatly reduce the amount of work the instructor has to do in the traditional way to explain to the students.

The motion path algorithm designed for the 3D robotic arm is a mechanism that is used to show the path taken by the robotic arm. For instance the initial 3D robotic arm is given a starting position ***s*** in a virtual environment as shown in Figure 2a, with a desired ending position ***t*** as shown in Figure 2b. The movement path of the robotic arm is based on the time input by the student, in this example say the time input is three seconds. Therefore the robotic arm should rotate in a curvilinear motion path from point ***s*** to ***t*** in the given time interval by plotting the motion path taken by the robotic arm. In this scenario the generated motion path on the screen can enforce visualization in the sense that it can clearly show the curve path taken by the robotic arm while moving from point ***s*** to point ***t***. This sort of motion could be difficult to explain in the traditional approach, for example from 2D drawings or static images.

Desktop Virtual Reality Environment Taps Package (Design Approach 2)

The extension work from previous research has been carried out to test the algorithm explained in the previous section. for another selected engineering problem. In this new TAPS package, progress was made to implement a problem that combines 2D, 3D and desktop virtual reality problem-solving model to help students understand relative motion analysis using rotating axes. To demonstrate this approach to students, a problem was taken from the textbook by Hibbeler (2001) as shown in Figure 3, and modeled in a 3D animation as shown in Figure 4. This type of analysis is useful for determining the motion of points on the same rigid body, or the motion of points located on several pin-connected rigid bodies. In some problems, however, rigid bodies (mechanisms) are constructed such that sliding will occur at their connections. The kinematics (motion of bodies) analysis for such cases is performed if the motion is analyzed using a coordinate system which both translates and rotates.

Figure 3. 2D mechanics dynamics problem of the DVR TAPS package

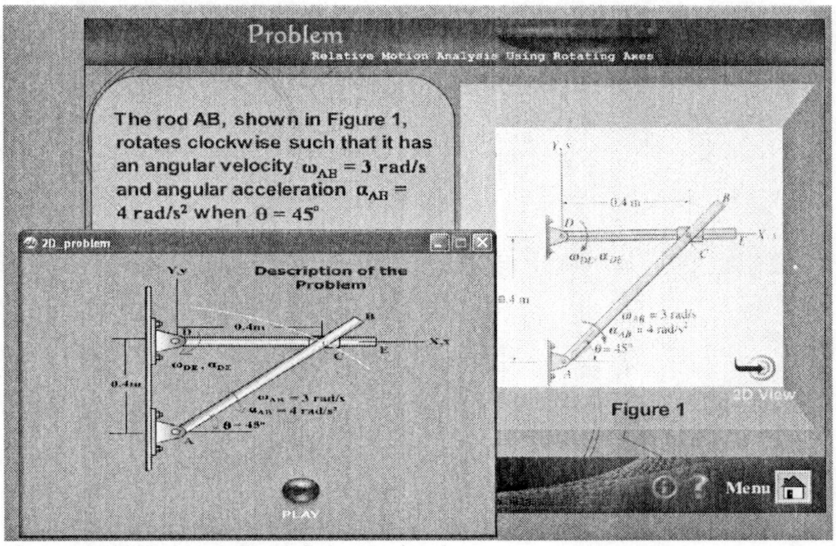

Copyright 2007, Manjit Singh Sidu. Used with permission.

Figure 4. 3D mechanics dynamics problem of the DVR TAPS package

Copyright 2007, Manjit Singh Sidu. Used with permission.

In the 3D TAPS package shown in Figure 4, students could click the trail button labeled as "T" to view the curvilinear path and motion of the rigid body and delete the trail by clicking the delete button labeled as "D". In addition, students could interact to gain a better visualization of the rigid body by clicking the "zoom in and out" buttons and rotate it by clicking the rotational buttons, that is, left, right, top and bottom. However, the most significant part of this TAPS package is that it could help students interact with the rigid body in a more natural and intuitive way as compared to the traditional approach. Similar explanation about solving the selected problem in the traditional approach could take a longer time for students to understand, particularly, students that have difficulties in understanding the engineering mechanics dynamics subject.

RESULTS AND EVALUATION OF THE PEDAGOGICAL EFFECTIVENESS

A group of 10 students who have taken the engineering mechanics dynamics subject was given a set of close-ended questionnaires in order to validate the problem solving technique adopted and the effectiveness of the TAPS packages. The evaluation of the TAPS package took place in one of the computing laboratories at the College of Engineering UNITEN. Each computer was installed with the TAPS package. The students were explained on the procedures using the TAPS package. The students were asked to go through the problem solving steps presented in the TAPS package and subsequently work on the problem given in the TAPS package. The session lasted approximately one hour under the observation of the researcher and the instructor teaching the engineering mechanics dynamics subject. Upon completion of the session, the students were given

Table 3. Student responses on the visualization, interaction and navigation of the TAPS package

Visualization			
Is the animated 2D model of the rotating axes helpful in visualization?	Easy to understand	Can understand	Quite difficult
Response:	100% (10)	0%	0%
Does the 3D model of the rotating axes provide better visualization?	Yes	No	No comments
Response:	60% (6)	20% (2)	20% (2)
What type of model do you think is more suitable to be used in the TAPS package?	Both 2D & 3D	3-D	2-D
Response:	60% (6)	0% (0)	40% (4)
How do you rate the narration (voice/sound) used in the TAPS package?	Very clear	Normal	Confusing
Response:	40% (4)	60% (6)	0% (0)
Interaction			
Does the step-by-step approach in the TAPS package provide a systematic way to solve the engineering problem?	Yes	No	No comments
Response:	80% (8)	20% (2)	0% (0)
Overall is the interactivity suitable for your level of understanding the problem presented in the TAPS package?	Yes	No	No comments
Response:	100% (10)	0% (0)	0% (0)
Navigation			
Do you think you could easily navigate while using the TAPS package?	Yes	No	No comments
Response:	80% (8)	20% (2)	0% (0)

close-ended questionnaires to evaluate the TAPS packages. The data collected from the study was used to investigate the students' views towards the TAPS packages in terms of visualization, interaction and navigation. The results of the statistical analysis are presented in Table 3.

The general outcomes of the statistical data collected from the students who used the TAPS package to visualize and solve the selected engineering problem can be summarized as follow:

- 60 percent of the students agreed that the animated 3D model could provide better visualization and it is easy to understand.
- The results about the interaction shows that 80 percent of the students agreed that the TAPS package provided a systematic way to solve the engineering problem, and 100 percent agreed that the interactivity is suitable for their level of understanding the problem as presented in the TAPS package. This result provides evidence that the TAPS package increased the interest of students in solving engineering problems by using multimedia technologies.
- From navigational point of view, the results indicated that 80 percent of the students found it easy to navigate while using the TAPS package. On the whole, the responses from the students feedback suggests that TAPS packages has the potential to guide them in solving the engineering problems better as compared to the traditional approach.

CONCLUSION

In this chapter two design approaches of TAPS packages for the study of multimedia patterns of interactions and visualization in engineering problem solution for undergraduates were described. These approaches can be employed and found to be suitable to aid learning. The TAPS packages targeted to integrate with, rather than replace, existing teaching methods, and efforts have been made to provide reinforcement of tutorial material, wherever possible. From the educational perspective, the TAPS packages benefited students experiencing difficulty in understanding the engineering mechanics dynamics subject in some situations, especially when it is properly implemented. The results of the study indicated that the TAPS packages has potential as an educational tool for promoting engineering education in terms of visualization, navigation and interactivity. In addition the TAPS packages could lead to better use of tutorial time, and an improved level of interactive help available to the student, which is seen as the primary deficiency in present teaching methods.

Although the traditional approach of teaching via lectures, tutorials and textbooks are essential part of any learning activities, it is believed that TAPS approach can be much more effective especially in the delivery of complex concepts such as those found in the engineering mechanics dynamics subject. In general, the TAPS packages were found to be effective in enhancing learning of engineering concepts and user-friendlier when compared to other computer based learning packages. One reason for this is because the existing computer based learning packages do not engage learners in solving the problems instead provides solutions that only encourages passive learning. Further work for this research includes extension of addressing students at different levels and the use of immersive virtual reality technology in the TAPS packages.

ACKNOWLEDGMENT

The authors would like to express their gratitude to UNITEN for the support provided.

REFERENCES

Bell, J. T., & Scott, H. F. (1995). The investigation and application of virtual reality as an educational tool. *Proceedings of the American Society for Engineering Education Annual Conference* (pp. 1718-1728).

Cairncross, S. (2002). *Interactive multimedia and learning: Realizing the benefits*. PhD Thesis. Napier University, Scotland.

Caramazza, A., McCloskey, M., & Green, B. (1981). Naive beliefs in 'sophisticated' subjects: Misconceptions about trajectories of motion. *Cognition, 9*, 117-123.

Champagne, A. L., Klopfer, J., & Anderson, M. (1980). Factors influencing the learning of classical mechanics. *American Journal of Physics, 48*, 1074-1079.

Clement, J. (1982). Students' preconceptions in introductory mechanics. *American Journal of Physics, 50*, 66-71.

Dede, C., Salzman, M., & Loftin, B. (1996). The development of a virtual world for learning newtonian mechanics. In P. Brusilovsky, P. Kommers, & N Streitz (Eds.), *Multimedia, hypermedia, and virtual reality*. Berlin: Springer (in press).

Fogler, H. S., Montgomery, S.M., & Zipp. R.P. (1992). Interactive computer modules for chemical engineering instruction. *Journal of Computer Applications in Engineering Education, 1*(1), 11-24.

Goldberg, F.M., & Anderson, J.M. (1989). Student difficulties with graphical representations of negative values of velocity. *The Physics Teacher, 27*, 254-260.

Gramoll. K. (2001). An internet portal for statics and dynamics engineering courses. *Proceedings of the International Conference of Engineering* (pp. 1-6).

Gunstone, R.F., & White, R. (1981). Understanding of gravity. *Science Education, 65*, 291-299.

Hibbeler, R. C. (2001). *Engineering mechanics: Dynamics* (9th ed.). Prentice-Hall.

Ismail, J. (2001). The design of an e-learning system: Beyond the hype. *The Internet and Higher Education, 4*(3-4), 329-345.

Janson, J. L. (1992). Computer-based training helps firms trim budgets. *PC Week, January 27*, 90-93.

Julia, M., Niall, S., & Su, W. (2002). Report on the British Council seminar and Asian association of open universities exhibition in Malaysia. Retrieved from http://cvu.strath.ac.uk/niall/malaysia/

Kahn, T. M., (1992). Multimedia literacy at Rowland: A good story, well told. *Technological Horizons in Education. 19*(7), 77-83.

Kim, J., Heon, P. S., Tae, Lee, H., & Yuk, K. C. (2001). Virtual reality simulations in physics education. *Interactive Multimedia Electronic Journal of Computer-Enhanced Learning, 3*(2). Retrieved from http://imej.wfu.edu/articles/2001/2/02/index.asp

Macpherson, C. (1998). Virtual reality: What is the state of play in education? *Australian Journal of Educational Technology, 14*(1), 60-74.

Manjit Sidhu, S., Ramesh, S., & Selvanathan, N. (2003). Using multimedia to minimize computational effort in engineering. *Proceedings of the Malaysian Scientific & Technology Congress (MSTC)* (pp. 811-815).

Manseur, R. (2005). Virtual reality in science and engineering education. In proceedings of the 35th ASEE/IEEE Frontiers in Education Conference (pp. 8-13).

McDermott, L.C., Rosenquist, M.L., &. Van, Zee, E. H. (1987). Student difficulties in connecting graphs and physics: Examples from kinematics. *American Journal of Physics, 55*, 503-513.

Norhayati, A. M., Halimah, B. Z., & Siew, P. H. (2001). Courseware development to motivate lifelong reading habits. Multimedia at work. *IEEE, 8*(4), 76-81.

Palmer, S. (2000). On-and off-campus computer usage in engineering education. Computers and education. *Elsevier Science, 33*(3), 141-154.

Ratan K., & Mitty, P. (1997). Using contemporary tools to teach dynamics in engineering technology. *International Journal of Engineering Education, 13*(6), 407-411.

Reif, F., & Sue, A. (1992). Cognition for interpreting scientific concepts: A study of acceleration. *Cognition and Instruction, 9*, 1-44.

Scott, N. W. (1996). *A study of the introduction of educational technology into a course in engineering dynamics: classroom environment and learning outcomes.* PhD thesis.

Squires, R. G. Andersen, P. K. Reklaitis, G. V. Jayakumar, S. & Carmichael, D. S. (1992). Multimedia-based applications of computer simulations of chemical engineering processes. *Journal of Computer Applications in Engineering Education, 1*(1), 25-30.

Steif, P. S., & Naples, L. M. (2003). Design and evaluation of problem solving courseware modules for mechanics of materials. *Journal of Engineering Education*. Retrieved from http://findarticles.com/p/articles/mi_qa3886/is_200307

Vallim, B. R. (2006). Practicing engineering in a freshman engineering course. *IEEE Transactions on Education, 49*(1), 74-79.

Chapter VII
Perceptions of Laptop Initiatives:
Examining Determinant Factors of University Students for Successful Implementation

Chuleeporn Changchit
Texas A&M University–Corpus Christi, USA

Robert Cutshall
Texas A&M University–Corpus Christi, USA

Susan Elwood
Texas A&M University–Corpus Christi, USA

ABSTRACT

Parallel to advancements in information technology usage, there are increasing demands for basic computer skills at minimum from today's college graduates. As a consequence, many colleges and universities have chosen to stimulate campus laptop initiatives as a way to provide their students opportunities to grow their computer skills and experiences. However, the success of laptop programs is very much dependent on the degree in which students and faculty are accepting a laptop environment and are willing to implement such programs. Defining which conception factors are necessary is essential for successful implementation. This study examines such factors by focusing on university student perceptions of required laptop programs in order to distinguish which factors they perceive as important. In understanding what factors encourage student support of laptop initiatives, such programs can be made more useful to students as well as more beneficial to universities.

INTRODUCTION

No longer are basic education supplies comprised simply of books, papers, and writing utensils. At an increasing number of universities, laptop computers have become one of the requisite technologies for incoming students. Many such laptop initiatives have sprung up in higher education institutions as they see a growing demand for technology awareness and skills in their students', society's, and many companies' expectations. The demand for technology-enhanced learning environments no doubt will continue to grow substantially as society, academic communities, and students continue to expect the educational process to employ technology comparable to that found in the real world (Brown & Petitto, 2003; Hall & Elliott, 2003; Weiser, 1998). In the real world, more and more, companies are requiring familiarity with technology as part of job requirements and daily work life. The need for technology know-how also extends to basic daily activities, such as familiarity with Internet functions, to view important personal information. Being able to adapt to an evolved model of communication, as well as the other capabilities available with technology, has broadened the scope of what individuals are required to be knowledgeable about in order to function in an increasingly high-tech society. Institutions of higher education have followed suit by offering services and creating programs centered on life's technology factor in order to adapt to and prepare students for these situations (Brown, 2003; Brown & Petitto, 2003; Weiser, 1998).

On campuses where all students are expected to have and use computers, laptops appear to be the popular choice. Laptop computers provide unsurpassed flexibility and convenience for students in the modern academic environment (Bazillion & Braun, 2001; Vaughan & Burnes, 2002). Compared to their desktop counterparts, the portability of laptops allows students to have easy and ready access to necessary technology and information with them in classes, libraries, and on trips. It has been argued that the laptop's flexibility and consequential ubiquity also bring benefits to colleges and universities by making it possible for the institutions to offer computer-enhanced classroom instruction, decrease the need for labs, and consequently lower the institution's computer budget (Badamas, 2001; Brown, Burg, & Dominick, 1998).

For higher education, providing experiences with computer tools tends to be one of the prerequisites to professional success as employers value extensive experiences with information technology in today's modern workforce (Brown et al., 1998; Rola, 2002; Tomek & Muldner, 1999). It has been found that prior exposure to and experiences with technology can have a significant impact on adapting to newer forms of technology, which can be vital in workplace success (Agarwal & Prasad, 1999). Previous researchers have shown that besides providing the ready convenience of technology access, integrating laptop computers into the classroom can lead to positive educational outcomes (Barak, Lipson, & Lerman, 2006; Finn & Inman, 2004; Fouts & Stuen, 1997; Gottfried & McFeely, 1998; Varvel & Thurston, 2002) which provides another incentive for universities to examine adopting their own laptop program.

New learning environments are being developed related to pioneering laptop programs. The studio environment, a careful blend of mini-lectures, recitations and hands-on laboratory experiences, mutually reinforce one another in large lecture hall settings (Dori et al., 2003). Major advantages to such learning environments include the facilitation of (1) procedural understanding through hands-on problem solving; (2) immediate feedback and appropriate responses in real-time interactions; (3) concrete learning of abstract concepts through visualization and simulation applications; and (4) collaborative work among learners and instructors through shared understandings (Barak et al., 2006).

Students' use of laptop computers is becoming more prevalent in today's universities. This more ubiquitous use of technology has forced several universities to discover and manage new perceptual issues in addition to the more familiar issues stemming from the primary use of university labs. Thus, defining the conception factors necessary to effectively implement a laptop initiative becomes a critical issue to the success of the program.

Literature Review

As laptop computers have become widely used in many workplaces and schools, they have developed into the largest growth area within the personal computer market (Berkhout, Hendriksson-Larséna, & Bongers, 2004). The need to access information technology on a daily basis continues to grow in greater amounts, and the laptop computer's functional advantages of being portable, lightweight and space saving to enable users to work anywhere and at anytime, have increased its popularity among personal computer users (Changchit, Cutshall, & Elwood, 2006). The time and effort required to locate and use a stationary desk computer can be burdensome in a mobile society. With laptops, the combined benefits of having most, if not all, of the capabilities of desk computers and of adaptable mobility ease the limits and burdens of stationary, personal computers.

In keeping with societal transitions, the decision to require student ownership of computers is not unusual among higher education institutions. Several universities are initiating laptop programs where all students are required to purchase laptop computers. These universities also offer their students computing and networking facilities that enable them to use many kinds of Web-based resources, from library catalogs to complete graduate degree programs (Brown, 2003; Brown & Petitto, 2003; Brown, Burg, & Dominick, 1998; Lehner, Nosekabel, & Kehmann, 2003). These resources are intended to help the integration of mobile technology into students' daily academic life, and therefore giving students a foundation of technological experience that they can further utilize in other areas. It has been pointed out that prior experiences with technology and how readily available technology is for use, are significant factors in further technology use, and the more positive the experience, the better outlook on further use (Al-Khaldi & Al-Jabri, 1998; Liaw, 2002). Laptop initiatives can aid in familiarizing students with technology. Institutions of higher education have the unique position to integrate technology familiarity into students' daily life with instructing students in their specific learning careers. Within an astonishingly short time, higher education has achieved a ubiquitous electronic presence. It is not uncommon now to see students and faculty carrying laptops and other mobile technology devices on their person as well as various technological advances on the campuses themselves.

The laptop initiative has not only given students a better education with ready access to more information at their fingertips, but it has made them more competent in using technology (Cutshall, Changchit, & Elwood, 2006). Several issues comprise a students' perceptual base regarding a laptop initiative. A study reported that key themes related to these issues include (1) academic and social use of laptops, (2) e-mail and instant-messaging, (3) faculty utilization, (4) Web uses, (5) comparisons with desktops, (6) cost, (7) library use, (8) problems, (9) family utilization, (10) service and help, (11) convenience, (12) network access, (13) worry, and (14) hardware and software (Demb, Erickson, & Hawkins-Wilding, 2004).

However, despite the benefits and societal issues with technology familiarity, there can be resistance to changes that are perceived as too rapid or overwhelming (Elwood, Changchit, & Cutshall, 2006). University laptop initiatives are still in their infancy, and many people may choose not to support such initiatives (Finn & Inman, 2004). Unless it is imperative for the university that students are willing to support the program, it is quite risky to start requiring all students to purchase a laptop. Not everyone exposed to an innovation will adopt the new technology at the same rate. The speed at which technology diffuses throughout a social system, such as higher education, is believed to be heavily dependent on several factors. The success of the laptop program relies heavily on the extent to which the laptop environment is accepted and wholeheartedly implemented by students and faculty. As with any new technology adoption, the acceptance of a laptop initiative is likely to be quite uneven. It is, in part, dependent on their attitudes towards technology and how well individuals are willing to accept what might be a change in their environment and fully utilize the technology by absorbing the implications of laptop use (Argawal & Karahanna, 2000; Mitra, 1998; Mitra & Hullet, 1997; Mitra & Steffensmeier, 2000). Some will readily accept it, whereas others will actively resist change.

This research was centered on the needs and attitudes of students. Typically, students are still skeptical when it comes to actually requiring them to purchase and use a laptop in higher education. Required materials are expected to be fully utilized and effective so that their expense is justified. Perceptions of faulty programs would hamper adaptation to these initiatives as participant willingness is needed as part of any program's success. The willingness to acquire additional programs has a dependency on perceptions of use. If individuals do not perceive a useful purpose for a newer technology or program, they then have a higher resistance to adaptation, regardless of usefulness in reality and perceptions of ease of use (Chau & Hu, 2001; Davis, 1989; Davis, Bagozzi, & Warshaw, 1989). This study attempts to investigate what factors are perceived as important to students. It is important for universities and other institutions to understand what factors can influence students' decision on the laptop program as part of development and initiation. By understanding what factors encourage students to support a laptop initiative, such a program can be made more useful to students as well as more beneficial to universities.

METHODOLOGY

A direct survey was used to collect the data for this study. The survey questions were compiled from previous study questions pertaining to information technology innovation as well as suggestions from researchers and students (Demb et al., 2004; Luarn & Lin, 2005; Moore & Benbasat, 1991). These questions were designed to gather data on students' perceptions on the prerequisite factors necessary to implement a laptop initiative, as well as their demographics. To validate the clarity of these questions, three professors and three students were asked to read through the survey questions. Revisions to the survey were made based on the feedback received.

A total of 54 items were used as five-point Likert scaled questions with end points rating from "strongly disagree" to "strongly agree." Survey items Q1 to Q28 collected demographic data. Survey items Q29 to Q53 measured students' perceptions on the prerequisite factors necessary to implement a laptop initiative. Survey item Q54 measured students' willingness to support a laptop initiative.

Data Collection

Surveys were distributed to 515 students enrolled in a mid-sized four-year university. The participants were given a 54 item survey and allowed class time to complete the survey. All participants were informed that participation in the study was voluntary and that all individual responses would be kept anonymous. The students were asked to rate each of the survey items on a Likert-scale from 1 to 5 with 1 being "strongly disagree" and 5 being "strongly agree." Five hundred and four participants completed and returned the survey instruments.

The data revealed that 20.44 percent of the respondents agreed or strongly agreed with requiring all students to purchase a laptop computer for use in their education. Approximately 52.58 percent of the respondents disagreed or strongly disagreed with a laptop computer initiative. The remaining 26.98 percent of the respondents were neutral on a laptop computer initiative. Table 1 summarizes additional demographic characteristics of the respondents.

Table 1. Demographic characteristics

Gender				
Female			**Male**	
293 (56.89%)			222 (43.11%)	
Ethnicity				
African	**Anglo**	**Asian**	**Hispanic**	**Native American**
26 (5.05%)	271 (52.62%)	21 (4.08%)	186 (36.12%)	11 (2.13%)
First Generation College Student				
Yes			**No**	
228 (44.27%)			287 (55.73%)	
Classification				
Freshman	**Sophomore**	**Junior**	**Senior**	**Graduate**
12 (2.33%)	56 (10.87%)	242 (46.99%)	165 (32.04%)	40 (7.77%)
College				
Arts & Humanity	**Business**	**Education**	**Nursing**	**Science & Technology**
6 (1.17%)	385 (74.76%)	116 (22.52%)	1 (0.19%)	7 7 (1.36%)
Own a Computer**				
Desktop			**Laptop**	
414 (62.92%)			244 (37.08%)	
Using Laptop in High School per Week (times)				
Never	One time		2-3 times	> 3 times
376 (73.01%)	45 (8.74%)		40 (7.76%)	54 (10.49%)
Using Laptop for class assignment per Week (times)				
Never	One time		2-3 times	> 3 times
238 (46.21%)	44 (8.54%)		79 (15.34%)	154 (29.90%)
Using Laptop for Leisure per Week (times)				
Never	One time		2-3 times	> 3 times
262 (50.87%)	35 (6.80%)		46 (8.93%)	172 (33.40%)
Annual Income (dollars)				
Under 20,000	**20,000-39,999**	**40,000-59,999**	**60,000-79,999**	**80,000 and Over**
355 (68.93%)	90 (17.48%)	31 (6.02%)	17 (3.30%)	22 (4.27%)

ANALYSIS AND DISCUSSION

The research data showed an odd-even reliability score of 0.943, suggesting internal consistency of the data. In addition, a Cronbach's alpha score of 0.924 was calculated as a second measure of reliability. It should be noted that these high levels of reliability relate to the data resulting from the measurement, not the instrument itself.

Factors Perceived as Critical

To determine which factors were deemed as critical to the successful implementation of a laptop computer initiative, the mean responses to each question were calculated and examined. Ten items with the highest mean scores were identified. The threshold for these factors was at a relative consensus point of 85 percent agreement. The 10 factors perceived as more critical by students are presented in the Chart 1.

Students believe that access to a wireless network is the most important factor for the success of a laptop initiative. This factor had a mean score of 4.53 out of 5 with 89.3 percent agreeing or strongly agreeing with this factor. Students are following the industry trend in ranking access to a wireless network as very important. For a laptop program to be successful, students believe that it is critical for the university to provide a wireless network for them to access stored information at various points on campus and to access the Internet.

The notoriously power-hungry laptop computers drove the students to rate the need for sufficient power outlets in the class room as the second most critical factor. Students rated the importance of access to sufficient power outlets in the classroom with a mean score of 4.52 out of 5, just slightly below the mean score for access to a wireless network. Of all the respondents, 89.9 percent agreed or strongly agreed with the need for sufficient power outlets inside the class room.

In addition, students believe that access to printers is a crucial factor for the success of a laptop initiative. This factor had a mean score of 4.49 out of 5 with 88.1 percent agreeing or strongly agreeing with this factor. This finding is consistent with observed student behavior in physical computer labs. Many students complete their assignments on computers off-campus and bring their work to the computer lab to print hard copies.

Chart 1. Critical success factors

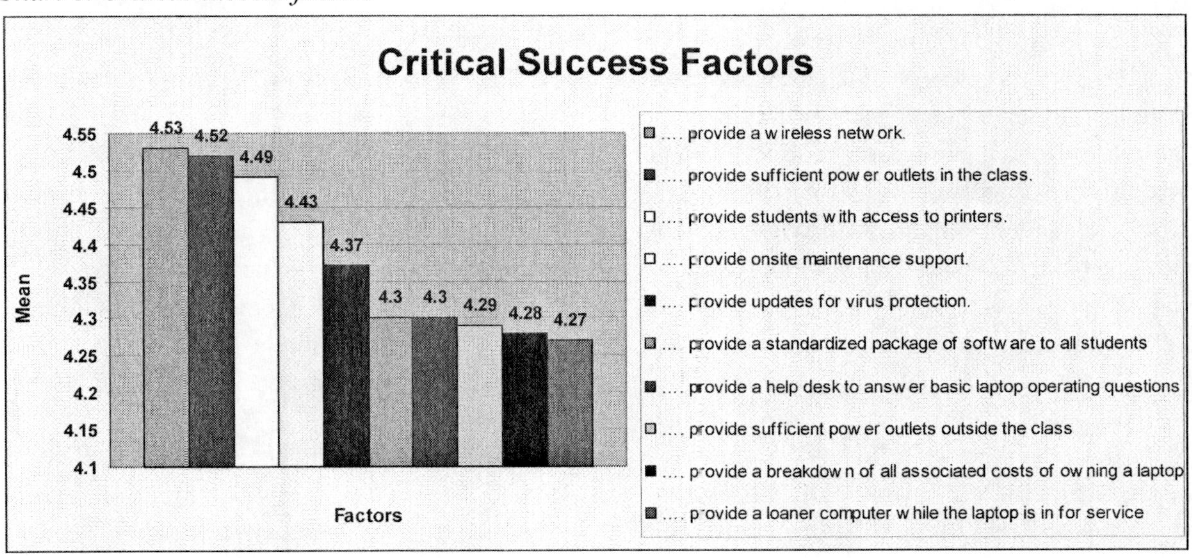

Dropping down to a mean score of 4.43 out of 5, students also believe that it is critical for the University to provide onsite maintenance for the laptop computers. The majority of the students, 86.4 percent, agreed or strongly agreed with the necessity of onsite maintenance. If the students are expected to use their laptops as a learning tool, they must be able to have their laptop quickly serviced when needed.

Another factor rated as critical by students was the issue of providing updates for virus protection. The mean response for this item was 4.37 out of 5 with 82.8 percent agreeing or strongly agreeing with this factor. This result points to student perceptions of the types of services needed in order to fully utilize laptops on campus.

With mean scores of 4.30 out of 5, students ranked the provision of a standardized software package and a help desk to answer basic laptop operating questions as equally important factors. Approximately 83 percent of the students agreed or strongly agreed with these two factors. This is evidence that students can see the benefit of using laptops in their education but only if there is a standardized starting set of applications that the faculty can expect the students to have and thus utilize in class.

In agreement with the need of sufficient power outlets inside of the classroom, students also realize that much of the learning takes place outside of the classroom. Thus students rated the need for sufficient power outlets outside of the classroom in student common areas such as the library as significant for laptop programs. The respondents rated this factor with a mean score of 4.29 out of 5 with 81.3 percent agreeing or strongly agreeing with this being a crucial factor in the success of a laptop program.

With the cost of higher education on the rise, students expressed the need for a breakdown of all of the associated costs required for laptop ownership. The mean score of 4.28 out of 5 shows that the students want to be aware of the total cost of ownership and not just the initial purchase price. This factor was seen as important by 81.5 percent of the respondents.

To round out the top 10 crucial factors associated with a university laptop program, students know that access to the technology needs to be available at all times with minimal downtime. Therefore students raked the need to have loaner laptops available when their own laptop has to be sent away for service. This factor had a mean score of 4.27 out of 5 with 81.3 percent agreeing or strongly agreeing with this factor.

Factors Perceived as Not So Critical

To determine which factors were deemed as not so critical to the successful implementation of a laptop computer initiative, the mean responses to each question were calculated and examined. Five items with the lowest mean scores were identified as the not so critical factors. The threshold for these factors was at a relative consensus point of 75 percent agreement. The five factors perceived as the not so critical factors with limited impact on the success of a laptop initiative are presented in Chart 2.

While students may see the benefits of using laptop computers in their education, they do not believe that it should be a requirement. This factor had a mean score of 2.76 out of 5 with only 31.7 percent agreeing or strongly agreeing with this factor.

Students believe that requiring all students to purchase a backup battery is not a critical issue. This factor had a mean score of 3.06 out of 5 with only 37.9 percent agreeing or strongly agreeing with this factor. This finding is consistent with the observed critical factor of providing sufficient power outlets both inside and outside of the classroom. With sufficient power outlets available, the need for a backup battery becomes a non-critical issue.

The majority of students also believe that it is not a critical factor to require a hardware update after two years. The mean score of this survey

Chart 2. The not-so-critical success factors

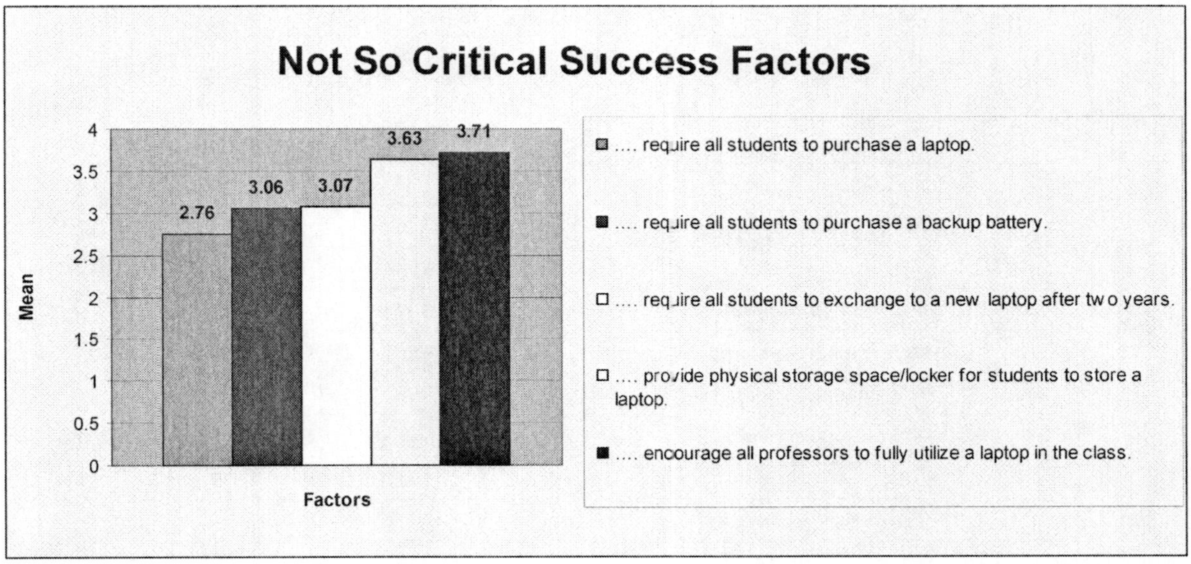

item was 3.07 out of 5. Only 36.2 percent agreed or strongly agreed with the requirement to exchange to a new laptop after two years.

Students also perceived that the necessity of providing physical storage space, for the laptop computer when not in use, was a non-critical issue. This factor had a mean score of 3.63 with 56 percent agreeing or strongly agreeing with this factor. This finding is consistent with the portability concept of the laptop computer. Laptop and notebook computers are designed to be smaller and lighter which make them easy to carry around when not in use. Hence, physical storage space on campus is not a necessity.

It does not seem to be of critical importance to students that professors should utilize a laptop computer in the classroom. This factor had a mean score of 3.71 with 62.5 percent agreeing or strongly agreeing with this factor. This finding is consistent with students' attitudes against requiring all students to purchase a laptop. It will be difficult for a professor to utilize a laptop computer in the classroom if not all students possess a laptop computer.

Differences Between Groups

To determine if there were significant differences, on the critical factors, between the group of students who support a laptop initiative and those students who do not support a laptop initiative, t-tests on the means were conducted. The responses from participants were divided into two groups based on their responses to survey item Q54. The two groups were those who favored the laptop initiative (support group) and those who did not support the initiative (reject group). The students who were uncertain on the laptop initiative were excluded from the t-tests. Table 2 shows the factors exhibiting a significant difference between the two groups at a *p-value* < 0.05.

The results of the t-tests revealed a significant difference between the two groups on eighteen factors: (1) provide a wireless network, (2) provide power outlets in the classroom, (3) provide access to printers, (4) provide updates for virus protection, (5) provide standardized software, (6) provide a helpdesk to answer questions, (7) provide a loaner computer, (8) provide network storage

Table 2. Significant between group differences

Factors	Support Group	Reject Group	p-value
Provide a wireless network	4.68	4.45	*
Provide power outlets in the class	4.68	4.46	**
Provide access to printers	4.64	4.40	**
Provide update for virus protection	4.52	4.27	*
Provide standardized software	4.50	4.20	*
Provide a help desk to answer questions	4.48	4.24	*
Provide a loaner computer	4.46	4.17	**
Provide network storage space	4.45	4.20	*
Provide email Account	4.40	4.10	**
Provide a !ease option	4.38	4.16	*
Provide a breakdown of costs	4.36	4.21	NS
Demonstrate laptop benefits	4.30	3.54	**
Provide onsite maintenance support	4.29	4.53	NS
Demonstrate how to fully utilize laptop	4.28	3.70	**
Encourage professors to utilize laptop	4.26	3.35	**
Provide power outlets outside the class	4.20	3.95	*
Provide basic training	4.20	3.95	*
Provide a physical storage space	3.83	3.58	NS
Purchase backup Battery	3.74	2.75	**
Require all students to purchase laptop	3.70	2.28	**
Exchange the laptop after two years	3.59	2.74	**

** Significant at the $p < 0.05$ level, ** Significant at the $p < 0.01$ level, NS – not significant*

space, (9) provide an e-mail account, (10) provide a lease option, (11) demonstrate laptop benefits, (12) demonstrate how to fully utilize the laptop, (13) encourage professors to utilize laptops, (14) provide power outlets outside of the classroom, (15) provide basic training on laptop use, (16) require the purchase of a backup battery, (17) require all students to purchase a laptop, and (18) exchange the laptop for a new one after two years.

It is also interesting to note that the support group rated all factors higher than the reject group. These results demonstrated that the support group tends to pay more attention to the details of program implementation. The findings suggest that the university may want to consider these factors before implementing a laptop program.

CONCLUSION

The initiative to use laptop computers in higher education is viewed as advantageous by many. Due to the ever increasing use of technology in primary education and the increasing demand, by industry, for more computer savvy graduates, the use of technology in higher education will continue to grow. Nevertheless in order to smooth the transition, the factors critical to a successful laptop program must be identified and addressed.

This study has provided an empirical glimpse into the minds of students as to what they perceive as critical factors in a laptop initiative. The findings revealed that students, both those who support and those who do not support laptop initiatives, place a critical level of importance on the following factors: (1) provide a wireless network, (2) provide power outlets in the classroom, (3) provide access to printers, (4) provide onsite maintenance support, (5) provide updates for virus protection, (6) provide standardized software, (7) provide a helpdesk to answer questions, (8) provide power outlets outside of the classroom, (9) provide a breakdown of all costs associated with laptop ownership, and (10) provide a loaner computer.

Both groups also perceived that the following five factors have a lower degree of importance in the success of a laptop program: (1) requiring all students to purchase a laptop, (2) requiring all students to purchase a backup battery, (3) requiring all students to exchange to a new laptop after two years, (4) providing physical storage space/locker for students to store a laptop when not in use, and (5) encourage all professors to fully utilize a laptop in the class.

The results also revealed that eighteen factors were perceived differently between the groups who support and do not support the laptop program. These factors were: (1) provide a wireless network, (2) provide power outlets in the classroom, (3) provide access to printers, (4) provide updates for virus protection, (5) provide standardized software, (6) provide a helpdesk to answer questions, (7) provide a loaner computer, (8) provide network storage space, (9) provide an e-mail account, (10) provide a lease option, (11) demonstrate laptop benefits, (12) demonstrate how to fully utilize the laptop, (13) encourage professors to utilize laptops, (14) provide power outlets outside of the classroom, (15) provide basic training on laptop use, (16) require the purchase of a backup battery, (17) require all students to purchase a laptop, and (18) exchange the laptop for a new one after two years. These findings suggest that the university may want to carefully consider these factors before implementing the laptop program.

The results in this study reveal the factors which are perceived by students as important or not important if the university would like to implement a laptop initiative. Determining such factors may allow educational institutions a base level awareness of students' perceptions. This awareness could provide insights into what needs to be done towards an effective laptop program. These initial findings warrant further investigation. To achieve a better understanding of all of the critical factors in a laptop program, future research should also include the perceptions of faculty, administrators, and staff as well as those of students.

REFERENCES

Agarwal, R., & Prasad, J. (1999). Are individual differences germance to the acceptance of new information technologies? *Decision Sciences, 30*(2), 361-391.

Al-Khaldi, M. A., & Al-Jabri, I. M. (1998). The relationship of attitudes to computer utilization: New evidence from a developing nation. *Computers in Human Behavior, 14*(1), 23-42.

Argawal, R., & Karahana, E. (2000). Time flies when you're having fun: Cognitive absorption and beliefs about information usage. *MIS Quarterly, 24*(4), 665-694.

Badamas, M. A. (2001). Mobile computer systems—security considerations. *Information Management & Computer Security, 9*(2/3), 134-136.

Barak, M., Lipson, A., & Lerman, S. (2006). Wireless laptops as means for promoting active learning in large lecture halls. *Journal of Research on Technology in Education, 38*(3), 245-263.

Bazillion, R. J., & Braun, C. L. (2001). Classroom, library and campus culture in a networked environment. *Campus-Wide Information Systems, 18*(2), 61-67.

Berkhout, A. L., Hendriksson-Larséna, K., & Bongers, P. (2004). The effect of using a laptop station compared to using a standard laptop PC on the cervical spine torque, perceived strain and productivity. *Applied Ergonomics, 35*(2), 147-152.

Brown, D. G. (2003). *Ubiquitous computing: The universal use of computers on college campuses.* Bolton, MA: Anker Publishing Company.

Brown, D.G., Burg, J.J., & Dominick, J.L. (1998). A strategic plan for ubiquitous laptop computing. *Communications of the ACM, 41*(1), 26-35.

Brown, D.G., & Petitto, K. R. (2003). The status of ubiquitous computing. *Educase Review, 38*, 25-33.

Changchit, C., Cutshall, R., & Elwood, S. (2006). Students' perceptions of the laptop program: what factors should be considered before implementing the program? *International Journal of Information & Communication Technology Education, 2*(2), 53-61.

Chau, P. Y. K., & Hu, P. J. H. (2001). Information technology acceptance by individual professionals: A model comparison approach. *Decision Sciences, 32*(4), 699-719.

Cutshall, R., Changchit, C., & Elwood, S. (2006). Campus laptops: Logistical and technological factors perceived critical to students. *Journal of Educational Technology & Society, 9*(3), 112-121.

Davis, F. D. (1989). Perceived usefulness, perceived ease of use, and user acceptance of information technology. *MIS Quarterly, 13*(3), 318-339.

Davis, F. D., Bagozzi, R. P., & Warshaw, P. R. (1989). User acceptance of computer technology: A comparison of two theoretical models. *Management Science, 35*(8), 982-1003.

Demb, A., Erickson, D., & Hawkins-Wilding, S. (2004). The laptop alternative: student reactions and strategic implications. *Computers & Education, 43*(4), 383-401.

Dori, Y., Belcher, J., Bessette, M., Danziger, M., McKinney, A., & Hult, E. (2003). Technology for active learning. *Materials Today, 6*(12), 44-49.

Elwood, S., Changchit, C., & Cutshall, R. (2006). Investigating students' perceptions on laptop initiative in higher education: an extension of the technology acceptance model. *Campus-Wide Information Systems, 23*(5), 336-349.

Finn, S., & Inman, J. (2004). Digital unity and digital divide: Surveying alumni to study effects of a campus laptop initiative. *Journal of Research on Technology in Education, 36*(3), 297-317.

Fouts, J., & Stuen, C. (1997). Copernicus Project: Learning with laptops: year 1 evaluation report. (ERIC No. ED 416 847)

Gottfried, J., & McFeely, M. (1998). Learning all over the place: Integrating laptop computers into the classroom. *Learning & Leading with Technology, 24*(4), 6-12.

Hall, M., & Elliott, K. M. (2003). Diffusion of technology into the teaching process: strategies to encourage faculty members to embrace the laptop environment. *Journal of Education for Business, 78*(6), 301.

Lehner, F., Nosekabel, H., & Lehmann, H. (2003). Wireless e-learning and communication environment: welcome at the University of Rosenburg. *E-Service Journal, 2*(3) 23-41.

Liaw, S. (2002). An internet survey for perceptions of computer and the world wide web: Relationship, prediction, and difference. *Computers in Human Behavior, 18*(1), 17-35.

Luarn, P., & Lin, H.H. (2005). Toward an understanding of the behavioral intention to use mobile banking. *Computers in Human Behavior, 21*(6), 873-891.

Mitra, A. (1998). Categories of computer use and their relationship with attitudes toward computers. *Journal of Research on Computing in Education, 30*(3), 281-295.

Mitra, A., & Hullet, C. R. (1997). Toward evaluating computer aided instruction: Attitudes, demographics, context. *Evaluation and Program Planning, 20*(4), 379-391.

Mitra, A., & Steffensmeier, T. (2000). Changes in student attitudes and student computer use in a computer-enriched environment. *Journal of Research on Computing in Education, 32*(3), 192-222.

Moore, G.C., & Benbasat, I. (1991). Development of instrument to measure the perceptions of adopting an information technology innovation. *Information Systems Research, 2*(3), 192-222.

Rola, M. (2002). Building IT into lesson plans. *Computing Canada, 28*(22), 28.

Tomek, I., & Muldner, T. (1999). Acadia advantage—evolution and experiences. *Interactive Learning Environments, 7*(2-3), 175-194.

Varvel Jr., V. E. & Thurston, C. (2002). Perceptions of a wireless network. *Journal of research on Technology in Education, 34*(4), 487-501.

Vaughan, J., & Burnes, B. (2002). Bringing them in and checking then out: laptop use in the modern academic library. *Information Technology and Libraries, 21*(2), 52-62.

Weiser, M. (1998). The future of ubiquitous computing on campus. *Communications of the ACM, 41*, 41-42.

Chapter VIII
Incorporating Geographic Information Systems for Business in Higher Education

David Gadish
California State University, USA

ABSTRACT

Schools of business can benefit from adoption of geographic information systems (GIS). A brief overview of GIS is presented along with an example of showcasing how it can be presented in a business school. Benefits for business schools, their students, and faculty are discussed. A comprehensive approach for promoting such spatial thinking is presented. The goal is to empower faculty to adopt GIS for their research and teaching, producing a large number of business school graduates that can promote spatial thinking in their organizations.

INTRODUCTION

This chapter discusses the introduction of GIS for business higher education in departments including management, marketing, economics, finance, and information systems. Although the value of geographic information systems (GIS) technologies is recognized by practitioners and educators alike, GIS instruction has yet to make significant inroads into business curricula (Miller, 2006). The goal is to have business school students, faculty, and administrators thinking about location and time issues relating to their research, teachings and business decisions using GIS technology to illustrate and implement business ideas in terms of location and time. We begin with a discussion of GIS in general and its benefits to business sectors and focus on business education. A detailed discussion of the approach ensues. It consists of an awareness campaign where business school faculty, administration, and students are made aware of the benefits of thinking about business, and business education in terms of location and time. Once an agreement is reached regarding the importance of spatial business thinking, resources must be secured to complement the approach. These include the purchasing and setup of GIS software, hardware and data. Faculty must be trained in use of GIS technology as well as in incorporating spatial thinking in the classroom. Faculty should also be introduced to the benefits of GIS in their research activities. The chapter concludes with lessons learned from the authors experience including the impact of spatial thinking and GIS technology on course curriculum.

California State University Los Angeles (CSULA) is reaching out to introduce spatial thinking in business education to build more academic-business bridges in the world. The school of business at CSULA, an AACSB accredited institute, is working to promote the use of GIS technology in business education, as well as in the multi-cultural business community that it services.

OVERVIEW OF GIS

GIS is an integrated computer system capable of capturing, storing, retrieving, analyzing and explaining spatial information that provides the user with knowledge of the location information in the context of time, about the world, a business, a project, or an objective. GIS is also a decision making tool that helps produce useful information in a cost-effective manner. The ability of GIS to analyze spatial data is frequently seen as a key element in its definition, and has often been used as a characteristic which distinguishes GIS from other systems. GIS facilitates spatial analysis which is a set of analytical methods. It requires access to both attributes of objects under study, and to their locational information and allows referencing traditional data sets to maps. Geographic information systems consist of a number of key components. These include computer hardware, software, data, procedures and people. GIS data consists of spatial or mapping objects as well as non-spatial attribute data. Spatial data includes points, lines, polygons, other graphical representations, as well as text that represent buildings, customers, roads and other real-world entities. GIS can help answer different types of questions. It can help you find what is at a particular location, where something specific is located, what has changed, which is the best way to get somewhere, what the pattern is, "what if" certain conditions arise.

GIS technology originated from computer aided design and drafting (CADD) systems initially used for engineering purposes. CADD systems were adapted, mainly by geographers to manage geographic and environmental data about earth.

Higher education institutions have largely focused on training a select number of GIS specialists in certificate and master's programs. Many of these specialists were hired by governmental organizations. This has resulted in a penetration of the technology in government and some private organizations. GIS is currently heavily entrenched in all levels of government across different departments and is making a substantial impact where it is used, including planning and coordination, and monitoring activities.

BENEFITS OF GIS FOR BUSINESS

Business knowledge is power and it can be increased by looking at the business data in terms of location and time. GIS enables viewing business information graphically, sharing information with others as well as making appropriate business decisions. GIS can be used for managing information about a business, a business sector, business activity in a region, country or worldwide (Grimshaw, 1999).

GIS can be used by businesses at a number of scales. GIS can be incorporated into individual projects, it can be used at the departmental level, at the enterprise level responding to overall organizational needs, or it can be used as a means for collaboration among multiple organizations. GIS allows organizations to make sense of large quantities of information that are prevalent in today's business environment.

In the last few years private business organizations started to see the benefits that GIS provides to the public sector and began adopting GIS technology for their own business needs. Table 1 lists a few of business sectors and how GIS technology can benefit them (Boyles, 2002; Harder, 1997). This serves to demonstrate the wide ranging applicability of GIS technology for business.

The GIS adoption process has proven to be lengthy and complex as there are no sufficient human resources that understand and appreciate spatial thinking and its benefits to business organizations. This gap in existing business and education appears more and more clearly as more and more businesses realize the benefits the implementation of the technology and their inability to cope with it. This provides business schools with an opportunity to fill an important educational gap and business need. Business schools should therefore begin and enhance their efforts of incorporating spatial thinking in their business education curriculum.

Table 1. GIS technology benefits to business organizations

Business Sector	GIS Contribution
Publishing (Newspapers)	Increase newspaper readership buy targeting new subscribers, Mapping courier routes
Banking	Measuring market potential
Retailing	Mapping customers, and providing custom advertising
Health (Gyms)	Evaluating suitability of sites for new gyms.
Dental supply	Realigning sales territories of sales people
Healthcare	Evaluating healthcare resources, analysis of demand for specific treatments by location to better serve the public
Real Estate	Determining where to locate commercial real estate—new shopping centers, new stores, by analysis of demographics and competition (Longley, 1996).
Food (Supermarkets)	Efficient delivery methods for food purchased via the web or by phone order to homes.
Insurance	Establish the value of real estate property to be insured.

BENEFITS OF GIS FOR BUSINESS EDUCATION

Benefits to Business Schools

Business schools are able to address business needs by educating future business leaders to think in terms of location and time. Business schools that will be early adopters of spatial thinking and GIS technology will have a competitive edge over other higher education institutions in satisfying the needs of business, promoting closer ties with these organizations. Business schools will also be able to attract those students that are interested in having a competitive edge once they enter the job market.

Business schools can also create new collaborations with industry, hosting workshops and extension classes for managers and executives.

Benefits to Students

Business students benefit from being exposed to a new way of thinking about business problems. This gives students a competitive edge over those that have not been exposed to such business thinking. Business students also benefit from exposure to GIS technology and having gained such a hands-on experience will be better able to manage real-world situations once they have started their careers.

The use of GIS technology allows for visualization of business issues, and its incorporation in classroom, lab and home study, would provide a more fun learning environment for students. This is likely to result in continued and possibly increased interest in pursuing business education.

This approach is different than that used in geography and environmental studies departments that train a select number of GIS experts. A complementary approach that involves educating a far larger number of students in spatial thinking is encouraged. The focus should be on training visionaries that are familiar with basic GIS functionality and can promote spatial thinking in the private sector.

Benefits to Faculty

Business faculty stands to benefit from spatial thinking complemented by the use of GIS technology in their teachings and research. Faculty will be better equipped to handle the analysis of real-world business problems by incorporating spatial thinking. Faculty can use GIS technology in the classroom to illustrate business concepts in terms of space and time by utilizing the visualization power of GIS technology to map business data. Business faculty is also able to use spatial thinking and GIS technology for their research objectives. GIS technology should help faculty explore new ideas discover new business patterns.

Benefits to the Business Community

Businesses will be able to hire business school graduates that are able to think spatially and have base knowledge of GIS technology. These graduates can look at the existing business processes and suggest new ways in which their organizations can leverage on their spatial thinking (Tomlinson, 2003; Wayne, 2002).

PROPOSED METHODOLOGY

The author of chapter has been working since 2002 to introduce GIS technology. This is a multi-step approach that begins with an awareness campaign lead by a champion and supported by one or more sponsors.

Introducing GIS requires a champion, which is a visionary within the school that believes in the approach and preferably is an expert in GIS technology. This person can be a faculty member in the business school if such expertise is available, be hired for this job, or come from a different college of the same higher education institution. The champion must lead an awareness campaign which should result in certain level of interest.

An Awareness Campaign

The proposed approach begins with an awareness campaign where the benefits of the use of GIS technology are presented to business school administrators, faculty and students. For an effective and comprehensive campaign both benefits and costs must be considered. Top school administration including the dean and associate deans must be clearly shown the benefits of the technology. It is important to bring top college administrators including the dean and associate deans to sponsor the vision early on in the process. This sponsorship does not need to include a financial backing in an environment of strained resources. Department chairs and their faculty must be behind the campaign as they will be the ones implementing it at the grass-roots level. Faculty must therefore believe in the benefits of GIS technology as it will benefit their students as well as their own research activities.

Incorporating GIS into the Curriculum

Incorporating GIS can be achieved on a number of levels. Initially one specialized GIS course can be introduced into the curriculum. The course should provide an overview of the theoretical aspects of spatial and temporal thinking, and will be complemented by an introduction to GIS technology and its use. The course should be focused on introductory business GIS applications. Business students will be exposed to ways in which spatial thinking can be incorporated into business decision making, business operations processes, and business management. The course would also show how GIS technology is used to solve specific business problems in the number of business sectors.

At CSULA, no GIS courses were offered at the business school prior to 2002. The author introduced GIS through the computer information systems (CIS) department of the business school. The author's course outline used at CUSLA's business school is presented in Table 2.

Table 2. Business GIS course: Topic outline

Topic ID	Description
1	What is GIS?
2	A Survey of the GIS Industry
3	Foundations of GIS: Data, Hardware, Software, Workflows, Science
4	Spatial Data: Collection, Build, and Maintenance
5	Spatial Data: Additional Sources of GIS data
6	Analysis of Spatial Data: Buffer Analysis, Network Analysis
7	Business Applications of GIS: Telecommunications, Utilities, Banking and Finance, Retail/Wholesale, Government Sectors
8	GIS Data Display
9	Introduction to the Global Position System (GPS)
10	Applications of GPS Technology to Business
11	Future of GIS for Business: Internet GIS, Intranet GIS, Field GIS, GIS Database Integration
12	Case Studies

This business GIS course is offered at both undergraduate and graduate levels simultaneously, and is open to all students at the school of business. The course complements the existing programs in the CIS department, as well as those students from other non-CIS majors at the school of business. Some of programs and benefits the course provides them are listed in Table 3.

Course work is complemented with individual student projects. Each student selects an area (business sector or specific organization) that interests them, and pursues a theoretical application of the ideas discussed in the course to the topic. The result is a paper discussing the existing state of GIS technology in the student's business area of interest. Each student shares their findings with the class during the last two weeks of the course. The paper provides students with an opportunity to see what others are doing in industry through Web and literature search, as well as provides them with the ability to dream up new and more effective ways of incorporating spatial thinking in their area of interest.

Having such a business course can serve as a springboard for students use their theoretical and practical experience in their studies in other courses offered in the business school.

In addition to offering such a new GIS specific courses, faculty can incorporate the use of GIS into a large number of business courses. This can be limited to assignments that ask students to start thinking terms of space and time to solve business problems, or assignments that explicitly require the use of GIS technology to analyze business data. An introductory course about business information systems could be modified to include a lecture introducing GIS technology. An introductory management course could be modified to include a lecture introducing the benefits of spatial thinking to the management of an organization.

Table 3. GIS technology benefits to the business school's programs

Department	Program	Benefits
CIS	B.A. Business Track	Provides spatial thinking geared towards business applications
	B.A. Technical Track	Discussion of possible ways to apply software development practices to spatial problems using GIS technology
	MSIS (Graduate)	Provides spatial thinking geared towards effective management of business organizations
Management	Undergraduate and Graduate	Provides spatial thinking geared towards business applications and business management
Marketing	Undergraduate and Graduate	Provides spatial thinking geared towards marketing applications
Economics and Statistics	Undergraduate and Graduate	Spatial thinking about local, regional and global economic issues including labor, transportation, international trade, banking, and regulatory agencies
Finance and Law	Undergraduate and Graduate	Spatial thinking about regional and global financial/monetary patterns

Purchase and Setup of GIS Software, Hardware, and Data

Incorporating GIS technology requires the purchase and setup of hardware and software. For the most part, existing hardware resources can be leveraged for GIS teachings and research. GIS requires more hardware resources than traditional computer programs such as word processors and spreadsheets. GIS software needs more disk space, and computer memory. Most modern desktop computers that are two or three years old have sufficient resources for GIS applications and therefore do not require any special or costly upgrades.

In the past, the purchase of GIS software was an expensive proposition. This however is no longer the case, as GIS software prices have considerably fallen in recent years. Falling GIS prices are attributed to increased competition in the GIS software market. This increased competition is derived from simplified tools for creating GIS software applications as well as a larger number of programmers and analysts that create GIS software. The result of this increasingly competitive GIS software market is the ability to purchase GIS software licenses at a low cost. Major software vendors are interested in having education institutions adapting their technologies and having their capabilities known to future business leaders. This has created an environment where GIS software can be licensed for GIS educational institutions at minimal costs.

Once licensing agreements have been established with GIS software vendors, the software needs to be installed on computer hardware in classroom environments, in computer laboratories, and in faculty personal computers. This can be done with minimal training by the system administrator of the business school.

Having set up the hardware and software capabilities for GIS teachings and research, attention must be turned to acquiring appropriate spatial data (mapping data) that would complement existing business data for analysis in the classroom as well as for faculty research purposes. Spatial data can be acquired from a number of sources in a variety of means. Some GIS data can be downloaded at no cost from the Internet. Business data can be purchased at an educational discount from private sector spatial data vendors. GIS data can be licensed from various levels of government, many times at no cost. In cases when spatial data cannot be located by an existing means it is possible to purchase or rent a hand held GPS/GIS device and use it to create a digital map that could then be used for GIS business analysis.

Training Faculty

Training faculty is fundamental to the success of the proposed approach. It is important for faculty members to feel comfortable in the use of the technology. It is also very important for faculty to be comfortable in explaining the benefits of the technology to themselves as well as to their students. It is important for faculty to be comfortable with the software and the thinking processes to an extent that they can convey the appropriate message and teach the technology to their own students at ease. Arriving at such a level of comfort with GIS technology may take time and resources. This may include sending faculty for training on the use of GIS software by GIS provided by GIS vendors. A better approach would be to bring GIS experts, preferably from within the business school, or if not available, from the Department of Geography of the same university, to teach business faculty in the use of the technology. Initial training must be followed up with regular events (possibly coinciding with the annual World GIS Day), whose goal is to reiterate and enhance the notion and the benefits of the use of the technology for business education, as well as refresh faculty with new GIS capabilities and tools and their benefit for business teachings and research. This ongoing process would be beneficial for existing faculty as well as new faculty that join the business school. Such meetings could be used to provide faculty and students an opportunity to share their research as pertaining to spatial thinking and GIS technology as well as their teaching experiences.

DISCUSSION

The introduction of GIS technology to a business school requires a leader, and must be done in a methodical way realizing that the process would take time. Some faculty may resist change due to fear of technology in general, and the perceived complications that are added with the adoption of GIS technology in particular. These concerns can be mitigated with time through a systematic and thorough campaign off educating faculty to the benefits of the technology and its used for their own teachings and research needs, and showcasing the benefits gained by those faculty that have adapted the technology.

Introducing GIS technology will and should have an impact on the course curriculum of the business school. As indicated earlier, one or more courses specializing in GIS would be added to the curriculum. Numerous other courses could be affected and their curriculum adjusted.

The proposed approach will, with time and effort, allow for increased awareness of spatial thinking and use of GIS technology to the benefit of business school administrators, faculty, and students. The benefits are expected to further propagate to the business sector and the community.

In the case of CUSLA's business school, the results are positive and promising. Students from various departments of the business school have taken the course and have commented that is provides them with a new way of seeing the business world. Student retention in the course is very high, and close to 20 percent of the students have indicated their intent to pursue business GIS learning beyond the existing framework. Many of these students have maintained are working on various research activities which incorporate

GIS and spatial thinking. Spatial thinking has been positively received by faculty in the school of business. The author has had an ongoing relation with a number of faculty from different departments request to present an overview of business GIS as part of their courses. The school of business continues to support these efforts in terms of financial and technical resources.

The author recommends the outlined approach to other business schools. The effort should be gradual, taking one small step at a time.

REFERENCES

Boyles, D. (2002). *GIS means business* (Vol. 2). ESRI Press.

Grimshaw, J.D. (1999). *Bringing geographical information systems into business.* Wiley, John & Sons, Incorporated.

Harder, C. (1997). *GIS means business.* ESRI Press.

Longley, P., &Clarke, G. (1996).*GIS for business and service planning.* John Wiley & Sons.

Miller, F., Mangold, W.G., & Holmes, T. (2006). Integrating geographic information systems (GIS) applications into business courses using online business geographics modules. *Journal of Education for Business, 82*(2), 74-79.

Tomlinson, R. (2003). *Thinking about GIS: Geographic information system planning for managers.* ESRI Press.

Wayne, R. (2002, March). Location, location, location. *SDMAGAZINE.*

Chapter IX
Programming Drills with a Decision Trees Workbench

Dimitris Kalles
Hellenic Open University, Greece

Athanasios Papagelis
University of Patras, Greece

ABSTRACT

Decision trees are one of the most successful machine learning paradigms. This chapter presents a library of decision tree algorithms in Java that was eventually used as a programming laboratory workbench. The initial design focus was, as regards the non-expert user, to conduct experiments with decision trees using components and visual tools that facilitate tree construction and manipulation and as regards the expert user, to be able to focus on algorithm design and comparison with few implementation details. The system has been built over a number of years and over various development contexts and has been successfully used as a workbench in a programming laboratory for junior computer science students. The underlying philosophy was to achieve a solid introduction to object-oriented concepts and practices based on a fundamental machine learning paradigm.

INTRODUCTION

A decision tree is a graphical representation of a procedure for classifying or evaluating an item of interest. It represents a function that maps each element of a domain to a value from a set of values; this value is typically a symbolic class label or a numerical value. Decision trees are excellent tools for supporting decisions, when a lot of complex information must be taken into account and the reasoning must be supplied for alternative paths (Mitchell, 1997).

Decision trees have two key merits when compared to other concept learners. First, they can manipulate a large amount of information due to the small computational power that is needed for the creation of the model of the underlying hypothesis (furthermore, the representation of the model does not demand excessive memory). Second, by providing classifications and predictions that can be argued about, they advance our insight in the problem domain.

Their success has motivated many researchers to attempt to improve decision tree learners. Efforts have mostly focused on pre-processing data (Musick et al., 1993; Quinlan, 1993), selecting splitting attributes (Breiman, 1984; Mingers, 1989) and tree pruning (Breiman, 1984; Quinlan, 1987).

Mundane but important tasks take up, usually, a large portion of the programming effort, when trying out a new idea. Such tasks include parsing input, creating data structures and statistics, printing and classifying. However, most of these typical components remain unchanged, even when a researcher wants to create a new tree algorithm. This observation necessitates the effort towards reusing the most flexible components, which can be easily adapted to each researcher's requirements.

The library described in this chapter addresses directly the problem of focusing research effort where it is mostly needed when one designs or implements a decision tree algorithm, which is the algorithm itself. The library is organized in components, each one corresponding to a clearly distinct stage of the tree building process. This architecture reduces the time of creating a tree algorithm by providing building blocks of the algorithm that do not need to be changed.

Two popular similar libraries are MLC++ (Kohavi et al., 1994) and WEKA (Witten & Frank, 2000), with WEKA being today the *de facto* choice and MLC++ being effectively sidelined. Both of them contain common induction algorithms (i.e., C4.5 (Quinlan; 1993), Naïve Bayes, ID3 (Quinlan; 1986)) under a unified framework. Moreover, they contain wrappers to wrap around algorithms including feature selection, discretization filters and bagging/combining classifiers and they provide a means for testing classifier accuracy.

This work will not replace those established and global machine-learning tools. Moreover, it does not compete with focused applications (Quinlan; 1993). Our library has a more limited scope: it focuses on providing the necessary infrastructure for creating and manipulating binary decision trees. Reducing the scope provides a more solid framework for the specific problem and results in a more attractive learning curve. Moreover, by limiting our attention to one domain, we can use the standard steps of decision tree learning as a pre-defined backbone, to which all new components must conform. Specificity, in this sense, allows for a tighter definition of which interface criteria components must satisfy and, eventually, results in a more structured (and easier) way of designing new algorithms.

The library is an open growing system (Christodoulou, 2001; Christodoulou et al., 2004; Christodoulopoulou, 2006; Drossos et al., 2000; Mpekou, 2006; Papagelis & Drosos, 1999) that supports the addition of algorithms and components, yet it is component and not algorithm oriented. The added architectural complexity it creates, from the software engineering perspective, is efficiently managed through a GUI, which provides an easy way to interchange "building blocks" between different tree implementations and to compare competing designs. In this respect, even if MLC++ and WEKA provide a good base of existing algorithms, they squarely trail our approach when the focus is on capability and usability to enhance the repertoire of algorithms.

The rest of the chapter is organized in six sections. In the next section, the basic characteristics of a decision tree algorithm are presented and, then, the library is described and the mapping of each component of a decision tree algorithm to each component of the library is explained. We then describe the shell used for creating and manipulating new objects and follow with the description of the GUI and assorted utility tools. We then present some observation on the software engineering perspectives of the system along with a historical recount of events that lead to its development, and we conclude with our observation and some issues for future research and development.

DECISION TREE BUILDING AND SYSTEM ARCHITECTURE

The process of building a decision tree can be divided in a number of sequential steps.

First, the training data must be acquired. The next step involves the pre-processing of the training data to transform the initial data to a suitable instance set format, which is then used to build the decision tree. The pre-processing step may involve a selection of any of the following: discretization of continuous features, handling of missing values, reordering of the instances in the instance set, exclusion of some attributes and similar steps. The instance set, after pre-processing, is used for the building of the decision tree, usually by recursive splitting on test values. The recursion terminates when a pre-specified criterion is reached. Optionally, one can prune the decision tree or use it to classify data, either labeled (accuracy testing) or unlabeled (actual classification).

These steps are shown in Figure 1 (dotted shapes indicate optional parts).

The Development Environment

A key issue was the selection of the development environment. An object-oriented language seemed necessary since the library demanded

Figure 1. Basic steps in decision tree building

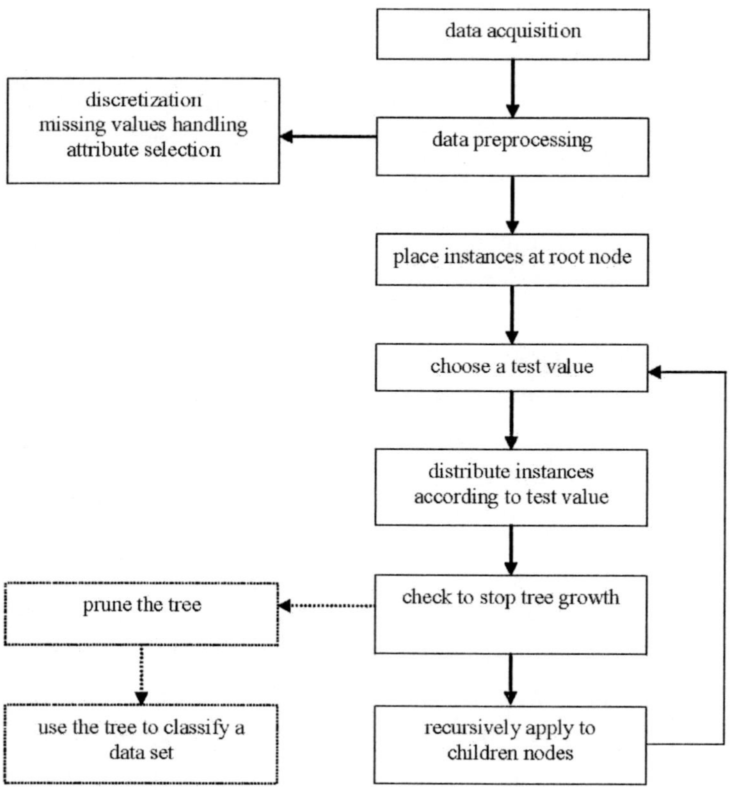

polymorphism and inheritance to accommodate the design of algorithms based on pre-specified templates. Furthermore, the system should support new algorithms and conventional ones, without the need to know its inner organization in detail.

These considerations led to the selection of Java as the coding language. Admittedly, from a requirements perspective, homing on the development platform so early may seem premature. In retrospect the decision was not so risky, as the Java language itself has been built around dynamically extendible libraries and this was a remarkably close fit to our architectural requirements.

Basic Building Blocks

After selecting the programming platform, the appropriate data structures were designed. First, the relatively stable parts of the decision tree lifecycle were identified, then they were transformed to primitive components with specific functionality and, finally, components were interconnected into a central framework. This framework acts as a communication protocol between the library components and is shown in Table 1.

The library comes with data classes, containers for the creation of data collections and an extended set of methods to manipulate them. A generic wrapper combines the decision tree components to derive a new decision tree. This wrapper contains methods for specifying all the necessary algorithms (building, splitting, pruning, classifying) to be used for creating the decision tree.

The Class Explorer

A significant problem that arose after the development of the library was the difficulty to test the combination of algorithms and to inspect the objects created. Of course, it was much easier than having to test everything from scratch, but it was felt that the overall development effort would not deliver an order-of-magnitude faster way of putting things together and observing them in action. The only way to perform some sort of testing was with the help of Java coding, by creating and executing Java programs that called the new algorithms.

This problem motivated the creation of class explorer, a command line tool to manipulate the library objects. Class explorer enables the user to construct objects "on the fly," even from classes that could have been unknown during library compilation time (this is based on the Java Reflection mechanism). Furthermore, it allows the organization of a series of actions in terms of command scripts that can run as batch programs.

Table 1. Components of a decision tree algorithm

Component	Role in decision tree framework
Parsing	Acquisition of data from various sources
Pre-Processing	Discretization, handing of missing values, ...
Splitting	Select test-values to be used during decision tree expansion
Distribution	Statistics about values' distribution among problem classes
Pre-Pruning	Stop the expansion of the decision tree
Post-Pruning	Prune the tree after it has been built
Tree Build	Batch mode, incremental mode builders
Classify	Classify instances using a decision tree

Scripts are an absolutely essential productivity tool: one can create algorithms that combine library components in an ad-hoc way, thereby being able to create command sequences that allow very enhanced functionality (for example, cross-validation testing). These features are key factors that differentiate our approach from established alternatives (MLC ++, WEKA) and, with class explorer, have been standard since the first release (Papagelis & Drosos, 1999).

The User Interface

Class explorer is a powerful tool but it targets a relatively seasoned researcher who is not apprehensive of learning a few new directives. This, predictably, motivated the development of a graphical user interface (Christodoulou, 2001) to facilitate user interaction with the library through class explorer. The GUI can be used to support a number of different actions, the most significant being object creation, script running, tree building (wizard assisted), tree visualization, and workspace management.

From an architectural point of view, the GUI is a layer above the class explorer. This particular decision may limit direct access to the library objects, but it lets the GUI access them in a more principled way and ensures that, if they change, one need only focus on the interaction between them and the class explorer, leaving the GUI intact.

Figure 2. The object explorer

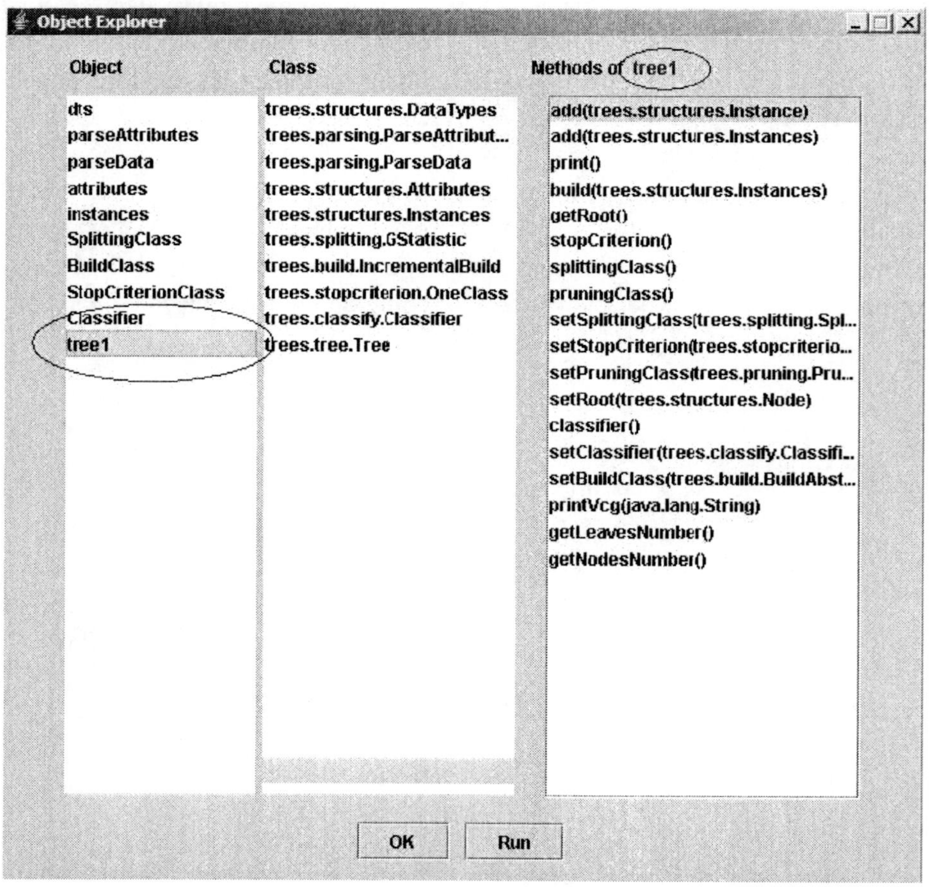

Figure 3. The tree creation wizard

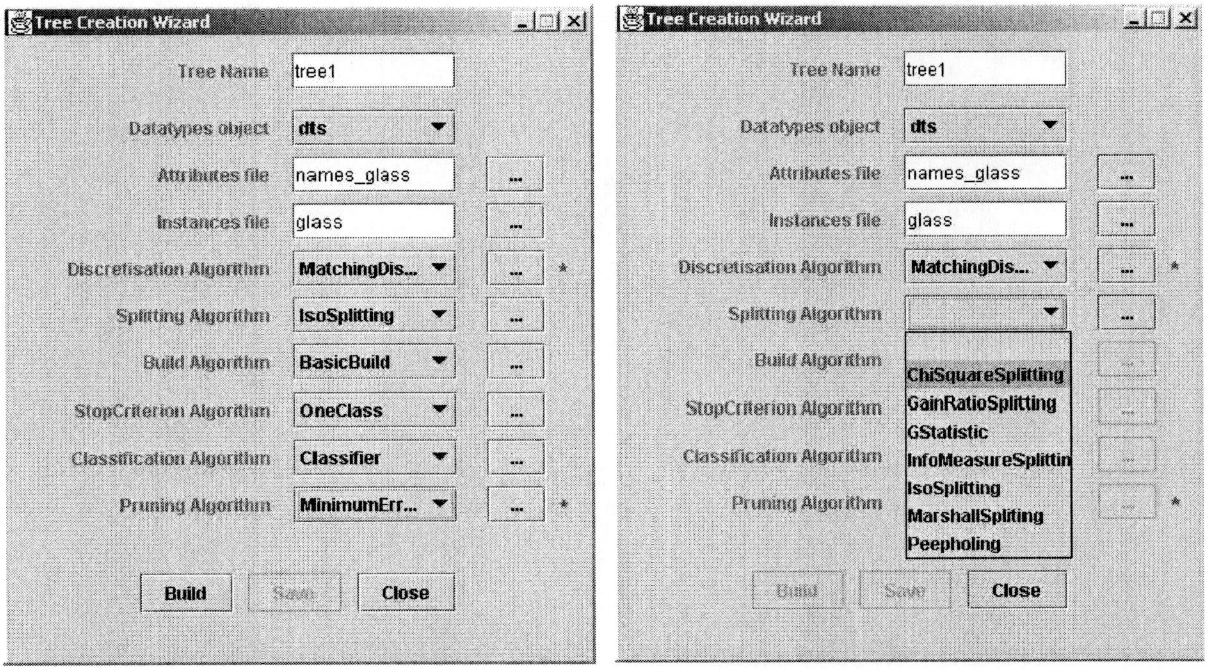

Figure 4. The tree visualizer

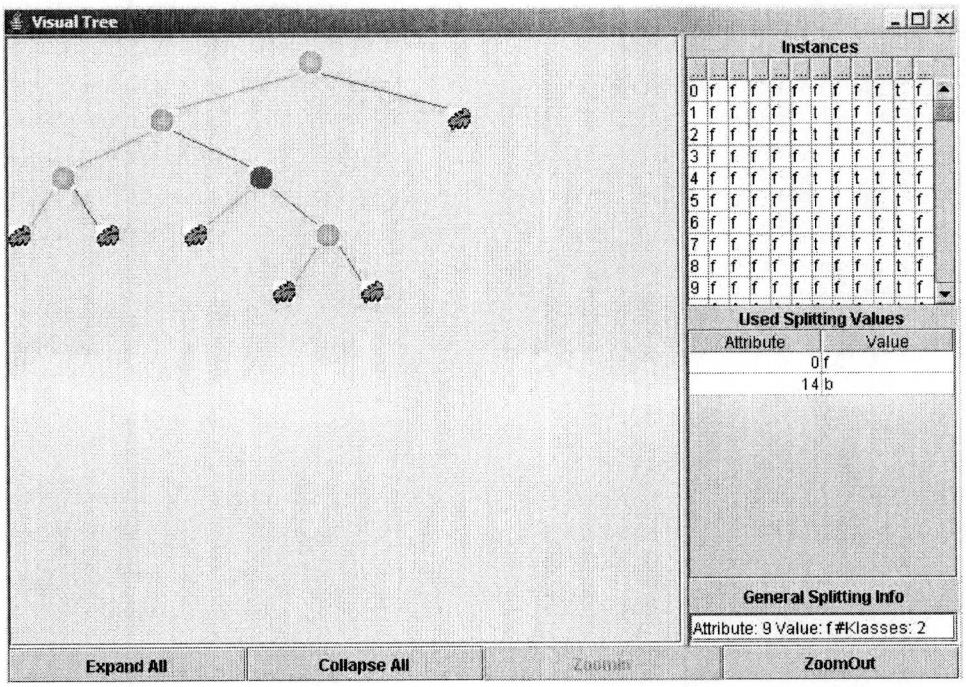

A key tool of the GUI is the object explorer (Figure 2), which is a fully visual tool for exploring all objects created during a system run. Furthermore, the user can create a new object or run a method of any live object. The system locates all possible inputs for any chosen method and prompts the user to choose among inputs, thus minimizing the risk of programming conflicts.

The tree creation wizard (Figure 3) manages the steps to build a decision tree. Specifying the instances and attributes files, choosing discretization, splitting and pruning criteria, and specifying a classification method are among these steps. After all required inputs have been specified the tree is built and can be subsequently saved.

The GUI also contains a tree visualizer (Figure 4). The tree visualizer constructs a visual representation of any (user-selected) sub-tree.

The GUI houses three further productivity tools: the script editor, the workspace manager and the tree generator. A script is defined as a sequence of commands of class explorer. The script editor may be used for the modification of an already existing script of the library or for the creation of a new one.

All the actions performed, either through the class explorer or the GUI, are stored in the workspace, which is a virtual repository of actions. One action performed by using the GUI may correspond to more than one action of the script language. Such actions are first decoded and then saved in the workspace. The user can browse through the actions that have been performed and fully edit the workspace. The idea of the workspace is to store actions and commands that are often executed by a user (i.e., instance and attributes parsing). These actions can de retrieved through a simple file opening, which is preferable to executing all actions one by one. One can think of workspace management as a close relative to action recording mechanisms that allow the construction of macro-commands.

The tree generator is a tool for converting a Java program that uses structures and components of the library (for example, an algorithm designed by a user of the library) to a Java package containing all the necessary information concerning the fraction of the library that deals with the specific experiment. The output of the tree generator may be used as a stand-alone program without requiring the whole library. The tree generator works mainly by automatically concatenating code excerpts that together implement a desired, pre-specified, functionality. The source for the generator can be either a Java program or a class explorer script. In a sense, the generator is a close relative of conventional linkers and even though it is the least developed of the library's features, it holds a significant potential as a tool that will facilitate the integration of algorithms designed with the toolkit into third-party systems.

THE ALGORITHM REPOSITORY

To date, the library embodies algorithms referring to the entire set of basic steps in building a decision tree. Taking the components of a decision tree building algorithm from the beginning, the step of preprocessing the values of the data set is now implemented using a chi-square based discretization algorithm (Mingers, 1989). We do not yet deal with missing values. As far as splitting is concerned, the user may choose among a chi-square based algorithm (Mingers, 1987), the G-Statistic (Mingers, 1989), the Information measure (Quinlan, 1983), the gain ratio (Quinlan, 1986), a variation of peep-holing (Catlett, 1992; Musick et al., 1993), the Marshall correction (Mingers, 1989) and Iso-Splitting, a simple benchmark algorithm (Papagelis & Drosos, 1999). Trivial incremental learning is supported.

The expansion of the decision trees is controlled by a simple pre-pruning algorithm, which terminates the expansion of the tree on the appearance of a node that contains instances of one and only class.

Quite some effort has been expended on the implementation of post-pruning algorithms. Currently, the post-pruning algorithms available are error-complexity (Breiman et al., 1984), a two-steps procedure which is based on the creation of multiple trees and the selection of the one that presents the fewer classification errors, minimum-error (Quinlan & Rivest, 1989), a method for finding the tree that theoretically would present the lowest error rate on classifying independent data sets, and reduced-error (Quinlan, 1987).

We assert that the architecture of the overall system, which allows for both easier insertion of new algorithms concerning any of the building blocks and "on the fly" combination of algorithms while building a new tree, should not only facilitate but also motivate users to contribute to the fulfillment of an integrated research platform.

It is also important to note that, even though the library itself does not provide constructs for comparing different algorithms, this can be done through scripting at the class explorer level, where evaluations and comparisons based on standard methodologies can be easily defined and edited.

ON ARTIFICIAL INTELLIGENCE AND ITS SOFTWARE ENGINEERING EDUCATION PERSPECTIVE

Getting AI technologies to work has usually been a difficult task. At one point, AI acceptance will depend on seeing it actually work on a commodity basis (for example, fraud detection, load default prediction, predictive maintenance, etc.). On the other hand this apparent success does much to remove many of the original difficult expectations of the AI concept in the sense that when technical intricacies are being explained, applications tend to be described as mainstream IT as opposed to real AI exhibits.

Having observed this fundamental relationship, we strongly believe that creating interfaces to AI technologies is a must for AI professionals and scientists. Professionals need to understand and trust research results to use them. Academics need to cross-fertilize ideas across technology and research domains without having to be proficient in all of them. And, students have to be taught the essential AI concepts but at the same time be able to appreciate the demonstration of the software engineering and architectural disciplines in AI products.

The viewpoint of this section is to lead to an appreciation of both the technical and the non-technical tasks that were laid before the authors, which have influenced the design and development of the library, and which have been a key influence to the way it is still being developed.

The original library prototype appeared as the result of a final year project (Papagelis & Drosos, 1999), where one of the authors acted as a student contributing to the project and the other author acted as a project supervisor. That project was carried out in the Department of Computer Engineering and Informatics, at the University of Patras, Greece. While the initial design goals were exactly the ones that were documented in the opening section of this chapter, the implementation scope was much more limited: the library featured just an extensive API for Java programming and an interpreted shell-like scripting language for rapid algorithm prototyping (the class explorer).

After the conclusion of the project, the supervising author was assigned with running the programming laboratory at the same department. The typical programming laboratory is organized in a set of programming exercises of increasing difficulty. The theory goes that students will be exposed to increased levels of technical sophistication and, therefore, obtain mastery of the language and the associated techniques. While this approach is valuable to create a technical appreciation of a programming language, it usually ends up delivering just that: a better appreciation of language constructs, with little or no insight as to the type of problems where this knowledge is applicable.

We decided to offer students a brief exposure to decision trees as a problem domain and to exploit the newly-delivered final year project as the test-bed for the assignments. Even though both the software and the documentation were well above the average expected of a final year project, they were practically untested. However, we also wanted to explore whether a conventional course could be used to sharpen some of the soft skills demanded of programmers: code inspection with sometimes inadequate documentation produced by others, teamwork and integration requirements.

The programming laboratory was run for three consecutive academic years. The overwhelming task for the first year of the course was to massively engage the student population in library extension work, by specifying exercises that demanded a basic understanding of the library to carry out the assignment (development of a visual browser of decision trees) and by allowing open-ended approaches for high-caliber students who were encouraged to specify and add features that made use of advanced understanding of either algorithmic or programming concepts. The key was to let the students define for themselves what constituted a worthwhile feature addition. One team with ample development talent undertook the code generator project that demanded a very strong mastering of the Java architecture and introductory compiler topics.

The second year of the course was different: high-caliber teams (self-assessed, again) were given integration tasks that demanded the understanding of both the initial library and the extensions built during the first year, whereas the rest of the class were given programming assignments that created new algorithms. The second-year procedure introduced a significant element of management risk. High-caliber teams were asked to rush through their assignments so that the rest of the class would be able to use the integrated software for their own assignments. The key issue here was to convince the class that the understanding of the library demanded code-inspection and research work that did not rely on the high-caliber teams being on time.

By the start of the third year development ceased; we took stock of library modifications and of a further final-year project providing new extensions (Christodoulou, 2001), and we altered the approach. All student teams were given modified pieces of obfuscated code and were asked to fit the missing pieces as well as understand to which parts of the library the code belongs to. The approach delivered larger but easier projects and trained the students in code inspection. No new software was produced.

Table 2. Library project development and class work timeline

Stage	Class Activities	Non-class Activities	Year
1	Prototype (Papagelis and Drosos, 1999)		
2	Extensions I (visual browser of decision trees, code generator)		1
3a	Integration of Extensions I	GUI development (Christodoulou, 2001)	2
3b	Extensions II (algorithm development)		
4	Code Inspection and Validation (of Stage 2)		3
5a	Distance learning material for object orientation (Mpekou, 2006)		
5b	Certification and validation of algorithms (Christodoulopoulou, 2006)		

Today, the whole package has been reviewed from two points of view, each one in the context of a final year project. One such project dealt with developing material for supporting the teaching of object oriented concepts (analysis, design and programming) in the possible context of a distance learning course (Mpekou, 2006). That aspect was already explored as a possible development direction in our earlier work (Kalles & Papagelis, 2006) and the development of the educational content was a necessary step towards that goal. The other project dealt with quality assurance aspects by establishing and carrying out a methodology for validating the algorithms developed by students and incorporated in the system, attempting to ensure that no stray code is left in the package and that the system will retain its application quality besides its educational value (Christodoulopoulou, 2006).

The above timeline can be briefly summarized in Table 2 (the year annotations only refer to work carried out in the laboratory course context).

The educational goals of achieving hands-on experience on large systems, teamwork, programming drill and commanding the attention of high-caliber and modest-effort students were all achieved with surprisingly lower than anticipated demand on tutors' time. However, every year recorded a different experience from the previous one and this has also significantly added to the quality of the software, since the development was always overseen by the authors and the student projects were steered accordingly. For the authors, and from the software engineering perspective, the development history itself is a significant lesson.

We now believe that our experience with self-styled high-caliber groups seems aligned with views about the success of agile methods (Armitage, 2004; Hedin et al., 2003) as well as the fact that students should be given the tools to judge their own progress (Edwards, 2003). When big or difficult extensions were sought for the software, group working seemed to be essential. We stress "seemed" as this estimation is based on what the student contributors themselves reported. Although the difficult extensions were considerably fewer when compared to the more predictable ones, it turned out that those teams were not dominated by a group member. It is also interesting to note that no high-caliber group decided to downgrade its assignment even though this was an option.

There is a caveat, however: we are talking about extensions undertaken by talented people, yet these very people have limited exposure to large systems. The same extensions would probably be better handled by any single good final year student. However, even this finding and this speculation suggest that peer "pressure" can be a significant driver in achieving group understanding of a difficult concept.

In contrast, less demanding extensions demonstrated a large portion of them being effectively an individual-led team of students. Therein, the group leader was typically responsible for the design and implementation of the extensions. This finding has more often-than-not been corroborated by academic colleagues and it is seems to be an inherent limitation of group projects. In that context, the group members tend to cluster around some strong-performing individual who sometimes also acts as their tutor throughout the project. The strong-performing member could sometimes be a possible member of a high-caliber team, however, when students are left to their own when asked to come up with groups for projects, social pressures may also influence the formation of groups.

Still, the process led to a much deeper appreciation of the need to design learning experiences where peer learning must be actively designed into the course (Barak et al., 1999). Thus, we now believe that there are some further learning enhancements one could make to the course in the direction of software engineering training. The most promising one, which would also take account of the need to accommodate a community of groups at various performing levels, is the introduction of open-source development concepts (Augustin, Bressler, & Smith, 2002), where the requirement for distant development, cross-fertilization of ideas and group software maintenance is most pronounced. This should indeed be an excellent training ground for future software engineers.

CONCLUSIONS AND FUTURE DIRECTIONS

Researchers and specialist business consultants alike could be prime users of the described toolkit, since it provides an efficient way to build new or modify existing algorithms, without having to worry about essential or rather standard parts of a decision tree algorithm. Students should also benefit from using the toolkit, as it constitutes an integrated environment for practicing machine learning and software engineering skills. Still, polishing today's version of the software is still essential before attempting a large-scale diffusion. It should then come as no surprise that we aim to keep pursuing the development agenda on the same track, namely by involving students, then consolidating the contributions and moving forward.

We have learnt a key technical lesson from an educational point of view: Java is an excellent language for educational purposes if one manages to teach it in an object-oriented context. The object-oriented context is easier said than done, since it demands a command of fundamental concepts that evade many junior undergraduates and their tutors. Our approach has been successful as a teaching paradigm only because the library was conceived and implemented as a Java-based object-oriented system itself. That the students had to first understand the problem domain, before embarking on coding experiments, meant that they were exposed to a system that applied basic object-orientation concepts in a non-trivial fashion. We thus claim that this top down approach in programming drill is essential if students are to be exposed to sophisticated concepts; these concepts do not exist in a vacuum and setting the surrounding context is a key success element.

Another key observation of the development-*cum*-teaching exercise has been that the process has been able to support several hundreds of students with surprisingly little resources in terms of physical presence, beyond lectures. The establishment and active management of a discussion forum provided the background for the successful collaboration of the most active teams, resulting in significant knowledge dissemination towards other teams as well. This was the type of knowledge that can only flow through peer learning, since it refers to technical details and a knowledge acquisition momentum that can be mastered by active contribution to group efforts. This experience has given rise to the expectation that the software, after this collective development effort, can be re-situated to teach programming concepts in a distance learning context. We aim to design a course to test exactly this direction, as we have already taken the step of developing content for that purpose (Mpekou, 2006).

We now elaborate on the educational implications of some technical extensions that we are considering. As described, the GUI takes advantage of the script language of Class Explorer. The script language may be enriched to allow more extended flow control in the tree building process. This would be a motivation for the creation of new GUI features shifting it explicitly towards a visual programming context. This particular extension, however, also has to be judged against the possibility of diluting the system's educational value.

We have initially decided to only handle binary trees; this reflects the original research oriented focus of the project. We do not plan to provide support for generalized trees, because the research potential does not compensate for the expected programming workload, however, we are keen to explore that direction for programming projects.

Summarizing, we first thought our work would be successful if researchers would associate quickly with the philosophy of components, accepting that they should focus on core research as opposed to trying to work out every implementation detail, and deciding to take a primer in Java that will allow them to become productive straightaway. Trying to develop the system to a standard and to a scope where this would be possible, we had no option but to revert to large scale development. This interleaved rather nicely with a software engineering education component and it was intriguing to observe that our initial assumptions tend, more often than not, to be satisfied today in most industrial and academic environments. Hence, we believe we are still on the correct track.

ACKNOWLEDGMENT

We acknowledge the contribution of all colleagues whose theses have contributed to our project and of Katerina Hatzara who assisted us in system maintenance. We acknowledge the contribution of all course students but we have also praised the extraordinary motivation or contribution of some of them. We also thank the postgraduate students who have assisted in running the laboratory.

Special credit is due to our teacher, Professor Paul Spyrakis, who introduced us in a successful and exemplary way to the concept of students programming for students. His series of programming projects on Operating Systems, assigned to undergraduates at the host department, ran for nearly 20 years.

Finally, we acknowledge the work of anonymous reviewers who have that has improved the presentation of our work. Earlier versions of this work have appeared in various venues (Christodoulou et al., 2004; Kalles & Papagelis, 2006) and a discussion on the educational perspective appears in a separate paper (Kalles, 2007).

REFERENCES

Armitage, J. (2004). Design: Are agile methods good for design? *ACM interactions, 10*(1), 14-23.

Augustin L., Bressler D., & Smith, G. (2002). Accelerating software development through collaboration. In *Proceedings of the International Conference on Software Engineering* (pp. 559-566). Orlando, Florida.

Barak, M., Maymon T., & Harel, G. (1999). Teamwork in modern organizations: Implications for technology education. *International Journal of Technology and Design Education, 9*(1), 85-101.

Breiman, L., Friedman, J. H., Olshen, R. A., & Stone, C. J. (1984). *Classification and re-gression trees*. Belmont, CA: Wadsworth.

Catlett, J. (1992). Peepholing: Choosing attributes effectively for megainduction. In *Proceedings of the 9th International Workshop on Machine Learning* (pp. 49-54). Aberdeen, Scotland.

Christodoulou, S. (2001). *A development environment for programming libraries for decision trees*. Diploma Thesis at Computer Engineering and Informatics Department, University of Patras, Greece (in Greek).

Christodoulou, S., Hatzara, K., Kalles, D., & Papagelis, A. (2004). Building decision trees with components. In *Proceedings of the 4th Panhellenic Conference on Artificial Intelligence* (pp. 42-51). Samos, Greece.

Christodoulopoulou, E. (2006). *Certification and validation of algorithms for decision trees*. Degree Thesis at Hellenic Open University, Patras, Greece (in Greek).

Drossos, N., Papagelis, A., & Kalles, D. (2000). Decision tree toolkit: A component-based library of decision tree algorithms. *In proceedings of the 4th European Conference on Principles and Practice of Knowledge Discovery in Databases*. Lyon, France.

Edwards, S.H. (2003). Improving student performance by evaluating how well students test their own programs. *ACM Journal on Educational Resources in Computing, 3*(3). Retrieved from http://portal.acm.org/citation.cfm?doid=1029994.1029995

Hedin G., Bendix L., & Magnusson, B. (2003). Introducing Software Engineering by means of Extreme Programming. In *Proceedings of the 25th International Conference on Software Engineering* (pp. 586-593). Portland, Oregon.

Kalles, D. (2007). Students working for students on programming courses. *Computers and Education (in press)*.

Kalles, D., & Papagelis, A. (2006). Managing the decision tree life-cycle with components. *International Journal of Information and Communications Technology Education, 2*(3), 1-13.

Kohavi, R., John, G., Long, R., Manley, D., & Pfleger, K. (1994). MLC++: A Machine Learning Library in C++. *Proceedings of IEEE Conference on Tools with Artificial Intelligence* (pp. 740-743). New Orleans.

Mingers, J. (1987). Experts Systems—rule induction with statistical data. *Journal of the Operational Research Society, 38*, 39-47.

Mingers, J. (1989). An empirical comparison of selection measures for decision tree induction. *Machine Learning, 3*, 319-342.

Mitchell, T. (1997). *Machine Learning*. McGraw Hill.

Mpekou, P. (2006). *Algorithmic library for decision trees: Development of educational content for decision trees and for object oriented designing and programming*. Degree Thesis at Hellenic Open University, Patras, Greece (in Greek).

Musick, R., Catlett, J., & Russel, S. (1993). Decision theoretic subsampling for induction on large databases. In *Proceedings of the 10th International Conference on Machine Learning* (pp. 212-219). Amherst, MA..

Papagelis A., Drosos N. (1999). *A decision trees components library*. Diploma Thesis at Computer Engineering and Informatics Department, University of Patras, Greece (in Greek).

Quinlan, J. R. (1983). Learning efficient classification procedures and their application on chess endgames (p.469). In R. S. Michalski, J. R. Carbonell, & T. M. Mitchell (Eds.), *Machine learning: An artificial intelligence approach* (pp. 463-82). Los Altos, CA: Morgan Kaufmann.

Quinlan, J. R. (1986). Induction of decision trees. *Machine Learning, 1*(1), 81-106.

Quinlan, J. R. (1987). Simplifying decision trees. *International Journal of Man-Machine Studies, 27*, 221-234.

Quinlan, J. R., & Rivest, R. L. (1989). Inferring decision trees using the minimum description length principle. *Information and Computation, 80*(3), 227-248.

Quinlan, J. R. (1993). *C4.5: Programs for machine learning*. San Mateo, CA: Morgan Kaufmann Publishers.

Witten, I., & Frank, E. (2000). *Data mining: Practical machine learning tools and techniques with java implementations*. San Mateo, CA: Morgan Kaufmann Publishers.

Chapter X
CareerQuesting Revisited:
A Protocol for Increasing Girls' Interest in STEM Careers

Karen S. White
Purdue University, USA

Mara H. Wasburn
Purdue University, USA

ABSTRACT

This chapter develops an educational strategy to foster the interest and persistence of middle school girls in science, technology, engineering, and mathematics (STEM) careers, using existing Web sites. Criteria are specified that enable middle school teachers to evaluate Web sites as supplemental learning activities within prescribed curricula. In particular, the evaluative criteria help evaluate sites that provide materials appealing to boys and girls, allowing teachers to adopt them without concern that they are providing an unfair advantage to girls.

INTRODUCTION

In September, 2001, the Council on Competitiveness, a group of industrial, university, and labor leaders whose mission is elevating national competitiveness to the forefront of national consciousness, launched the building engineering and science talent (BEST) initiative. It is described on the Council Web site as a public-private partnership "to identify the most effective strategies for building a more diverse science, engineering, and technical workforce and to bring best practices to communities nationwide" (Council on Competitiveness, 2004). One of the first reports to emerge from the BEST program begins:

There is a quiet crisis *building in the United States—a crisis that could jeopardize the nation's pre-eminence and well-being. The crisis has been mounting gradually, but inexorably, over several decades. If permitted to continue unmitigated, it could reverse the global leadership Americans currently enjoy.* (Jackson, 2004)

This report goes on to cite "the gap between the nation's growing need for scientists, engineers, and other technically skilled workers, and its production of them."

The literature refers to a "leaky" pipeline of female talent, leading to "under-representation" of women in the STEM (science, technology, engineering, and math) professions. Much research documents this problem, tracing attitudes and behaviors of girls from elementary school through graduate studies and employment. The studies reflect a clear gender distinction at all ages, (Freeman, 2004; Jones, Howe, & Rua, 2000) showing that as girls grow older, there is steady decline in the number expressing interest in STEM subjects, and a corresponding decline in the numbers of women entering higher levels of study. Numerous programs proposed, and implemented over the past decade to improve gender equity in STEM education and workforce. These programs aim to provide information and other support for women and girls, allowing them to make better informed decisions with respect to their educational activities and career planning (AAUW, 2004; National Science Foundation, 2003). Given the proliferation of such efforts, some measurable effect on "entry and persistence" of women into these professions should be expected. However, data do not indicate substantial gains (Freeman, 2004; Huang, Taddese, & Walter, 2000).

The apparent failure of these programs may be due to a failure to implement or adopt them broadly. In light of recent political pressure to improve standardized testing scores, teachers must devote most of their time to the specific curriculum requirements. They may lack the time to locate and evaluate additional resources to address such gender equity issues, or may lack the budget to adopt commercially available resources. Even when appropriate materials are available for little or no cost, such as those on Web sites for many gender-equity programs, teachers may not understand how or when to use them (Furuta et al., 1999; March, 1997).

Web site-based resources are often developed as part of informal experience to be used as a supplement to regular classroom activities. However, it is unlikely that students will seek out these resources on their own. Additionally, research suggests that only a small percentage of teachers believe it is appropriate to address gender-equity concerns in the typical classroom (Bullock, 1997). Those who do may not have support of the school administration to adopt programs perceived to benefit only girls (Posnick-Goodwin, 2005).

The goal of this paper is to identify a set of criteria for use by teachers in evaluating existing Web sites with the potential for increasing interest and persistence of middle school girls in STEM fields. In particular, the criteria emphasize materials that might appeal to both boys and girls, allowing teachers to adopt them without concern that they are providing an unfair advantage to girls.

REVIEW OF THE LITERATURE

There are many alternative explanations for why women continue to avoid STEM professions. Early studies cited difficulties in mathematics and science, or avoidance of these subjects by girls as the main concern. Girls entered college level studies without adequate preparation for STEM fields, and so avoided them. Programs designed to stimulate interest in mathematics and science,

(GirlTech.org, 2004) to make them entertaining and appealing to girls, were offered as a means of addressing this problem. Other research suggested that it is the very nature of scientific disciplines and the research questions they ask that makes them less appealing to women (Blickenstaff, 2005; Rosser, S. V., 1997). Some recent research has offered within-child differences as the explanatory variable (Waite, S. J., Wheeler, S., Bromfield, C., 2007).

Other research indicates that skills development programs have, in fact, been successful, and that girls are participating equally in middle school and high school studies in science, mathematics, and technology, then entering college studies without the skill gap observed in the early 1990's (Clewell & Campbell, 2002; Freeman, 2004; Mead, 2006; Watt, Eccles,& Durik, 2006). If this is true, then another influence must be responsible for the continued absence of females involved in college degree programs, or seeking employment in STEM fields.

Often, studies report that girls' and women are choosing other fields of study because they are "not interested" in STEM subjects (McDonnell, F., 2005; Morgan, Isaac, & Sansone, 2001). The problem may be explained as being linked to gender stereotypes, (Seymour, 1998) leading girls and young women to dismiss certain jobs as too "masculine." Educational materials, books and Web sites, typically address such stereotypical thinking by providing "virtual" role models for girls through biographies of female scientists and engineers. Alternately, there are Web sites devoted to increasing interest in, or raising awareness of, STEM careers as exciting and challenging, with the intention that young girls will then aspire to these in spite of the stereotypes. These resources highlight scientific and engineering specialties thought to appeal more strongly to females. Many such programs include a substantial Web site component, providing a mechanism for distribution of educational resources for use by students, teachers, and even parents.

Informal vs. Formal Educational Strategies for Gender Equity in STEM

While Web sites may have some beneficial effects, some research indicates that girls would benefit more from increased *curricular integration* of science, technology, and math (BEST, 2004). For Web-based materials to be effective, the target audience must be directed toward to the resources, or alternatively, exhibit a strong interest in the subject in order to seek them out (Millar & Shevlin, 2003). Given that one goal shared by most of the Web sites is to "develop" interest, it follows that the target audience in this case is not inclined to find these sites independently. The medium itself has connotations of being too technical, and girls are assumed to have no interest in computers (Hanson, 2002).

The gender & science digital library (GSDL) project has addressed the needs of teachers seeking to provide an "interactive collection of high-quality, gender-equitable science, technology, engineering and mathematics (STEM) resources for K-12, higher education (community college and university), women's studies, teacher preparation programs, and informal learning environments" (EDC, 2004). Materials accessed on the GSDL Web site have been reviewed for quality, providing the user with confidence in their use. The site serves as a clearinghouse for materials, providing a description referencing the basic content and aims of the material, appropriate age groups, and other relevant information (Hanson, 2002). However, the primary goal of the GDSL project is to make available this type of resource. The site does not clearly provide guidance as to how or when to access these resources, or in what context, except to the extent that the referenced materials do so.

Also, the GSDL project attempts to learn more specifics regarding what teachers are looking for in Web-based resources, as part of their *Digital Libraries: Effective Access* project for STEM educators. A preliminary report available online (Sucher, 2003) describes survey results showing that a significant number of the STEM educators surveyed (86 percent) had not received training in equity in STEM education. Further, one-half of the teachers surveyed had not received training and were not interested in such training (Sucher, 2003, p. 38).

The overwhelming number of projects and Web sites available produces some confusion concerning which would be best for a particular need. Teachers noted that they relied on some of their regular professional reading (educational journals or newsletters) or colleagues to reference good Web sites (Sucher, 2003). The GDSL study focused on "effective access," and participants were asked to "select the top three challenges faced in seeking and using Web-based resources" 71 percent responded that the time it takes to locate resources was the primary concern. When participants were asked, "How long do you typically spend looking for educational resources on the Web?" Sucher (2003) reports slightly more than two-thirds replied with some amount up to 10 hours per month.

PROCEDURES

The AAUW (2004) report provided a key insight with respect to the development of this project, in terms of how Web sites and electronic resources might retain the advantages of "informal" education, yet maintain the accessibility and structure afforded by the classroom. The report summarized 416 projects in terms of subject content (science, technology, engineering, or math related, or some combined subject elements).

Of the four "STEM subjects," science received the greatest attention, with 196 of the 416 programs citing science content as a primary subject matter. The lowest representation was for engineering programs, with only 64 programs claiming engineering as a focus. In addition, in spite of the funded programs' fundamental commitment to gender equity, a significant number of programs (approximately 40 percent) served both boys and girls

The AAUW report further highlighted the characteristics of each subject emphasis. In this, technology focused programs were singled out for three unique characteristics. First, every technology-focused program specified at least some kind of goal for the program, while some programs in science, engineering, and mathematics omitted this. More than two-thirds of these had a goal to "increase student engagement with technology" and one-third including goals of career awareness or gender equity awareness. Second, one-third of these were school-based (which was the highest percentage of school-based programs of all the STEM subjects). Finally, in comparison to the other STEM subjects, technology projects seemingly provided a unique opportunity for interdisciplinary focus, with 71 percent combining technology activities with science, engineering, or mathematics content.

Taken together, these findings indicate that "technology" as the focus of program subject matter provides a vehicle compatible with achievement of other learning goals. Further, 10 of the 17 Web-based programs were in the subject area of technology. Thus, technology was selected as a subject focus uniquely suitable to the purpose of the directed project, to facilitate the use of existing Web-based resources in the classroom setting.

In moving forward with the second stage of the project, and the final analysis to identify appropriate Web sites or Internet tools for inclusion in the Web site, it was necessary to provide some criteria for evaluation of the candidates in terms of the potential learning experience and ease of incorporation into the regular curriculum. A review and analysis of three previous studies on evaluation of gender-equitable software and programs, each focusing on a different aspect of "assessment issues" with respect to gender equity programs, was undertaken:

- Gender & Science Digital Library (GSDL)
- Building Engineering & Science Talent (BEST)
- Girls Tech (from The Douglass Project)

The gender & science digital library (GSDL) project addressed the needs of teachers seeking to provide an "interactive collection of high-quality, gender-equitable science, technology, engineering and mathematics (STEM) resources for K-12, higher education (community college and university), women's studies, teacher preparation programs, and informal learning environments" (EDC, 2004). Materials accessed on the GSDL Web site have been reviewed for quality, providing the user with confidence in their use. The site serves as a clearinghouse for materials, providing a description referencing the basic content and aims of the material, appropriate age groups, and other relevant information (Hanson, 2002). However, it does not clearly provide guidance as to how, when, or in what context to access these resources.

The Web site for the BEST initiative specifically notes its emphasis on programmatic results: "BEST sought to convene the nation's respected practitioners, researchers and policymakers and identify "what's working" across the country to develop the technical talent of under-represented groups in pre-K through 12, higher education, and the workplace" (BEST, 2002).The BEST program goal is to encourage adoption of programs that have extensive research studies supporting effectiveness. However, the end list of resources did not provide any that would be readily accessible to a teacher involved in preparing lesson plans.

The Girls Tech investigators created a framework for evaluating the appeal of Web sites, CD-ROMs, and other electronic information resources to young women. This framework, called "The GirlsTech Model," was developed by analyzing library and information science and gender studies research, and through original theoretical work. Consequently, resources selected using the GirlsTech Model can encourage young women to increase their use of Web sites and related technology, thereby increasing their computer experience and confidence and making computer and technology professions more appealing career options. (The Douglas Project, 2003). This template addresses how to interact with technology, and is more concerned with "attractiveness" and improving attitudes of girls toward the STEM subject by making the experience more enjoyable.

Given budgetary constraints of most schools, it is necessary to focus on Web site resources that are free of charge, and yet provide high quality materials. Therefore, some of the government funded projects were considered as primary candidates for inclusion (AAUW, 2004; National Science Foundation, 2003). These sites will provide free or low cost materials, with associated government financial support to ensure continued access and maintenance.

Evaluation of *CareerQuesting* Resources

None of the three evaluation protocols reviewed actually would serve the purpose of this project as originally designed. The GSDL tool is perhaps the most useful, but it is overly detailed to facilitate curriculum development. The Girls Tech approach to "gender specific" criteria attempts to generalize "female" preferences in a categorical manner to enhance "enjoyment" or engagement but does not necessarily address the effectiveness of resources. In practice, the Girls Tech tool also failed to identify a substantial pool of resources that were more than "average" with respect to some of the key "female friendly" factors. Finally, the BEST template is the only one that strongly emphasizes programmatic effectiveness, and attempts to impose a strict definition of success (using rigorous investigative assessment of student outcomes). However, this approach fails to provide a sufficient pool of accessible (and affordable) materials, unlike the GSDL, for which one criterion of any resource selected for inclusion is that it be readily available via the Internet at no cost to users.

The literature review on gender-specific issues in career selection also identified four key elements that the chosen resources or activities should emphasize:

- **Career information and exploration:** Career specific information offered at a point where the girls have not internalized a negative perception of STEM subjects
- **Personal identification and relevance:** Students may perceive that technology is not something that they need to learn more about because they have no personal need to do so. If they can find ways to use technology that benefit them directly in their daily lives, they are more motivated to learn and master the basic skills necessary to use it.
- **Real world application and context:** Allowing girls to acquire basic technical skills or a knowledge base to enhance their sense of competence in STEM-related activities.
- **Social interaction and teamwork:** STEM or technology education should emphasize the potential for working with others to solve problems, including application of technology to further communication and social relations.

Together, these should provide a foundation for students to exercise their talents and creativity in a scientific or technical outlet, and find the experience to be both enjoyable and rewarding.

It is important to emphasize that each of these elements will also benefit all students, not just girls, providing an opportunity to encourage less technically inclined boys to develop these same competencies, and further broaden the pool of skilled workers in technology and related fields. Levine argues that the classroom must focus more on work-life readiness, giving students the tools they need to be productive and effective in their careers (Levine, 2005, pp. 10-11).

The emphasis of the *CareerQuesting* resources is upon providing the students with the learning experiences that will engage them in the process of career exploration and selection. Unlike resources explicitly targeted to girls, teachers and administrators have a clear and compelling motivation to provide students with this kind of career preparation experience.

The final section of the tool included a two-stage review, with a preliminary screen that focused on the key elements of accessibility and content suitable for general classroom use. Each of the resources identified was evaluated using the pre-screening questions. In particular, the Web sites should be easy to use without excessive adaptation of materials, available at no cost, and the materials should be appropriate for a diverse student population. A summary of the evaluation criteria appears in Table 1.

The ideal final pre-screen score would be 22, with a preferred score of 15-21. If an electronic resource met the pre-screening criteria, it was further evaluated. The resource must have met at least "average" standards with respect to the listed criteria, with preference given to with highest scores. A template for evaluation of the CareerQuesting Web sites is presented in Table 2.

Web sites and WebQuests

The search results were divided into two broad resource categories, Web sites and WebQuests. WebQuests are Web-based constructivist lesson formats that present a problem or series of tasks to be addressed by students using suggested Web-based resources for research (Dodge, 2005). If a site passed the initial pre-screening, it was subjected to the final review process, as indicated by the Web site and WebQuest Scoring Sheet presented in Table 3.

Table 1. Summary of CareerQuesting evaluation elements

General Content and Media Criteria Pre-screen

- **Domain:** Contains subject matter in the domains of natural sciences, pure or applied, mathematics, engineering, or technology.
- **Grade level:** Middle school students, appropriate for classroom use.
- **Access/Availability:** Readily available, via the Internet, at no cost, with clear means of ongoing support and funding. Minimal use of advertising (may acknowledge corporate sponsors) Easy to use, clear directions and functionality, utilizes only content and tools available on the site or commonly available and accessible to the public.
- **Functionality:** Must meet minimum design and technical support criteria commonly expected of Web sites and multimedia, including good page layout and navigational design.

Final Evaluation elements

- *Gender Equity Criteria* *
  1. Consistent with general principles of gender equity/inclusivity
  2. No specific programmatic goals for STEM diversity
- *Gender-Specific Criteria for Media and Software* *
  1. Career Information a
  2. "Real world" application/Contextuality
  3. Relevance/ Personal Identification-
  4. Social Connectivity/ Collaboration
- Pedagogical Criteria
  1. Technology use
  2. Standards and credibility
  3. Versatile Use
  4. Flexibility
  5. Interdisciplinary

** If practical, the materials can be integrated with other resources or adapted to compensate for a lower score in these areas.*

The search results were clearly tracked into two broad resource categories, Web sites and WebQuests. The final selection of recommended resources proceeded to identify the "top ten." The rankings of those Web sites and WebQuests are shown in Table 4.

Best Overall Web site

KidzOnline—Tech Training at www.kidzonline.org/TechTraining best met the criteria specified. It includes extensive technical education resources on subjects ranging from animation to Webpage creation. There are lesson plans for nearly every subject from English and language arts to science and technology education. The materials are intended for students from 6th to 12th grade. Technology skills are highlighted in nearly every area.

Best Career Planning Web sites

In addition, two Web sites were chosen to provide alternatives for career exploration tools. They included USA TODAY Education CareerQuest at www.usatoday.com/educate/careers/careers.htm and a general career exploration resource, Career Voyages—Welcome at www.careervoyages.gov that contains valuable content but may be more suitable for slightly older students.

Table 2. Template for evaluation of CareerQuesting Web sites

General Content and Media Criteria Pre-screen
Domain: Natural sciences, pure or applied, mathematics, engineering, or technology.
1. Yes (+5) 2. No (-5)
Grade level: Middle school students, appropriate for classroom use.
1. Yes (+5) 2. No (-5)
Access/Availability: Readily available via the Internet at no cost; clear means of ongoing support and funding. Minimal use of advertising. Easy to use, clear directions and functionality, utilizes only content and tools available on the site or commonly available and accessible to the public.
1. Yes (+5) 2. No (-5)
Functionality: Meets minimum design and technical support criteria commonly expected of Web sites and multimedia, including good page layout and navigational design.

Ranking	Weighted Score
1= 1st quartile (site ranks in top quartile)	5
2= 2nd quartile	0
3= 3rd quartile	-2
4= 4th quartile ranking (site ranks in lowest quartile)	-5

Web sites or other Internet materials to be considered for inclusion in the "Career Questing" resource guide will be evaluated according to the following criteria. Each site will be rated with respect to the factors listed as follows:

Score	Evaluation Guidelines
"Poor"	Contains no aspects of desired criteria, or inadequate in functionality or content.
"Below Average"	Materials have few desired elements or factors listed. Materials considered for inclusion only if there is necessary functionality or content not available otherwise.
"Average"	Site consistent with similar resources. For gender-specific criteria, preference given to resources that can be directly adopted with minimal modification
"Above Average"	The resource or Web site conforms in most respects to the desired criteria, or does provide a sound foundation for adaptation to meeting the criteria.
"Excellent"	Materials are exemplary with respect to the desired criteria, and are acknowledged as such by experts and/or educators.

Score Ranking: Each of the relative scores was assigned a weighting value as follows:

Score	Weight
Excellent	+5
Above Average	+3
Average	0
Below Average	-3
Poor	-5

Score Range	High	Low
Gender Equity Criteria	10	-10
Gender Specific Criteria	20	-20

Table 3. Web site and WebQuest scoring sheet

Resource: ________________ URL: ________________________

Criteria	Description	SCORE
Gender Equity Criteria		
Gender equity/ Inclusivity	Content and activities suitable for both boys and girls. Resources should portray balanced images and content with respect to the participation of men and women, boys and girls in related activities, and be inclusive of other "underrepresented" groups.	
Programmatic goals for STEM diversity	No strict requirement for explicit gender equity focus, but the materials or Web sites should conform to best practices for engagement of student interest and building enthusiasm for understanding and application of STEM skills.	
Gender-Specific Criteria for Media and Software		
Career Information and/or "Real world" application	Content linked to inspirational content on related careers, specific career exploration component that addresses issues relevant to student interests, or which highlights the real-world applications of the technology or subject matter.	
Relevance/ Personal Identification	Web site should provide students with a chance to experiment and apply technology in ways that relate to their own interests or allow them to leverage additional talents and interests.	
Social Connectivity/ Collaboration	Will provide elements allowing for collaborative interactions between students, or can be integrated with other sources add opportunities for students working together.	
Contextuality	Content is provided along with background or history, supplemental information on application is provided. If this is not provided, within the site or resource, can the material be integrated with other content to frame and present the activity or experience?	
Pedagogical Criteria		
Technology use	Interactive content, increases student engagement with technology, development of fundamental technology use skills.	
Standards and credibility	Content will meet highest standards of scientific and technical creditability, and will be endorsed/ produced by a reputable source.	
Versatile Use	Materials or tools allow for customization of the activity or applications Can be adapted or customized for use in different class setting or curriculum content.	
Flexibility	Preferences will be given to sites that provide a general purpose experience, where skills can be applied in a variety of other contexts, including general application of technology to problem solving, etc. Provide opportunity for students to exercise basic skills and processes creatively.	
Interdisciplinary	Preference will be given to technology related activities that integrate other STEM related content, including mathematics, physical sciences, and engineering	

Table 4. Web site and WebQuest Rankings

Web sites	Pre-screen Ranking	Gender Equity Criteria	Gender specific Media Criteria	Pedagogical Criteria	Overall
Kidz Online	**1.00**	**1.00**	**0.90**	**1.00**	**0.97**
USA Today Education—Career Quest	**0.91**	**0.80**	**0.55**	**0.60**	**0.71**
Career Voyages	**0.91**	**0.80**	**0.45**	**0.60**	**0.68**
Project Cybercareers	**0.77**	**0.60**	**0.65**	**0.80**	**0.72**
Ohio Math Works	**1.00**	**0.80**	**0.90**	**1.00**	**0.94**
GM—GMability	**1. 00**	**0.80**	**0.65**	**0.80**	**0.82**
NASA Quest	**1.00**	**0.80**	**0.80**	**0.75**	**0.85**
Invention at Play	**1.00**	**1.00**	**0.70**	**1.00**	**0.92**
GetTech.org	**1.00**	**0.80**	**0.55**	**1.00**	**0.85**
eCybermission	**1.00**	**1.00**	**0.65**	**0.90**	**0.88**
WebQuests					
Who am I? Career Webquest	**0.45**	**0.80**	**0.65**	**0.70**	**0.63**
William's Career Webquest	**0.55**	**0.30**	**0.30**	**0.45**	**0.42**
Futurequest	**0.55**	**0.80**	**0.45**	**0.45**	**0.53**
CyberScience Magazine	**0.77**	**0.60**	**0.55**	**0.55**	**0.63**
NetForce Webquest	**0.77**	**0.50**	**0.75**	**0.09**	**0.76**
Dr. B's Atlantis Quest	**0.55**	**1.00**	**0.75**	**1.00**	**0.79**

Best Technical Sites

Three Web sites were chosen as the best sources of in-class technical assignments to foster interest in STEM subject matter. These Web sites are primarily considered as resources for technical assignments and application of technology, building of technical skills, but in some cases included elements of career exploration as well:

- **Ohio Math Works:** (http://www.ohiomathworks.org/) A high-quality resource focusing on development of quantitative skills and real-world application of mathematics, with elements of career exploration.
- **GM—GMability Education 5-8:** (http://www.gm.com/company/gmability/edu_k-12/5-8/index.html) Emphasizes technical subjects and engineering related careers.
- **Welcome to NASA Quest!:** (http://quest.arc.nasa.gov/index.html) Significant classroom and educational resource support.

Best General Education Resources.

Three Web sites were chosen as general-purpose Web sites for teachers to find lesson plans or ideas for class assignments. These sites have interactive content or game-like activities for students. Resources focus on more involved, longer-term activities:

- **Lemelson Center presents Invention at Play:** (http://inventionatplay.org/) Focuses on creativity and having fun with technology and basic science or engineering skills.
- **GetTech.org [also Getsmarter.org Home]:** (http://www.getsmarter.org/index.cfm) Provide online game-like activities, along with tutorial or quiz-applications
- **Welcome to eCYBERMISSION:** (http://www.ecybermission.com/index.cfm) A resource for a team-competition activity with more than average involvement and lesson plan development. Other competitions often focus on "math drills" or robot building, or similar activities that either more like trivia contests or are somewhat remote from daily life. Projects focus on "working to solve problems in your community." Also, the competition is directed toward for teams of 6th to 9th grade students, making it more likely that students will find the projects more interesting and relevant.

WebQuests were the subject of a separate search and evaluation process, to find specific examples that embody key elements of the *CareerQuesting* functionality. Since these are designed by teachers for their own use, they provide a very flexible resource for classroom use.

Best for Career Exploration

Each of the following is a well-developed career exploration activity. The FutureQuest activity includes a group task to allow for student collaboration on the project. There are links to several good interactive resources for career interest testing and assessment as well:

- **Who Am I? Career WebQuest:** (http://home.sullivan.k12.il.us/teachers/lawson/webcareers.htm)
- **Martin Willams's Career WebQuest:** (http://home.sullivan.k12.il.us/teachers/lawson/webcareers.htm)
- **FutureQuest:** (http://www.pvpusd.k12.ca.us/teachweb/twidwelll/FutureQuest.html)

Best for Technical Skills

Each of the following WebQuests focuses on assignments to learn Web page construction skills or use of the Internet for research.

- **Cyber Science Mag:** (http://projects.edtech.sandi.net/kearny/cybermag/index.html) Suitable for use in general English/Composition courses and journalism, or as part of science and technical education classes.
- **NetForce:** (http://www.geocities.com/lukasaurus_smith/) Group assignment structure makes this useful for teamwork and collaboration activities.
- **Dr. B's Internet Research Guide: Atlantis Quest:** (http://drb.lifestreamcenter.net/Lessons/Atlantis/index.htm) Group assignment structure makes this useful for teamwork and collaboration activities.

Web-Based Resources for Evaluation

The following are several Web-based resources for the reader to evaluate employing the protocol in this article: Each was available as of February 10, 2007, and appeared to have reliable funding sources:

- **http://www.engineergirl.org/**
- **http://www.niehs.nih.gov/kids/labcoat.htm**
- **http://www.girlpower.gov/girlarea/sciencetech/jobs/index.htm**

RECOMMENDATIONS AND CONCLUSION

The goals of this chapter were to identify criteria for evaluating Web site resources with potential for increasing interest and persistence of middle school girls in STEM fields, and to develop a means by which teachers might use and evaluate these existing Web sites and resources. In particular, the tool was used to identify general-purpose materials that might appeal to both boys and girls, allowing teachers to adopt them without concerns that they are providing an unfair advantage to girls.

Resource selection should be validated using an independent review process to establish quality standards for all resources. It may be necessary to make some modification of the tool, since in finalizing the list of Web sites for inclusion as recommended resources, it was apparent that the final selection process remained difficult. This was especially true in the decisions as to whether or not a particular Web site met the requirement that a resource be from a reputable source with evident means of funding or support. It is possible to argue that some of the resources selected as recommended resources did not fully meet this standard.

This was particularly the case with government and non-profit sources. Generally, both met the reputation requirement but the funding element was difficult to evaluate. Most of the non-profits, such as universities, were dependent on money from federal grants. Some of the resources selected were part of government initiated educational reform, and thus were subject to policy changes that would eliminate a particular program or Web site. For corporate resources, funding was not seen as an issue, but corporate outreach and community involvement policies could shift dramatically, and a particular Web site might be eliminated.

Finally, while the resources do focus on technical or STEM subjects, career exploration remains as the central concern. The intent of the *CareerQuesting* model is that students will be taught to be "mapmakers" as part of the regular curriculum. Levine summarizes the problem as follows:

A sizable hunk of a child's success is measured by her ability to comply, to learn what she is expected to learn, and to do what she is told to do. An adult must be able to chart her own road maps. The odyssey leading into adulthood can be a lonely and harsh voyage, especially if a startup adult is naive and uninformed, if he's never learned to be a mapmaker. (Levine, 2005, p.10)

A quest usually refers to a long journey of discovery and learning, with a person seeking an important goal or treasure at the end. Students need a map for this inevitable "career" quest each must embark upon. They need to know how to make such a map, and how to use the map. The goal of evaluating Web sites is to provide teachers with better access to some of the best tools and resources to help create a generation of mapmakers.

REFERENCES

AAUW. (2004). *Under the microscope: A decade of gender equity projects in the sciences.* Retrieved March 12, 2004, from http://www.aauw.org/research/underthemicroscope.pdf

Battle, A., & Wigfield, A. (2003). College women's value orientations toward family, career, and graduate school. *Journal of Vocational Behavior, 62*, 56-75.

BEST. (2002, April 2004). *BEST.* Retrieved September 28, 2004, from http://www.bestworkforce.org/

BEST. (2004, April 2004). *What it takes: Pre-K design principles to broaden participation in science, technology, engineering, and mathematics.* Retrieved February 11, 2007, from http://www.bestworkforce.org/PDFdocs/BESTPre-K-12Rep_part1_Apr2004.pdf

Blickenstaff, J. C. (2005). Women and science careers: Leaky pipeline or gender filter? *Gender and Education, 17*(4), 369-386.

Bullock, L. D. (1997). Efficacy of a gender and ethnic equity in science education curriculum for preservice teachers. *Journal of Research in Science Teaching, 34*(10), 1019-1038.

Butler, M. B. (1999). Factors associated with students' intentions to engage in science learning activities. *Journal of Research in Science Teaching, 36*(4), 455-473.

Clewell, B. C., & Campbell, P. B. (2002). Taking stock: Where we've been, where we are, where we're going. *Journal of Women and Minorities in Science and Engineering, 8*(3&4), 255-284.

Council on Competitiveness. (2004). *Council on competitiveness Web site: Worldclass workforce initiatives.* Retrieved November 1, 2004, from http://www.compete.org/

Dodge, B. (2005). *The WebQuest Page.* Retrieved January 15, 2007, from http://webquest.sdsu.edu/

EDC. (2004). *GSDL, FAQ.* Retrieved April 12, 2004, from http://gsdl.enc.org/ Education Development Center. (2004). *GSDL Web site.* Retrieved January 15, 2007, from http://gsdl.enc.org/

Freeman, C. E. (2004). *Trends in educational equity of girls & women: 2004* (No. NCES 2005-016). Washington, DC: U.S. Government Printing Office: U.S. Department of Education, National Center for Education Statistics.

Furuta, R., Shipman III, F., Hsieh, H., Francisco-Revilla, L., Karadkar, U., Rele, A., et al. (1999). *Using the Internet in the classroom: Variety in the use of Walden's Paths.* Paper presented at the ED-MEDIA 99: World Conference on Educational Multimedia, Hypermedia, and Telecommunications, Seattle, Washington.

Gaskell, P. J., Hepburn, G., & Robeck, E. (1998). Re/presenting a gender equity project: contrasting visions and versions. *Journal of Research in Science Teaching, 36*(8), 859-876.

GirlTech.org. (2004). *Girl Tech Web site.* Retrieved February 8, 2007, from http://www.girltech.org/

Hanson, K. (2002, October 3-6, 2002). *Space aliens? Women, ICTs, and gender-equitable electronic resources.* Paper presented at the Association of Women in Development(AWID) Conference, Guadalajara, Mexico.

Huang, G., Taddese, N., & Walter, E. (2000). *Entry and persistence of women and minorities in college science and engineering* (No. NCES 2000–601,). Washington, DC: U.S. Department of Education, National Center for Education Statistics.

Jackson, S. A. (2004). *The quite crisis: Falling short in producing American scientific and technical talent.* Retrieved February 10, 2007, from http://www.bestworkforce.org/PDFdocs/Quiet_Crisis.pdf

Jones, M. G., Howe, A., & Rua, M. J. (2000). Gender differences in students' experiences, interests and attitudes toward science and scientists. *Science Education, 84*(2), 180-192.

Levine, M. (2005). *Ready or not, here life comes.* New York: Simon & Schuster.

March, T. (1997, June 2001). *Working the Web for education.* Retrieved February 10, 2007, from http://www.ozline.com/learning/theory.html

McDonnell, F. (2006). Why so few choose physics; An alternative explanation for the leaky pipeline. *American Journal of Physics, 73*(7), 583-586).

Mead, S. (2006). *The truth about boys and girls.* Washington, DC: Education Sector.

MentorNet. (2004). *MentorNet Web site.* Retrieved January 27, 2007, from http://www.mentornet.net/

Millar, R., & Shevlin, M. (2003). Predicting career information-seeking behavior of school pupils using the theory of planned behavior. *Journal of Vocational Behavior, 62,* 26-42.

Morgan, C., Isaac, J. D., & Sansone, C. (2001). The role of interest in understanding the career choices of female and male college students. *Sex Roles, 44*(5-6), 295-320.

NASA. (2004). *Women of NASA Web site.* Retrieved February 10, 2007, from http://quest.arc.nasa.gov/women/intro.html

National Science Foundation. (2003). *New formulas for America's workforce: Girls in science and engineering* (No. nsf03207). Washington, D.C.: National Science Foundation.

Packard, B. W.-L., & Nguyen, D. (2003). Science career-related possible selves of adolescent girls: A longitudinal study. *Journal of Career Development, 29*(4), 251-263.

Posnick-Goodwin, S. (2005, March). In the march toward gender equity, girls surge forward, but boys fall back. *California Educator, 9.*

Rosser, S. V. (1997). *Re-engineering female friendly science.* New York: Teacher's College Press.

Seymour, E. (1998). *The role of socialization in shaping the career-related choices of undergraduate women in science, mathematics, and engineering majors.* Paper presented at the Choices and Successes: Women in Science and Engineering, New York.

Sucher, Y. (2003, March 2003). *Digital libraries: Effective access survey of STEM educators, preliminary summary of survey results.* Retrieved January 27, 2007, from http://www2.edc.org/GDI/homepage_SR/Prelim_Report_March2003.doc

The Douglas Project. (2003). *GirlsTech: Girls, science, and technology.* Retrieved January 27, 2004, from http://girlstech.douglass.rutgers.edu/

Waite, S. J., Wheeler, S., Bromfield, S. (2007). Our flexible friend: The implications of individual differences for information technology teaching. *Computers & Education, 48,* 80-99.

Watt, H. M. G., Eccles, J. S., & Durik, A. M. (2006). The leaky pipeline for girls; A motivational analysis of high school enrolments in Australia and the USA. *Equal Opportunities International, 25,* 642-663.

WITI. (2004). *Women in Technology International Web site.* Retrieved February 10, 2007, from http://www.witi.com/

Chapter XI
How to Use Vignettes in an Online Environment to Expand Higher Order Thinking in Adults

Maria H. Z. Kish
Duquesne University, USA

ABSTRACT

A challenge in teaching and providing any type of instruction in the online learning environment is to ensure that participants are engaged in the process and find meaning in their learning. This case study investigated the use of vignettes as a teaching strategy and learning activity of the Generative learning model in a hybrid online course. Vignettes are short and realistic stories that may help bridge participants' previous experiences to applying course material in relevant situations. The generative learning model, consisting of five main components: attention, motivation, knowledge, generation, and metacognition (Wittrock, 2000), was incorporated when requiring students to answer teacher-generated vignettes and to generate their own vignettes. Two outcomes were anticipated using vignettes within the generative learning model in a hybrid online course: (1) enhancement of academic achievement and (2) higher order thinking[1]. This study considered data from student work collected from the Instructional Techniques Course, GITED 631, taught in the Graduate School of Education at Duquesne University, Pittsburgh, Pennsylvania, in the fall of 2003. Eight participants responded to teacher-generated vignettes, created diagrams and rubrics, created their own vignettes, and recorded their observations concerning vignettes in reflective learning logs. The adult online learners in this study professionally focused on teaching children and adults. This study's participants all professionally focused on teaching children and adults. The research findings indicate that the use of teacher-generated vignettes can increase academic achievement, and that learner-generated vignettes can help students achieve higher order thinking. This article also discusses the methods that have been used to teach adult learners how to respond to and create vignettes for their own teaching and presentation purposes.

INTRODUCTION

You are interested in getting a new job. After thinking about the type of position you want, you develop your own portfolio. You realize this is the best way to really display what you have learned, accomplished, and what your current projects are. Because the job market is tight, you need a way to "stand out" from everyone else.

At one point you have an interview for the "perfect job." The interview goes well, and you are told the interviewer was impressed with your portfolio.

The following addresses what you had to do to develop a successful portfolio (as defined in *Professional Portfolios for Teachers*, by Tomei and Wilcox, 1999):

1. Name and describe the type of job you are pursuing. What type of portfolio (intelligent or smart) would you develop? After defining this type of portfolio, explain why you would develop this type.
2. What level (learner, expert, scholar) should you focus on? After defining this level of portfolio, explain why you would work with this level.
3. Define what collection points mean. Consider and state the problem you face in this vignette, and indicate eight different collections points you would include in your portfolio that could address this problem. Also indicate which of the folders (collecting, working, and showcase) you would use to store these particular collection points.
4. Name and define the two types of assessments that should be considered for your portfolio. How would you consider these when you assess your portfolio? Name three different people who would be a part of the assessment process and explain why they should be a part of the assessment process.

This is an example of a vignette, an incomplete short story "written to reflect, in a less complex way, real-life situations in order to encourage discussions and potential solutions to problems where multiple solutions are possible" (Jeffries & Maeder, 2004, p. 8).

Online learning plays a significant part of the future of education, as online learning continues to provide education to individuals facing time constraints. With the benefits of online learning, vignettes may be used to provide meaningful learning experiences to adult learners.

The primary purpose of this study is to show how adults achieve higher order thinking and academic achievement when provided with a supportive online learning environment. The generative learning model is used in this study because it has been shown to help adults meet their learning needs. The techniques used for writing vignettes and teaching others how to write their own vignettes are discussed.

BACKGROUND

The following are the major components of this study: adult learning needs, the generative learning model, vignettes, online learning, and teaching strategies and learning activities helpful for working with narratives.

Adult Learning Needs

The learning environment best suited for adults is supportive, focuses on course and individual goals, and considers active learning activities that assist adults in transferring their learning to their own situations. Three significant learning outcomes that matter to adults include: higher order thinking (Pepicello & Tice, 2000; Wojnar, 2000); academic achievement (Kim, 1999; McKeachie, 1999; Thompson, 1997; Wlodkowski, 1999); and finding the learning activities helpful in understanding and applying the knowledge they gain in their own lives (Knox, 1986; Moore & Bogotch, 1993; Knowles, 1995; Thompson, 1997).

Assisting Adult Learners in Achieving Higher Order Thinking

Higher order thinking includes the following levels of Bloom's taxonomy: application, analysis, synthesis, and evaluation. "A demonstration of 'application' shows that he *will* use it correctly, given an appropriate situation in which no mode of solution is specified" (Bloom et al., 1956, p. 120). Analysis "emphasizes the breakdown of the material into its constituent parts and detection of the relationships of the parts and the way they are organized" (Bloom et al., 1956, p. 144). Synthesis is "the putting together of elements and parts so as to form a whole" (Bloom et al., 1956, p. 162). The highest cognitive level, evaluation, is "the making of judgments about the value, for some purpose, of ideas, works, solution, material, etc" (Bloom et al., 1956, p. 185). Researchers have shown the significance of higher order thinking with adult learners (Odom, 1998; Van der Wal and Van der Wal, 2003 Wojnar, 2000).

Academic Achievement

One way of demonstrating academic achievement is through performance assessments. Performance assessments are assessments that require students to perform authentic tasks (such as working with narratives) to show what they have learned (Wiggins, 1998; Airasian, 2001). Robert Campbell (1999) and Sri Ananda (2000) discuss the significance of using narratives with adult learners. Guidelines for developing scoring guides and rubrics (Airasian, 2001; Linn, Baker, & Dunbar, 1991) were considered for evaluating the performance assessments in this study.

The Generative Learning Model

The generative learning model, developed by Merlin Wittrock (1974), focuses on learners actively generating meaningful relationships by linking prior learning experiences to new learning, thus creating a transfer of knowledge and skills to new situations and environments.

Since its inception, The generative learning model has been modified over the years to include five major components: (1) attention; (2) motivation; (3) knowledge; (4) generation; and (5) metacognition (Wittrock, 2000, p. 210). Attention is important in directing and sustaining activity (Wittrock, 1990). The motivation component focuses on assuring the learner knows his or her role in the learning process and understands that he or she has the responsibility and ability to achieve their learning goals (Wittrock, 1990). "The knowledge component of the generative learning model considers how the learner's memory works in the learning process. The generative process includes: '(a) organizational structures for storing and retrieving information; and (b) processes for relating new information to the stored information" (Wittrock, 1974, p. 182). Wittrock says, "Metacognition refers to knowledge about, awareness of, and control over one's cognition. Cognition includes thoughts, motivations, and feelings" (1994, p. 1).

Stories and Vignettes

The use of stories provides experiential learning and promotes critical thinking and problem solving skills (Kerka, 2001). Writing stories provides several benefits for the learner, such as the expression of individuality, self-expression of problems or feelings, independent thinking, confidence building (Zinkoski, 2001) and language skills, such as listening, reading, speaking, and reflecting (Campbell, Campell, & Dickinson, 2004).

Table 1. Comparison of vignettes to other narrative forms

Narrative Form	Description/Purpose	Similarities to Vignettes	Differences from Vignettes
Case Studies	• Help learners see different view points, bridge the gap between theory and practice, increase involvement in learning, and bring together the insights of several people (Ford, 1969, p. 19).	• Based on real-life situations	• "A second-hand account of something that did happen, or a first hand account of something that could happen" (Ford, 1969, p.14). • Include three components: a report, an analysis, and a critical discussion, (Marsick, 1998) • "Traditional cases can be static and dated [and] are read passively" (Kerka, 2001, p.3).
Case Stories	• Stories that simulate the real world. This approach combines the reflection required with traditional case studies and adds the creativeness of storytelling (Maslin-Ostrowski & Ackerman, 1998, p. 303).	• Based on realistic events and "help people learn problem solving techniques and analysis, and to link classroom learning with practice" (Maslin-Ostrowski & Ackerman, 1998, p. 303). • Advantages include (Maslin-Ostrowski & Ackerman, 1998, p. 312) (1) Helping people learn problem-solving techniques and analysis, fostering collaboration and collegiality; (2) Bridging the gap between action and thought	• Require oral and written descriptions given by individuals within the classroom • Include five essential steps (pp. 305-307): (1) free write activity; (2) writing case stories; (3) telling, listening to, and discussing case stories; (4) small group reflection; 5) whole group reflection • Disadvantages include: (Maslin-Ostrowski & Ackerman, 1998, p. 313) • Require more time (an entire process takes a minimum of three hours) • "War stories" or stories told repeatedly that are not subject to alteration may appear
Scenarios	• "Scenarios are a tool for helping us to take a long view in a world of great uncertainty" and they are "stories about the way the world might turn out tomorrow, stories that can help us recognize and adapt to changing aspects of our present environment" (Schwartz, 1991, pp. 3-4).	• Because scenarios are short it usually takes less time for students to understand, analyze and evaluate what is occurring or what should be occurring. • The term "vignettes" has been used interchangeably with the term "scenarios" (Dede, 1998; Fogarty, 1997)	• Students may or may not be able to relate to the scenarios presented • Students may or may not have direct instruction or guidance as to how the scenarios should be addressed. • Scenarios may or may not be written as a complete, stand-alone narrative.
Critical Incidents	"An incident is any observable human activity that is sufficiently complete in itself to permit inferences and predictions to be made about the person performing the act. To be critical, an incident must occur in a situation where the purpose or intent of the act seems fairly clear to the observer and where its consequences are sufficiently definite to leave little doubt concerning its effects" (Flanagan, 1954, p. 327).	• Meaningful narratives to which the reader can relate, and are followed by questions promoting reflection and further discussion, due to the presence of many different possibilities. • Allow for many different points of view.	• Are not based directly upon the background (or lack of experiences) of participants in the class • Are not purposely incomplete so that the participants are required to fill in the details. • Are usually followed by comments and may or may not be followed by questions.

Vignettes are used as "reflective, research, and modeling tool" (Jeffries & Maeder, 2004, p. 17), as an assessment tool (Jeffries & Maeder, 2004; Kish, 2004), and as a learning activity encouraging creativity and higher order thinking (Kish, 2004). Whether created by the teacher or by the student, the use of vignettes is considered a generative activity, because it requires students to "generate integrated relationships between the external stimuli and the memory components" (Grabowski, 2004, p. 738).

Vignettes have two parts: narrative and questions. Table 1 compares vignettes to other narratives used successfully with adults.

The questions that follow the vignette may be compared to essential questions and have the following characteristics:

- "Go to the heart of a discipline" (Wiggins, 1987, p. 12; 1998, p. 214)
- "Have no one, obvious, 'right' answer" (1987, p. 12; 1998, p. 214)
- Require students use higher order thinking skills (in relation to Bloom's Taxonomy) (1987; 1998)
- "Recur; they are raised naturally rather than asked throughout one's learning" (1998, p. 215)
- Framed to encourage personal interest among students (1987; 1998)
- "Link to other essential questions" (1998, p. 215)

Online Learning

Successful online learning provides adults with flexibility for learning material and engaging in meaningful discussions (McKenzie, 2001; Wojnar, 2000), freedom to participate as an individual with unique ideas and insights (Harrasim, 1987), and challenge adult learners in higher order thinking (McIsaac & Gunawardena, 2004; Pepicello & Tice, 2000, p. 54; Wojnar, 2000).

Generative learning can be successfully supported via computer-based instruction (Higginbotham-Wheat, 1991); requires active learning, a "major outcome of learning networks" (Harasim, Hiltz, Teles et al., 1995, p. 29); can be supported in asynchronous learning (Hiltz, 1997); can be used in an online learning environment to create meaningful artifacts showing a connection of "new ideas to existing knowledge structures" (Ryder, 1998, p. 4); and is a significant model to consider in the online learning environment (Grabowski & Koszalka, 1999; McGuire, 2003).

The American Distance Education Consortium (ADEC) (2003, pp. 1-3) documents principles for setting up a learning environment and characteristics of quality Web-based teaching and learning. Online course components that work well in meeting the needs of adult online learners include: course management system, asynchronous discussions, and e-mail correspondence.

Teaching Strategies and Learning Activities Used with Adult Learners

Several different teaching strategies and learning activities have been used successfully with adults when reading, discussing, and creating narratives (aee Table 2).

DESIGN OF THE STUDY

This case study investigated how vignettes affect academic achievement and enhance higher order thinking[2]. The following research questions pertain to a hybrid online course designed with the generative learning model:

1. Will the student completion of teacher-provided vignettes enhance academic achievement as measured in the following types of student work/activities:

Table 2. Teaching strategies and learning activities used with adult learners when working with narratives

Teaching Strategy/Learning Activity	Description	Purpose	Benefits
K-W-L Chart	• Created in 1986 by Donna Ogle as a reading and learning strategy. • The chart, comprised of in three columns, requires the users to complete what they know (in the "K" column), what they want to know (in the "W" column), and what they learned (in the "L" column).	• The K-W-L chart is a cognitive bridge between the participant's prior knowledge and understanding as to what participants are about to learn. • Used for reading and discussing narratives.	• The K-W-L chart can help students focus on what they need to know, as well as on their professional and personal goals as they relate to the course.
In-class Discussions	• Face-to-face meetings in which the instructor and adult learners converse.	• Provide dialog for adult learners, which is significant to their learning needs (Daloz, 1986; Vella, 1994). • Considered in reading and creating narratives.	• Allows for communication between the instructor and participants, as well as among the participants themselves. • Allows for informal, spontaneous discussions
Online Discussions (Asynchronous)	• A discussion in which "the participants may connect at any time around the clock, and from any location in the world accessible by the Internet or a reliable telephone system, rather than having to be online at the same time. The system stores the entries in a permanent, ordered transcript which keeps the equivalent of 'bookmarks' to separate anything that is 'new' for each individual from items that have already been seen" (Hiltz, 1997, p. 2).	• Provide dialog for adult learners, which is significant to their learning needs (Daloz, 1986; Vella, 1994). • Considered in reading and creating narratives.	• Allows for communication between the instructor and participants, as well as among the participants themselves. • Can be accessed at anytime, at the convenience of the adult learner (Hiltz, 1997)
Online Presentations	• PowerPoint presentations posted online	• Considered as a form of direct instruction. • Used in reading and creating narratives.	Direct instruction is helpful in teaching skills (Kauchak & Eggen, 1998)
In-class Presentations/Demonstrations	• Presentations and Demonstrations provided by the instructor in the classroom	• Considered as a form of direct instruction. • Used in reading and creating narratives.	Direct instruction is helpful in teaching skills (Kauchak & Eggen, 1998)
Learning Logs	• Journals that require learners to reflect upon learning experiences.	Used to consider experiences in reading and creating narratives.	Learning logs have been used successfully with adult learners to help them reflect upon what they have learned and how they can possibly use their learning in other circumstances (Brookfield, 1995; Wojnar, 2000).

continued on following page

Table 2. continued

Teaching Strategy/Learning Activity	Description	Purpose	Benefits
Modeling	Modeling was developed by Albert Bandura, which entails, "Taking an individual through a series of progressively more difficult behaviors" (1965, p. 311). Modeling is a generative activity, because it relies on the learner's perceptual abilities (DiVesta, 1989).	Modeling an activity shows the learner how something is done so that he or she can imitate the process.	Educational researchers cite modeling as a significant method that has been used widely (Knowles, 1978). Because the components of modeling are similar to components of cognitive theory, modeling has been described as a generative activity, and that Wittrock cited its importance in his writing (1978), its consideration and use in the generative learning model is appropriate.
Think Aloud	The "think aloud" technique, coupled with modeling, usually involves having teachers describe their thinking while working with examples, so that students understand how the skills work (Campbell et al., 2004, p. 21; Kauchak & Eggen, 1998, p. 278).	Used to show one's thought process.	Having subjects think aloud while composing has been able to provide more insight on the composing process (Scardamalia & Bereiter, 1986, pp. 780-781).
Advanced Organizer	A visual guide used to assist the learner in arranging new ideas or concepts. Developed by David Ausubel.	Provides a link for the learner from previous knowledge and understanding to new learning.	Helps learners incorporate and maintain meaningful learned material in three ways: (1) "new material is rendered more familiar and potentially meaningful" and "the most relevant ideational antecedents in cognitive structure are also selected and utilized in integrated fashion"; (2) advanced organizers can "promote initial learning" and help learners remember the material to be learned when developed at the appropriate level for learners; and (3) memorization becomes unnecessary (Ausubel, 1968, p. 137).

- Asynchronous discussions
- Diagrams
- Rubrics

2. Will the writing of learner-generated vignettes promote higher order thinking including application, analysis, synthesis, and evaluation as measured in three different asynchronous discussions?

Purpose

The purpose of this case study was to describe, both qualitatively and quantitatively, how vignettes enhanced academic achievement and higher order thinking in an adult online course. The case study involved master's level students enrolled in a graduate-level elective three-credit course in the School of Education at Duquesne University, *Instructional Techniques*, GITED 631, fall of 2003.

The major unit of analysis was a collection of student artifacts gathered during the course[3].

At the conclusion of the course, three readers were trained to score vignette assignments and rate discussions, using a group moderation process described by Gipps (1994). The readers were instructed how to make structured observations according to a set of guidelines provided by Goetz and LeCompte (1994), and used scoring guides and rubrics to record observed achievements and higher order thinking in participants' work.

Pilot Studies

Two pilot studies determined the reliability of the background and course-ending questionnaires as well as the scoring guides and rubrics.[4]

Participants

There were eight students consenting to be in the study, five females and three males[5]. Because vignettes were considered as a teaching strategy and learning activity, the researcher selected a participant population interested in teaching or training. This study used criterion sampling, a type of purposive sampling that is the most common form of sampling in qualitative studies (Merriam, 1988).

Instructional Considerations

Enhancing Academic Achievement with Vignettes

The first research question focused on the generative learning model and academic achievement as measured by participants' responses to teacher-generated vignettes and participant creation of diagrams and rubrics. All of the answers to the teacher-generated vignettes were sent to the instructor privately; after she had received all of the responses the instructor then posted them in the asynchronous discussion. The researcher required the participants to rate all of the other posted responses and to post to two of these responses. Finally, participants were instructed to record how they felt about writing vignettes in a learning log.

Promoting Higher Order Thinking with Vignettes

The second research question regarded the generative learning model and higher order thinking as measured by learner-generated vignettes in asynchronous discussions. To answer this question, the researcher showed participants how to write vignettes and then had them posted in three separate asynchronous discussions so that all of the participants could read the learner-generated (or participant-generated) vignettes. Participants were instructed to record how they felt about writing vignettes in a learning log.

Procedure

The researcher obtained permission from the participants, collected and evaluated student artifacts, and selected and trained the readers who were familiar with scoring guides, rubrics, higher order thinking, and teaching adult learners. The instructor evaluated all student work during the course according to the scoring guides and rubrics.

A reader-training program was developed to evaluate and analyze the artifacts. The three readers first evaluated each student response independently and then compared the scores. In cases of score differences, one reader, designated as the recorder, decided on the final score and recorded the final score in the appropriate scoring guide.

Instrumentation

Several instruments directly addressed the first question. The background questionnaire, diagnostic pretest, and K-W-L chart were used to gather information regarding the participant's educational background and experience. Both the diagnostic pretest and the K-W-L chart were provided at the beginning and the end of the course so that the researcher could determine what the participant had learned. Wojnar (2000), an international expert in teaching online courses, reviewed the background questionnaire to address construct validity. Construct and content validity were partially addressed by the diagnostic pretest and the K-W-L chart.

Other instruments used to address the first question included teacher-generated vignettes and scoring guides and rubrics for responding to the teacher-generated vignettes, and scoring guides and rubrics for participant-created diagrams and rubrics. Because the instructor wrote two sets of vignettes during the two pilot studies, the process of writing vignettes for the participants represented some evidence of construct and content validity. To address content validity, criterion validity, and construct validity, all of the scoring guides and rubrics were modified following the results of pilot study data and comments from other committee members.

Again, several instruments were considered for the second question. The first instrument used to determine frequency of higher order thinking when writing vignettes was Evaluating Bloom's Levels of Cognitive Activity in Student Work, derived from Wojnar's (2000) study. The second instrument, Evaluating Higher Order Thinking in Participant-Generated Vignettes (An Example) (Kish, 2004, Appendix I), outlined how participants could exhibit higher order thinking when writing vignettes. To address content validity, criterion validity, and construct validity, each table included the levels of higher order thinking from taxonomy of educational objectives and the characteristics of each level in participant work. The scoring guide and rubric for participant-generated vignettes were used to assist in measuring participant-generated vignettes. To address content validity, criterion validity, and construct validity, the scoring guide and corresponding rubric were modified following the results of pilot study data and comments from committee members.[6]

Triangulation

Triangulation of the measurement process was accomplished when investigating each research question. Question 1 involved methodology triangulation, and both questions involved investigator triangulation.

Data Collection

At the conclusion of the course, the readers were given the anonymous participants' coursework. The readers used the rubric to complete a scoring guide for each of the vignettes the participants answered using the table evaluating Bloom's levels of cognitive activity in student work to determine the number of occurrences of analysis, application, synthesis, and evaluation in each participant-generated vignette.

Data Analysis

First, the external readers were trained to score the vignette assignments to see if their scores agreed with the academic achievement scores that the instructor assigned. The researcher noted any score differences between those determined by the readers and the instructor for each subcategory and final total. Descriptive statistics (range, mean, median, frequencies, etc.) were compiled for each type of assignment including sub-categories. The method for analyzing the data was enumeration (Goetz & LeCompte, 1984, pp. 5-6).

RESULTS

Enhancing Academic Achievement

Responses to Teacher-Generated Vignettes

Participants responded to two sets of teacher-generated vignettes. There were five components used to determine the participant's final score: valid response, analysis, defense, discussion, and language (Kish, 2004, Appendices E and K).

Intercorrelations for the first set of participant vignette scores determined by the readers and the instructor were significant ($r = .80$, $p < .02$). Considering the low power of all statistical tests in this study, (since $n = 8$), this finding suggests a high level of inter-rater reliability between the readers and the instructor when scoring the first set of vignette responses.

Intercorrelations for the second set of participant vignette scores determined by the readers and the instructor were not significant ($r = .55$, $p < .16$). The apparent lack of inter-rater reliability between the readers and the instructor was most notable in the scores of participants 2, 4, and 6 but may be a statistical artifact since $n = 8$.

The instructor's scores indicated there was an increase in scores from the first set of vignettes to the second set of vignettes for six of the eight participants. The external readers' scores indicated that five of the eight participants' scores increased from the first to second set. Most of the participants evidently learned how to complete the vignette task more successfully.

Participant-Developed Diagrams

Participants chose a diagram from the course text *Up and Out* (Johnson, 2000) and applied it to a lesson they planned to teach. They were asked to provide instructions to their own students on how to fill in the diagram and indicate a follow-up activity. The participants were required to provide learning objectives for the activity, how the diagram would be used (including the subject and student age), a description of how the participant's students would complete the diagram, and how the participant would assess the students. All of the diagram scores were 93 percent or higher ($M = 97.9$), suggesting that the diagram task was successfully completed by all of the participants.

Participant-Developed Rubrics

The third part of the first research question concerned how well participants created their own rubrics. Participants chose a task they wanted their own students to complete and determine the assessment criteria they would use to grade their students. Participants were graded in six areas. They were required to identify the task at the top of the rubric and follow the rubric format provided in the *Up and Out* course text (Johnson, 2000). Participants were also required to include descriptors for each level of performance, three to six criteria related to the task, indicators of the criteria to be met at each level, and a scoring scale. Even though two rubric scores were 89, the fact that the rest of the rubric scores were 95 or higher ($M=95.9$) indicates that most of the participants successfully completed the rubric task.

Observing Higher Order Thinking with Vignettes

The participants generated three different vignettes. For each set, the researcher and the readers noted the participant number, level of higher order thinking obtained, explained why that level was obtained, and indicated the frequency. The researcher and the readers searched for higher order thinking in the writing of the vignette, its questions (tasks), and its explanation.

The intercorrelations for the higher order thinking frequency scores determined by the instructor and the readers for the first set of participant-generated vignettes were significant ($r = .80, p < .02$). This finding suggests a sufficient level of reliability between the instructor and readers when determining higher order thinking frequency for the first set of participant-generated vignettes. When the same comparison was conducted on the second set of participant-generated vignettes, intercorrelations were not significant ($r = .09, p = .83$). When this comparison was conducted on the third set of participant-generated vignettes, intercorrelations were significant ($r = .78, p < .02$). Taken together, these three findings suggest some, but not complete, inter-rater reliability when determining higher order thinking frequency for the three sets of participant-generated vignettes.

Based on instructor ratings, the total frequency of application was 31, analysis 23, synthesis 22, and evaluation 42. Based on external reader ratings, the total frequency of application was 39, analysis 20, synthesis 22, and evaluation 20. Overall, the instructor and the readers noted similar patterns of analysis and synthesis on the part of the participants across the three vignette assignments but were less similar in their observations of application and evaluation. In the case of evaluation, the fact that the instructor noted more than twice as many instances than did the readers (42 and 20, respectively) was due to the instructor's broader definition of the construct, viz., the instructor recorded instances of evaluation in the participants' vignette explanations as well as in the task questions they created whereas the readers looked for evaluation only in the participants' explanations.

STUDY CONCLUSION

Based on the data gathered and analyzed, the researcher has come to the following conclusions[7]:

1. Having students answer teacher-generated vignettes and having them respond to other responses to these vignettes in an asynchronous environment can help them in their academic achievement
2. Students achieved higher levels of the taxonomy of educational objectives when creating their own vignettes

Educational Implications

Based on the findings, the researcher suggests the following educational implications:

1. Teachers can successfully use vignettes with adult learners within the context of the generative learning model to encourage higher order thinking and enhance academic achievement
2. Completing teacher-generated vignettes and writing vignettes can enhance academic achievement, regardless of whether they are combined with other learning activities
3. Completing teacher-generated vignettes is a useful assessment tool and an effective instructional method
4. Completing teacher-generated vignettes and writing vignettes can measure different course content from more traditional forms of assessment.
5. Because they are text-based, vignettes can be used effectively in a hybrid online course, partly face-to-face and partly online
6. Teachers can learn how to write appropriate vignettes for students of any grade level
7. Age level is likely to be a factor when determining if students are ready to write their own vignettes

Application of Teaching with Vignettes

First, the instructor used vignettes as a teaching strategy, and then as a learning activity for her students. Because the use of vignettes was new to most of the participants, she first modeled the process of writing vignettes and then required the participants to answer certain vignettes. Then, she demonstrated in more detail to the participants how they should go about writing their own vignettes for their own students. The generative learning model was followed in both writing vignettes for the participants (teacher-generated vignettes) and in helping the participants write their own vignettes (learner-generated vignettes).

It should be noted that in her study (Kish, 2004) and in subsequent *Instructional Techniques* courses, the instructor wrote the vignettes so that they did relate directly to the background and experiences of the participants. This is slightly different from how Maeder and Jeffries write vignettes, which includes writing stories with which the participants have no experience. Based on comments and feedback from participants, vignettes that are based on their backgrounds and interests may help motivate participants to focus on the course material, to apply the course material so that it is the most useful or applicable to them, and to become more self-directed in their learning. That is, a number of adult learning needs are met.

Using Teacher-Generated Vignettes to Model the Writing of Vignettes

The *attention* component required that the vignettes written on the background questionnaire (Kish, 2004, Appendix A), data and textbook information. The instructor addressed participants' prior knowledge by having them complete a diagnostic pretest, fill in a K-W-L chart, and discuss previous experiences with instructional techniques in a classroom discussion. There were online presentations posted that summarized main textbook concepts. The five components of the generative learning model were highlighted as well.

The instructor addressed the *motivation* component by spending face-to-face sessions linking specific topics in the course syllabus that matched participants' backgrounds and areas of academic interest. The K-W-L chart, indicating what participants knew and wanted to know, was used to assure the participants how the course would address their professional needs.

Prior *knowledge* was considered by encouraging participants to incorporate their previous knowledge and current teaching experiences when completing vignettes. To promote online discussions, the instructor posted participants' vignette responses, required participants to follow "netiquette," or guidelines to communicate effectively when posting responses, and required each participant to rate two of their classmates' responses according to how helpful or informative they were with respect to the course topic. Individual feedback on all participant coursework was provided.

Having participants complete the vignettes required several types of generative activities: constructing main ideas, writing summaries, solving problems, and providing definitions, examples, and explanations. It was essential that participants solve the problem in the vignette by providing examples of techniques that would work within the vignette. The instructor assisted in their *generation* of knowledge by providing a detailed scoring guide and a rubric, both of which acted as advanced organizers so that the participants would be assisted in the process of answering these vignettes. There were five components considered in determining the participant's final score: valid response, analysis, defense, discussion, and language. All of these are explained in more detail in the updated scoring guide, located in Appendix A.

The *metacognition* component required participants to reflect upon and respond to questions in a learning log concerning their learning experience when completing the vignette assignments. Questions included what participants thought about the use of vignettes when they had to complete them and whether participants would consider using teacher-generated vignettes in their own courses.

Setting up Asynchronous Discussions for Answering Teacher-Generated Vignettes

The instructor required that participants send their vignettes responses privately via the digital drop box first, so that participants would not use others' ideas. After receiving all of the vignette responses from the participants, the vignettes were posted online in the asynchronous discussion so that the participants could refer back to them at any time. Each set of teacher-generated vignettes was posted in two different asynchronous discussion forums, in which each individual teacher-generated vignette had its own threaded discussion. This was done so that it was easier for participants to refer to one teacher-generated vignette and then read through the responses that pertained to that vignette.

The participants were required to read through all of the responses and to respond to at least two of them. Finally, participants were to rate all of these responses and send these ratings to the instructor via e-mail within one week. The instructor responded to everyone within the asynchronous discussion environment; however, scores for the vignette responses were sent to the participants via e-mail to maintain privacy.

Training Session for Participants Creating their own Vignettes

The instructor focused participant *attention* on the vignette creation process by reviewing the course objectives and the teacher-generated vignette. She provided general definitions of vignettes, types and examples, and an explanation of how vignettes are created and used with adult learners.

The *motivation* component was considered by encouraging participants to create and use their own vignettes in their own teaching. The instructor reminded them that by creating vignettes they can connect participants' background and interests with course material, hence possibly helping their own students more interested and motivated in their classes.

Prior *knowledge* was addressed by making no assumptions about the participants' creative ability or past writing experiences and therefore reviewed the components of a story: characters, setting, plot, and some type of problem or challenge. The participants were reminded to relate the process of creating the vignettes to their own understanding and background. The instructor also explained that there are two types of vignettes: truncated, where the "plot line stops at a critical juncture and participants complete the vignette" and abridged, where the "story's details are omitted so that multiple interpretations can be defended" (Jeffries & Maeder, 2004, p. 20).

The instructor modeled the vignette-creation process through the think-aloud technique so that the participants could understand the thought processes she was going through in creating the vignettes and so that they could go through the process with her as well.

Vignette-creation is an important generation activity: linking prior knowledge to the subject matter. To assist participants with the "building" or creation of their stories, the instructor directed them to complete the "vignette starter," a type of advanced organizer, before writing their vignettes (Appendix B). The "vignette starter" provided participants cues to generate their own vignettes and reminded participants to view and edit the vignette so that all of the requirements specified in the scoring guide and rubric were met. The instructor worked with each participant individually in class to be sure each participant had a preliminary idea and understood the process.

The scoring guide and the rubric were also reviewed with the participants so that they would understand how they would be assessed; an updated version of the scoring guide is located in Appendix C.

A table, developed by Wojnar (2000) determining elements of higher order thinking in the student work of adult learners, was used to illustrate how participants could write follow-up questions to their vignettes that would require higher order thinking in their own students. This table summarizes the level of higher-order thinking addressed and types of questions that would correspond to each level (Kish, 2004, Appendix H).

In an *Instructional Techniques* course recently taught to instructors, the instructor included an explanation of the revision of Bloom's Taxonomy of Educational Objectives (Anderson, Krathwohl, Airasian, Cruikshank, Mayer, Pintrich, Paths & Wittrock, 2001). The revised taxonomy is made up of two dimensions: four categories of knowledge and the different levels of cognitive activity. The instructor referred to Table 4.1, The Knowledge Dimension (2001, p. 46), which includes factual, conceptual procedural, and metacognitive knowledge. The instructor referred to Table 5.1, The Cognitive Process Dimension (Wittrock, 2001b, pp. 67-68), which includes remember, understand, apply, analyze, evaluate, and create. Because the focus is on higher order thinking, the instructor developed a table that described evidence in student work according to the higher levels of the revised taxonomy: apply, analyze, evaluate, and create.

To address *metacognition*, participants were required to explain how they wrote their vignettes and reflect in their learning logs about using vignettes as a teaching strategy or learning activity.

Asynchronous Discussions Considering Learner-Generated Vignettes

Separate asynchronous discussion forums were set up for each set of vignettes the participants generated. Participants were encouraged but not required to post responses to these forums.

CONCLUSION

Using vignettes within the generative learning model was beneficial with adult online learners because the teaching strategy and learning activity addressed their learning needs. The teaching techniques and learning activities used within the generative learning model focus upon significant adult learning needs such as attracting and sustaining the learner's attention, motivating the learner by showing how to be successful in answering or creating vignettes, connecting their prior knowledge, experiences, and concerns to the course content, having participants create or generate responses and their own vignettes, and having participants reflecting upon what and how they learned. As a teaching strategy, adult online learners were challenged to consider their prior knowledge, experience and interests and apply their understanding of course concepts to discuss solutions in a meaningful way. Through asynchronous discussions, participants were able to reflect on other responses and reflect and collaborate with other participants to address meaningful situations and solutions. As a learning activity, vignettes encouraged individuality, creativity, and reflection upon how particular situations can challenge and become meaningful to participant's own students. Having learner-generated vignettes posted in asynchronous discussions that consider "netiquette" can provide participants with supportive feedback and insight from other participants. Even though teaching the writing of vignettes has been done with graduate level students in the field of Education, Kish projects that other presenters or instructors from other fields could benefit from learning how to write vignettes as well.

REFERENCES

Airasian, P. (2001). *Classroom assessment: Concepts and applications* (4th ed.). New York: McGraw-Hill.

American Distance Education Consortium. (January, 2003). *ADEC Guiding Principles for Distance Teaching and Learning*. Retrieved January 14, 2004 from http://www.adec.edu/admin/papers/Distance-teaching_principles.html

Ananda, S. (2000, July). *Equipped for the Future Assessment Report: How Instructors Can Support Adult Learners Through Performance-Based Assessment.* (Equipped for the Future (EFF) Technical Report). Washington, DC: National Institute for Literacy. Retrieved August 19, 2002 from http://eff.cls.utk.edu/resources//ananda_eff.pdf

Anderson, L. W., Krathwohl, D. R., Airasian, P. W., Cruikshank, K. A., Mayer, R. E., Pintrich, P. R. Raths, J. & Wittrock, M. C. (Eds.). (2001). *Taxonomy for learning, teaching and assessing: A revision of Bloom's taxonomy of educational objectives.* New York: Addison Wesley Longman, Inc.

Ausubel, D.P. (1968). *Educational psychology: A cognitive view.* New York: Holt, Rinehart and Winston, Inc.

Bandura, A. (1965). Behavioral modifications through modeling procedures. In L. Krasner & L.P. Ullmann, L. (Eds.), *Research in behavior modification: New developments and implications* (pp. 310-340). Chicago: Holt, Rinehart and Winston, Inc.

Bloom, B.S., Engelhart, M.D., Furst, E.J. Hill, W.H., & Krathwohl, D.R. (1956). *Taxonomy of educational objectives: The classification of educational goals.* New York: Longmans, Green and Co.

Brookfield, S.D. (1995). *Becoming a critically reflective teacher.* San Francisco: Jossey-Bass Publishers.

Campbell, L., Campbell, B., & Dickinson, D. (2004). *Teaching and learning through multiple intelligences* (3rd ed.). Boston: Pearson Education, Inc.

Campbell, R. (1999). The development and validation of an instructional design model for creating problem based learning (Doctoral dissertation, University of Pittsburgh, 1999). *Dissertation Abstracts International, 61*(01), 67.

Daloz, L.A. (1986). *Effective teaching and mentoring: Realizing the transformational power of adult learning* San Francisco: Jossey-Bass Publishers.

Dede, C. (1998, June 5-6). Future visions of information technology in math education: Design heuristics for vignettes embedded in the standards. Technology and NCTM Standards 2000 Conference, Arlington, VA.

DiVesta, F. (1989). Applications of cognitive psychology to education. In M. C. Wittrock & F. Farley (Eds.), *The future of educational psychology* (pp. 37-73). Hillsdale, NJ: Lawrence Erlbaum Associates.

Ellner, C.L., & Barnes, C.P. (Eds.). (1983). *Studies of college teaching: Experimental results, theoretical interpretations, and new perspectives.* Lexington, MA: D.C. Heath & Company.

Fischer, C.G., & Grant, G.E. (1983). Intellectual levels in college classrooms. In C.L. Ellner &

C.P. Barnes (Eds.). *Studies of college teaching: Experimental results, theoretical interpretations, and new perspectives* (pp. 47-60). Lexington, MA: D.C. Heath & Company.

Flanagan, J. C. (1954, July). The Critical Incident Technique. *Psychological Bulletin, 51*(4), 327-358.

Fogarty, R. (1997). *Problem based learning and other curriculum models for the multiple intelligences classroom*. Arlington Heights, IL: IRI/Skylight Training and Publishing, Inc.

Ford, L. (1969). *Using the case study in teaching and training*. Nashville: Broadman Press.

Galbraith, M.W. (Ed.). (1998). *Adult learning methods: A guide for effective instruction*. (2nd ed.). Malabar, FL: Krieger Publishing Company.

Gipps, C.V. (1994). *Beyond testing: Towards a theory of educational assessment*. London: The Falmer Press.

Goetz, J.P., & LeCompte, M.D. (1984). *Ethnography and qualitative design in educational research*. Orlando, FL: Academic Press, Inc.

Grabowski, B. (2004). Generative learning: Past, present, and future. In D.H. Jonassen (Ed.). *Handbook of research on educational communications and technology* (2nd ed.), (pp. 719-743). Mahwah, NJ: Lawrence Erlbaum Associates, Publishers.

Grabowski, B. & Koszalka, T. (1999). *Web-enhanced learning environment strategies for classroom teachers*. Retrieved April 12, 2002 from http://www.comms.uab.es/inet99/inet98/4e/4e_3.htm

Harasim, L., Hiltz, S.R., Teles, L., & Turoff, M. (1995). *Learning networks: A field guide to teaching and learning online*. Cambridge, MA: The MIT Press.

Higginbotham-Wheat, N.L. (1991). Use of generative learning strategies in computer-based instruction: Effects of learner-generated visual and verbal elaborations (Doctoral Dissertation, Memphis State University, 1991). *Dissertation Abstracts International, A(52/12),* 4205.

Hiltz, S.R. (1997). Impacts of college-level courses via asynchronous learning networks: Some preliminary results. *Journal of Asynchronous Learning Networks (JALN), 1*(2), 1-19.

Jeffries, C., & Maeder, D.W. (2004). *Using vignettes to build and assess teacher understanding of instructional strategies*. Unpublished manuscript, University of California at Los Angeles.

Johnson, A.P. (2000). *Up and out: Using creative and critical thinking skills to enhance learning*. Needham Heights: Allyn & Bacon.

Jonassen, D.H. (Ed.). (2004). *Handbook of research on educational communications and technology* (2nd ed.). Mahwah, NJ: Lawrence Erlbaum Associates, Publishers.

Kauchak, D. P., & Eggen, P.D. (1998). *Teaching & learning: Research-based methods*. (3rd ed.). Boston: Allyn and Bacon.

Kerka, S. (2001). *Capstone experiences in career and technical education*. (ERIC Publications 071). Washington, DC: Office of Educational Research and Improvement (ED). (ERIC Reproduction Service No. ED456333).

Kim, K.A. (1999). A comparison of academic achievement between the predominantly adult classes and the predominantly traditional age student classes and their teachers' perceptions of teaching (Doctoral Dissertation, North Carolina State University at Raleigh, 1999). *Dissertation Abstracts International, A60/09,* 3235.

Kish, M.H.Z. (2004). *Using vignettes to develop higher order thinking and academic achievement in adult learners in an online environment* (Doctoral Dissertation, Duquesne University, Pittsburgh, 2004).

Knowles, M. (1978). *The adult learner: A neglected species* (2nd ed.). Houston: Gulf Publishing Company.

Knowles, M. (1995). *Designs for adult learning: Practical resources, exercises, and course outlines from the father of adult learning.* Alexandria, VA: American Society for Training and Development.

Knox, A.B. (1986). *Helping adults learn: A guide to planning, implementing, and conducting programs.* San Francisco: Jossey-Bass Inc., Publishers.

Krasner, L., & Ullmann, L.P. (Eds.). (1965). *Research in behavior modification: New developments and implications.* Chicago: Holt, Rinehart and Winston, Inc.

Linn, R.L., Baker, E.L., & Dunbar, S.B. (1991). Complex, performance-based assessment: expectations and validation criteria. *Educational Researcher, 20*(8), 15-21.

Maslin-Ostrowski, P., & Ackerman, R.H. (1997, March). *A Case for Stories: Toward Further Understanding of Situated Knowledge and Practice.* Paper presented at the Annual Meeting of the American Educational Research Association, Chicago, IL. (ERIC Reproduction Service Document No. ED408265).

Maslin-Ostrowski, P., & Ackerman, R.H. (1998). Case Story. In M.W. Galbraith (Ed.), *Adult learning methods: A guide for effective instruction* (2nd ed.), (pp. 303-316). Malabar, FL: Krieger Publishing Company.

Marsick, V. J. (1998). Case study. In M.W. Galbraith (Ed.), *Adult learning methods: A guide for effective instruction* (2nd ed.), (pp. 197-218). Malabar, FL: Krieger Publishing Company.

McGuire, K. (2003). *Some definitions and guidelines for distance education.* Retrieved February 3, 2003 from www.uscampus.com/get_help/library/library_edurelated/article_distance.htm

McIsaac, M.S., & Gunawardena, C.N. (2004). Distance education. In D.H. Jonassen (Ed.), *Handbook of research for educational communications and technology* (2nd ed.), (pp. 355-395). NJ: Lawrence Erlbaum Associates, Publishers.

McKeachie, W.J. (1999). Introduction. In W.J. McKeachie (Ed.), *McKeachie's teaching tips: Strategies, research, and theory for college and university teachers* (10th ed.), (pp. 1-8). Boston: Houghton Mifflin Company.

McKenzie, J. (2001). *Planning good change with technology and literacy.* Bellingham, WA: FNO Press.

Merriam, S.B. (1988). *Case study research in education: A qualitative approach.* San Francisco: Jossey-Bass Publishers.

Moore, R.C., & Bogotch, I.E. (1993, November). *No longer neglected: Adult learners in graduate learning programs.* Paper presented at the Annual Meeting of the Mid-South Educational Research Association, New Orleans, LA. (ERIC Document Reproduction Service No. ED367905).

Odom, E. (1998, Fall). Critical thinking skills. *Word's worth,* (pp. 5,10). (ERIC Reproduction Service No. ED434227).

Ogle, D. (1986). K-W-L: A teaching model that develops active reading of expository text. *The Reading Teacher, 39,* 564-570.

Olson, G.M., Malone, T.W., & Smith, J.B. (Eds.). (2001). *Coordination theory and collaboration technology.* Mawhah, NJ: Lawrence Erlbaum Associates.

Pepicello, B., & Tice, E. (2000). Reshaping and teaching and learning: The role of liberal arts in online education. In K.W. White & B.H. Weight (Eds.), *The online teaching guide: A handbook of attitudes, strategies, and techniques for the virtual classroom* (pp. 45-56). Needham Heights, MA: Allyn & Bacon.

Ryder, M. (1998, June 7-11). *The world wide web and the dialects of consciousness*. Paper presented at the Fourth Congress of the International Society for Cultural Research and Activity Theory, Aahus, Denmark. Retrieved October 15, 2003 from http://carbon.cudenver.edu/~mryder/iscrat_98.html

Scardamalia, M., & Bereiter, C. (1986). Research on written composition. In M.C. Wittrock (Ed.), *Handbook of research on teaching* (3rd ed.), (pp. 778-803). New York: Macmillan Publishing Company.

Schwartz, P. (1991). *The art of the long view*. New York: Double Day.

Thompson, L. (1997, May). *Education graduates' post-degree learning needs study*. (SIDRU Research Report No.19). Saskatchewan, Canada: Saskatchewan Instructional Development and Research Unit, Regina. (ERIC Document Reproduction Service No. ED434081).

Tomei, L., & Wilcox, B. (1999). *Professional portfolios for teachers: A guide for learners, experts, and scholars*. Norwood, MA: Christopher-Gordon Publishers, Inc.

Van der Wal, R.J., & Van der Wal, R. (2003). Assessing life skills in young working adults — Part 1: The development of an alternative instrument. *Education and Training, 45*(3), 139-151.

Vella, J. (1994). *Learning to listen, learning to teach: The power of dialogue in educating adults*. San Francisco: Jossey-Bass, Inc., Publishers.

White, K.W., & Weight, B.H. (2000). *The online teaching guide: A handbook of attitudes, strategies, and techniques for the virtual classroom*. Needham Heights, MA: Allyn & Bacon.

Wiggins, G. (1987, Winter). Creating a thought-provoking curriculum. *American Educator,* 11(4), 10-17.

Wiggins, G. (1998). *Educative assessment: Designing assessments to inform and improve student performance*. San Francisco: Jossey-Bass Publishers.

Wittrock, M.C. (1974). A generative model of mathematics learning, *5*(4), 181-196.

Wittrock, M.C. (Ed.). (1986). *Handbook of research on teaching* (3rd ed.). New York: Macmillan Publishing Company.

Wittrock, M.C. (1986). Students' thought processes. In M. C. Wittrock (Ed.), *Handbook of research on teaching* (3rd ed.), (pp. 297-314). New York: Macmillan Publishing Company.

Wittrock, M.C. (Nov-Dec, 1987). Teaching and student thinking. *Journal of Teacher Education, 38*(6), 30-33.

Wittrock, M.C. (1990). Generative processes of comprehension. *Educational Psychologist, 24*(4), 345-376.

Wittrock, M.C. (1991). Generative teaching of comprehension. *The Elementary School Journal, 92*(2), 169-184.

Wittrock, M.C. (1994). *Metacognition*. Unpublished manuscript, University of California at Los Angeles.

Wittrock, M.C. (2000, Winter and Spring). Knowledge acquisition and education. *The Journal of Mind and Behavior, 21*(1 and 2), 205-212.

Wittrock, M.C. (2001a). The knowledge dimension. In L.W. Anderson, D.R. Krathwohl, P.W. Airasian, K.A. Cruikshank, R.E. Mayer, P.R. Pintrich, J. Raths, & M.C. Wittrock (Eds.), *A taxonomy for learning, teaching, and assessing: A revision of bloom's taxonomy of educational objectives* (pp. 38-62). New York: Addison Wesley Longman, Inc.

Wittrock, M.C. (2001b). The cognitive process dimension. In L.W. Anderson, D.R. Krathwohl, P.W. Airasian, K.A. Cruikshank, R.E. Mayer, P.R. Pintrich, J. Raths, & M.C. Wittrock (Eds.), *A taxonomy for learning, teaching, and assessing: A revision of bloom's taxonomy of educational objectives* (pp. 63-92). New York: Addison Wesley Longman, Inc.

Wlodkowski, R.J. (1999). *Enhancing adult motivation to learn (Revised Edition): A comprehensive guide for teaching all adults* (2nd ed.). San Francisco: Jossey-Bass Inc., Publishers.

Wojnar, L. (2000). *Instructional design and implementation of a best practice model of online teaching and learning.* Unpublished doctoral dissertation, Duquesne University, USA.

Wojnar, L. (2002, August). Research summary of a best practice model of online teaching and learning. *English Leadership Quarterly, 25(1)*, 2-9.

Zinkoski, P. (August, 2001). The many benefits of writing. *Partnership for Learning.* Retrieved January 11, 2004 from http://www.partnershipforlearning.org/article.asp?ArticleID=319

ENDNOTES

1. In the original study (Kish, 2004), student preference of the use of vignettes as a teaching strategy and learning activity was also researched.
2. The original study also included the following question: Do students prefer vignettes to lectures, teacher demonstrations, student demonstrations, projects, online slide presentations or online discussions as shown in student reflective learning logs and a questionnaire (distributed at the end of the course).
3. To address the question regarding student preferences of teaching strategies and learning activities, an additional questionnaire was distributed at the end of the course.
4. In the original study, pilot study data were also analyzed for vignette preference.
5. For more information on the participants, please refer to the original study.
6. To address student preferences of teaching strategies and learning activities, content validity, criterion validity, and construct validity were addressed for the learning logs and for the preference questionnaire.
7. For conclusions concerning the third question studied, please see the original study.

APPENDICES

Appendix A. Scoring guide for evaluating participant-completed vignettes that are teacher-generated (in asynchronous discussions)

Criteria	Score
Valid Response	
Student answered all parts of the questions	5
Analyses and responses addressed the question	10
Section Total	**15**
Analysis	
Included appropriate contextualized resource material (use of book and notes)	5
Represented and analyzed at least 3 points of view (where appropriate) including clear and focused statement of agreement / disagreement	10
Section Total	**15**
Defense	
Appropriate references (including quotations and page numbers) were made to readings and research to justify answers (At least one was made per question)	10
Included relevant evidence in support for all 3 of the viewpoints	10
Accurate definitions and components of key terms were included for each question.	10
Provided appropriate examples of key terms and issues (where required)	10
Defined the problem and suggested viable resolutions (where required)	10
Section Total	**50**
Discussion	
The participant's response provided a thoughtful contribution that added to the understanding of others	6
Student responded appropriately to at least two other postings appropriately	4
Section Total	**10**
Language	
Language and phrasing were appropriate for the audience	2
Responses were clearly written	2
Spelling was correct	2
Punctuation was accurate	2
Grammar and usage were correct	2
Section Total	**10**
GRAND TOTAL	**100**

Appendix B. Vignette starter

DIRECTIONS:

I. **Topic**: After considering your students and the subject you teach, think of a lesson you teach in which you might be able to use a vignette. You might want to try to "web" your ideas by coming up with the topic of your lesson, and then determine different objectives you would want to meet.

II. **Try to fill in the answers for the following questions (in the space provided on the right):**

My lesson focuses on....	
My objectives for this lesson include....	
The materials or books I am considering include....	
This vignette will show me whether or not students know....	
The following is a setting or situation with which either all of my students or none of my students have experience with	
The setting I would like to use can be described as (remember to consider what you see, hear, feel, and the tone of the story)...	
Indicate one of the following: I will write a truncated vignette so that the students have to "fill in" the ending of the vignette. OR I will write an abridged vignette so that students have to "fill in" the middle" of the vignette.	
The plot or general sequence of events will be....	
Important characters in this vignette include...	
I want my students to be able to answer the following questions so that they have to either "fill in the ending" or "fill in the middle" of the vignette (remember to keep in mind what the objectives are)...	

III. **Follow-Up Activity:** Once the chart is complete, try to write out the vignette. Make sure that you include questions that require your students to fill in the appropriate parts of the vignette and that have your students meet the objectives. Go back and refer to the scoring guide and rubric to make sure that you have included all of the necessary elements. When it comes to the editing process and finalizing what you have, make sure that you can answer the following questions: "Have I provided enough information here for the participant to find the clues to complete my tasks?" and "Can I make it any shorter without losing anything critical to the vignette?"

Appendix C. Scoring guide for participant-generated vignettes

Criteria	Category	Score
Context	Vignette simplified a real-life situation	5
	Subject was clearly recognizable	5
	Problem was clearly identified	4
	Section Total	**14**
Story Elements	Included one or more main characters	3
	All of the participant's students either have experience with the setting or have not	2
	The vignette had a plot that at least has a beginning and middle (if truncated vignette) or beginning, middle (left incomplete) and end (if abridged vignette)	5
	The plot told the events in logical order	5
	Section Total	**15**
Content	Vignette was either purposely left incomplete or was vaguely written so that multiple solutions (to the questions) can be defended	5
	There were at least 4 questions that require the participant's students to answer the vignette	3
	The questions encouraged independent thinking and unique responses	5
	The questions had the participant's students do one or more of the following: provide examples, read through and critique a situation, explain a point of view, recall an experience, or solve a problem	5
	The vignette was fictional and original	5
	The vignette was written as a narrative and is short (1–3 paragraphs, <250 words)	7
	Section Total	**30**
Language	Words selection made it is easy to understand just what the writer meant.	1
	Language and phrasing were natural, effective and appropriate for the audience	2
	Spelling was correct	1
	Punctuation was accurate	1
	Grammar and usage were correct	1
	Section Total	**6**
Explanation	Included what the participant was trying to teach	5
	Included how the participant came up with the vignette	5
	Included the objectives that were considered when the vignette was written	10
	Explained how situation presented in the vignette was either familiar or not familiar to the participant's students.	5
	Explained the types of responses expected or anticipated	10
	Section Total	**35**
	GRAND TOTAL	**100**

Chapter XII
Business-Plan Anchored E-Commerce Courses at the MBA-Level

C. Derrick Huang
Florida Atlantic University, USA

ABSTRACT

The diversity and currency of subjects covered in e-commerce courses at the MBA level present a challenge to educators. In this chapter, we analyze and recapitulate our experience in using the business plan to anchor the e-commerce course to address those challenges. Business plan requirements can link the various subjects together, afford students with a real-life experience learning process, and, with proper curriculum design and course delivery, give students an opportunity to be "reflective practitioners." Results showed that students' learning and interests for the e-commerce subjects were high with the business plan requirement.

INTRODUCTION

The introduction and adoption of electronic commerce courses in the MBA curriculum have been a rapid, dynamic, yet volatile process for many universities. From the early days of "jubilation" about anything e-commerce, to the crash and burn of the dotcoms, to the relative stabilization of Internet businesses in recent years, e-commerce courses have undergone dramatic changes in terms of content, teaching orientation, and interest level. But even with the burst of the Internet bubble, the popularity of e-commerce courses and programs actually grew (Fusilier & Durlabhji, 2003). In less than 10 years, they have become a mainstay in many MBA programs, especially in the MIS or IT track.

However, to successfully offer e-commerce courses at the MBA level poses continuing challenges to educators for several reasons. First, the concept and practice of e-commerce are still new, dynamic, and evolving constantly. Materials are often ill-defined, and topics become out of date quickly. For instance, applications service provider and independent business-to-business intermediaries, two of the most important Internet applications in 1999, have been reduced to minor topics in the general areas of e-commerce operations. Conversely, popular and important current Internet activities, such as social networking, did not exist in the 1990s boom. Such fluid nature of the course requires instructors to renew, sometimes redo, the course content and coverage from year to year, or as often as every semester.

Second, an e-commerce course covers topics that are often diverse and seemingly unrelated. Based on three widely adopted graduate-level e-commerce textbooks (Awad, 2007; Laudon & Traver, 2007; Turban et al., 2006), the topics range from the technology infrastructure and Website construction, to marketplace structure and economics, to marketing and operations, to various issues regarding security and social impact. It is difficult, if not impossible, to cover all these relevant topics in a semester's time while maintaining a consistent tone for the course. At worst, students come away with a collection of materials without a central theme or integrated idea of what e-commerce really is about. We therefore see many different "flavors" of e-commerce courses—for example, technology or programming-oriented, e-marketing focused, entrepreneurship—often based on the interest and specialty of the instructors.

Last, but not least, MBA students come with a variety of backgrounds, and they take e-commerce courses with differing objectives. Some, for instance, already have strong hands-on knowledge in IT, and they hope to complement that with a more management-orientation e-commerce program. Others, however, are not familiar with the technical side and expect to learn more about e-commerce technology and how they can apply it in business. Offering interdisciplinary choices as motivators to all students and balancing all subjects carefully to satisfy individual preferences become an important task in making an e-commerce course successful.

How does one deliver an e-commerce course that is dynamic and consistently updated? How does one paint the whole picture of e-commerce with those diverse topics as tightly knitted components? Perhaps, more importantly how does one integrate the technology components with all the business subjects to equip the students with real-world applications of e-commerce? There are a number of proposed ways to address these concerns. One approach, arguably passive, is to adapt the lectures, linking them together while citing real-world applications, and supplementing lectures with cases (Hackney, McMaster, & Harris, 2003). While this approach exposes students to critical subject matter, it does not engage them in the process of integrating the materials. Unfortunately, quality e-commerce cases are not that common, and few cover more than one topic adequately. Another approach is to require students to compose a final paper on a practical topic that involves more than one area of e-commerce. This requirement improves their comprehension of the concepts and ability to apply them. Regrettably the impact is limited, because students only engage in the integrated exercise toward the end of the semester, rather than throughout the whole course. Yet, another is to use small business consulting projects as a hands-on exercise in the actual implementation of e-commerce applications. Because of, and limited by, the nature of available small business sites, this approach is best suited for undergraduate e-commerce courses with a technical focus (Tabor, 2005). Finally, a "student-driven" approach is proposed, where students, divided into groups, develop and learn e-commerce materials with the help of tutors based on narratives and tasks issued to them for each section. This seems to be an excellent approach for a graduate-level e-commerce course, provided that the class is large enough and the resources (e.g., the availability of tutors) are available (McBride, 2005).

In this chapter, we discuss a different approach to addressing those challenges, based on our experience in using business plans as the anchor for the course. In the following sections, we describe the pedagogical framework of this approach, curriculum design, and course delivery. We further discuss the results and evaluations of such a course, as well as important success factors and issues to address.

PEDAGOGICAL FRAMEWORK

To design an e-commerce curriculum for MBA students, we started by defining the goals and guidelines for the course. After reviewing available textbooks, other e-commerce course offerings, and MBA school guidelines, we determined the following academic goals for the student:

- An in-depth knowledge of the Internet in business: its operations, benefits, limitations, and interdependencies; and
- The ability to create, plan for, and critique a new (line of) business based on or heavily involved in use of the Internet and/or Internet technologies.

These two curricular goals acted as the basis for the course design. Along with the goals, we adopted the following guidelines to govern our approach to the course: First, integration of e-commerce with business and management subjects. We do not believe e-commerce is its own subject; rather, we adopt a "two-way model:" e-commerce permeates all areas of business—operations, marketing, finance, information technology—and all subjects of management influence how e-commerce is implemented. This is true whether e-commerce is conducted as an independent business—the Amazon.com model—or as an extension of an existing business—the Wal-mart.com model. It is, therefore, imperative that e-commerce touches on all areas of business and management *in an integrated fashion.*

Second, experiential learning, or problem-based learning, that requires planning, analysis, and reflection (Gackowski, 2003; Nuldén & Schepers, 2002). We feel that it is crucial for MBA students to learn how to create an e-business or apply e-commerce concepts to their own business. A well-designed experiential learning project, where students actually get to practice such an application in a controlled environment with instructor feedback and self-reflection, would be an effective approach.

Third, team work is necessary. We have found, at the MBA level in particular, that students learn from each other as much as they learn from course materials and instructors. This mutual learning can be reinforced in many ways, such as case discussions, student presentations, and team projects. We encourage students to form teams with others from different backgrounds to maximize their mutual learning experience.

With these goals and guidelines in mind, we intended to design the e-commerce course with a hands-on, team-based learning tool that "anchors" the various components of the course—textbook, cases, readings, lectures, and assessments. Students must learn from doing and applying all the topics covered in the course in order to fulfill the requirements of the anchor. The following section describes why we chose business plans as the anchor, and how such a course was designed and delivered.

BUSINESS-PLAN ANCHORED E-COMMERCE COURSE

Business Plans

A business plan "describes all critical internal and external elements and strategies for guiding the direction of [a] venture's first several years as well as giving potential investors an idea of the venture's structure, objectives, and future plans.

It communicates important entrepreneurial management practices, such as how the venture will mitigate risk … [and] manage uncertainty" (Price, 2004, p. 7). A typical business plan includes the following elements:

- **Executive summary:** Highlights of the business plan.
- **Company description:** Legal establishment, history, and start-up plans.
- **Product or service:** What the venture sells and how it benefits the customers.
- **Market analysis:** Customer needs, where they are, and how to reach them.
- **Strategy:** Venture's competitive advantage and how to achieve it.
- **Implementation:** Specific management responsibilities with dates and budgets, and how the results may be tracked.
- **Management team:** The organization and the key management team members.
- **Financial analysis:** At the minimum, a current and *pro forma* balance sheet, an income statement, and a cash flow analysis.

The writing of a business plan is an activity extensively endorsed in the academic literature in entrepreneurship (Fry & Stoner, 1985; McKenna & Oritt, 1981; Timmons, 1980) and has been a major component in such courses. The main reason, of course, is that it is a crucial product of initial entrepreneurial activities: A business plan records and captures an entrepreneur's visions of how a start-up business should initiate and operate, and it is often the basis for potential financiers to make funding decisions (Sexton & Bowman-Upton, 1991). In addition, as a teaching tool, the business plan serves several important purposes: It brings realism into the classroom and allows the students to learn by doing; it helps students develop high-order reasoning and organization skills; and it is a natural setting for major team work.

The business plan requirement satisfies our goals and guidelines for the e-commerce course. It addresses our concerns for the diversity of subjects covered in such a course and presents a multidisciplinary approach to learning: The business plan exercise encompasses many, if not all, of the aspects of an e-business—strategy, marketing, operations, management, financing, human resources, and organization—and "glues" them together. In order to write a good business plan, students need to integrate all those aspects and address them equally well. Further, MBA students in e-commerce can benefit by gaining a hands-on experience working on a near-real-life business opportunity of their own choice. Composing a business plan affords them an opportunity to practice what they learn in the classroom, gaining crucial skills—analytic ability to apply concepts to practical problems, organization and presentation of business ideas, and outlining details of running a successful e-business—in a real-world setting. Lastly, an e-commerce course has a strong link to entrepreneurship by nature. The majority of the e-commerce coursework is concerned with starting up a new e-business, creating a new e-commerce division within traditional companies, or evaluating and reacting to the impacts of e-commerce by other start-ups. A business plan helps students understand and "experience" the various aspects of starting and building a new e-business, as well as learning how to take it from planning to the next stage, such as getting financing, finding partners, and sustaining growth.

To maximize the benefits of such a tool, we made the business plan requirement not just an assignment, but the "anchor" of the course, and class delivery was scheduled accordingly. The design is based on a reflection of the "aligned teaching" philosophy (Biggs, 2003). The course is taught in such a way that students can learn and apply key concepts to their business plan exercise gradually. At the end, the students practice all the key points learned in the course, have an opportunity to revise and critique their plans, and learn how to improve their skills after their business plans are completed and submitted.

Course Design

The course, named "E-Commerce and Internet Business Applications," is divided into two modules, with the business plan as the common thread. The topic syllabus for the Spring 2007 class is presented in Appendix 1. The first module, spanning one half to two-thirds of the sessions, covers the *traditional* topics of e-commerce, focusing on those subjects and concepts of e-business that they will need for their business plans. The first three weeks concentrate on the basics of e-commerce, with an introduction to the concept of business planning and how to write a business plan, as well as a discussion of the e-business strategies and business models. Although this course is not technically oriented, a survey of e-commerce technologies is included not only to familiarize the students with various jargons and buzzwords, but also provide them with the knowledge of how technologies can be used as a tool for gaining business advantage. It is then followed by a series of key elements of launching and running an e-business, including the managerial, marketing, and financing aspects of the business operation. Finally, we discuss how the effectiveness of e-business can be analyzed and measured to complete the knowledge core for e-business planning in this module.

The second module focuses on *current* and *special* topics of e-commerce, such as social networking, media transformation brought about by the wide use of Internet, security, mobile and ubiquitous commerce, and ethical issues. These topics are applications of the subjects in the first module and can be chosen based on their importance and currency.

The transition from the first to the second module is marked by the completion of students' business plans. Instead of a final, the business plan is used as a midterm project to anchor the course. During the first module, students work on developing and completing their business plans, making use of the materials that they learn in class. At the end of the module, they present and submit their business plans and receive comments from peers and the instructor. This gives students time to internalize the experience and reflect on their business plans in class discussions during the second module, maximizing the benefits of this exercise.

Course Delivery

An e-commerce course is intrinsically current; to cover the topics, we adopt the use of a combination of textbooks (for the concepts) and cases (for the applications). We had considerable difficulty in identifying an appropriate textbook; in fact, we adopted a different one each semester when the course was given. The textbook we use for the Spring, 2007 term is *Creating a Winning E-Business,* 2nd edition (Napier et al., 2006), supplemented by 12 teaching cases from a variety of sources.

The textbook and cases cover the course materials, while the business plan constitutes the main avenue for internalization of those materials. Students are prepared from the first day and told that the entire course will "evolve around" the business plan requirement. Within the first three weeks, the students are required to form teams of up to three people, research potential ideas, and select appropriate e-commerce topics for their business plans. By the end of the third class, each team turns in a proposal that outlines their business ideas with possible references and information sources. They can only proceed after the proposal is approved. This step is designed to be a screening process to insure that the business proposals are appropriate and doable, and that the students have a clear idea what they intend to do from start to finish. The proposal helps the instructor diagnose problems that students may encounter when they start working on their business plans.

The students have one-and-a-half to two months to work on their business plans. In each class during the first module, they are encouraged to discuss their plan-in-progress, based on the topic of that session, so they can get maximum exposure to the various areas of their plans. They are required to meet with the instructor at least twice after proposal approval and before final plan submission to ensure their progress. The deliverables of this requirement, at the end of the first module, consist of a written business plan as well as a 20-minute presentation in front of the class. The presentation setting is such that they are presenting to a group of venture capitalists—if their business plan is for a new venture—or the board of directors—if their business plan is for an e-business extension of an existing company—to get funding or approval. After the presentation, the presenters have to answer questions from the audience related to their plans. And they have the opportunity to incorporate all the comments from fellow students and the instructor before they submit the final version of their business plans.

During the presentation session the students must evaluate the other teams' plans. Each is required to fill out an evaluation form for each business plan presented, based on pre-determined criteria, such as the perceived accuracy of the market analysis, sustainability of the competitive advantage, and the potential of the business (see Appendix 2 for the evaluation form). They must "score" each business plan—marking whether they would invest as lead investor, invest as junior investor, require more information, or decline—and explain why they reached that decision.

The second module begins after all business plans are submitted. Having "created" an e-business from ground up, the students now have a better understanding of the issues around those special topics of e-commerce. Their business plans are often used as case examples in class discussions of those topics. For example, the instructor might ask the team that proposed a business of an online tracking device to elaborate on the potential ethical issues associate with their proposed business, and the whole class benefits from a discussion on that topic.

Two important sets of events are arranged to complement the business plan requirement. Every year, during the spring semester, The Center for Entrepreneurship and Venture Capital at the case university sponsors a business plan competition for university affiliates. Students in the e-commerce course are highly encouraged to enter the contest. To facilitate the students' entry, the business plan presentation session is scheduled one week before the competition deadline. They benefit from the feedback from the instructor and fellow students before they finalize their competition submission, as well as practice their business plan presentations, which may be required later if their plans make it to the final round.

This type of course links classroom learning with real-world experience directly, and we take advantage of this setting by inviting guest speakers (NOT guest lecturers) to appear in four to six sessions. The guest speakers are industry practitioners who speak about their past or present e-commerce experience. To maximize the learning value, in the first module we invite these practitioners to address topical issues—for instance, online marketing channels and their effectiveness—that can help the students' writing of the business plan. In the second module, after their business plans are done, we invite e-business entrepreneurs to discuss their start-up experience and lessons. Listening to those entrepreneurs after the completion of the business plan gives the students a better understanding of the real-life entrepreneurial experience. They can absorb the information, reflect on what they have done in their business plans by comparing and contrasting with the shared stories of successful or failed experiences, and internalize the process of writing a business plan.

Course Evaluation

The e-commerce course with a business plan as anchor has been taught five times at the case university since 2003. A selection of business plans created in the past can be found in Appendix 3. Sources of the business plan ideas varied: Some were extensions of businesses in development; others were original; whereas still others were creative or futuristic. But all students worked through the planning process and turned their ideas into real business proposals successfully.

The business plan competition has been a great motivation, as evidenced by the fact that teams from this course have won prizes in the competition three years in a row. The course evaluations and surveys administered at the end of the semesters indicate that students consistently agreed that the business plan was the most effective way to learn various subjects of e-commerce. (Appendix 4 lists some of the comments students provided in the surveys.) Students particularly liked the fact that real-life e-business entrepreneurs were invited to speak after the completion of their business plans, so that they could validate and/or improve their limited experience. The combination of the survey result, the result from the business plan competition, and the feedback from class discussions led this author to conclude that the business plan requirement accomplished what it was intended to do, namely be a tool to help the students learn, internalize, integrate, and apply the diverse subjects of e-commerce.

There are some caveats that must be addressed. Students might not take the business plan seriously. Although the instructor makes it clear at the beginning of the course that it is the central, or anchoring, learning tool, students might merely regard the business plan requirement as one of grading components and treat it accordingly. Another potential problem is that students might perceive the business plan as one single assignment, not an overall pivotal project that integrates all the subjects covered in the course. The most effective way to address these risks is with the course delivery itself. Throughout the course, lectures on many subjects are regularly mapped to the various components of the business plan, and students' plans are frequently used as examples in class discussions. This active involvement helps them learn the subjects through their own business plan efforts and reinforces the role of this requirement. The business plan competition has proven to be an enormous incentive for students to put forth their best efforts.

We found that, occasionally, some students had difficulties in completing such a project that required self-initiation, independent research and thinking, and strong closure. Others did not recognize how hard and time-consuming a business plan could be, as reflected in some survey feedbacks. They tend to be those who excel in more well-defined, structured teaching environments characterized by lectures, topical papers, and final exams. To help those students make the transition and maximize their learning through the business plan requirement, a considerable amount of supervision and "hand-holding" was required. Students should be encouraged to consult with the instructor when they encounter difficulties. They will feel more comfortable to do so when they realize that the business plan project is designed for learning, not grading, purposes.

CONCLUSION

Business plan teaching requires a considerable amount of planning then tweaking during delivery. The course materials and the business plan requirement need to fit so that the students can learn and digest the maximum amount of information. The level of emphasis placed on the instruction of business plans is an important balancing act. Too little emphasis will reduce the business plan to a regular classroom assignment, while too much may enable the writing of the business plan to take over the whole course, obscuring the true mission of the course. The two-module approach appears to strike the balance fairly well.

Slotting the business plan in between the two modules is an important scheduling feature of the course. To achieve the objective of "reflection on action" (Schön, 1987) in this business plan exercise, feedback and follow-up learning are almost as important as the composition phase. If it is due at the end of the course, students have a tendency to finish the final project assignment and completely forget it after the semester is over, making feedback wasted. By requiring the business plan submission at the end of the first module instead, the instructor can use the second module to reiterate and emphasize many of the concepts that the students encountered during the business plan exercise. And such an experience should help them absorb and digest those concepts better.

REFERENCES

Awad, E.M. (2007). *Electronic commerce: From vision to fulfillment* (3rd ed.). Upper Saddle River, NJ: Pearson Prentice Hall.

Biggs, J.B. (2003). *Teaching for quality at university: What the students does.* Society for Research into Higher Education Open University Press, Buckingham, Pennsylvania.

Fusilier, M., & Durlabhji, S. (2003). No downturn here: Tracking e-business programs in higher dducation. *Decision Sciences Journal of Innovative Education, 1*(1), 73-98.

Fry, F.L., & Stoner, C.R. (1985) Business plans: Two major types. *Journal of Small Business Management, 23,* 1-6.

Gackowski, Z.J. (2003). Case/real-life problem-based learning with information systems projects. *Journal of Information Technology Education, 2,* 357-365.

Hackney, R., McMaster, T., & Harris, A. (2003). Using cases as a teaching tool in IS education. *Journal of Information Systems Education, 14*(3), 229-234.

Laudon, K.C., & Traver, C.G. (2007). *E-commerce: Business, technology, society* (3rd ed.). Upper Saddle River, NJ: Pearson Prentice Hall.

McBride, N.K. (2005). A student driven approach to teaching e-commerce. *Journal of Information Systems Education, 16*(1), 75-83.

McKenna, J.F., & Oritt, P.L. (1981). Growth planning for small business. *American Journal of Small Business, 5*(4), 19-29.

Napier, H.A., Rivers, O., Wagner, S., & Napier, J.B. (2006). *Creating a winning e-business.* Boston: Thomson Course Technology.

Nuldén, U., & Schepers, H. (2002). Increasing student interaction in learning activities: Using a simulation to learn about project failure and escalation. *Journal of Information Systems Education, 12*(4), 223-232.

Price, R.W. (2004). *Roadmap to entrepreneurial success: Powerful strategies for building a high-profit business.* New York: American Management Association.

Sexton, D.L., & Bowman-Upton, N.B. (1991). *Entrepreneurship: Creativity and growth.* New York: McMillan Publishing Company.

Schön, D.A. (1987). *Educating the reflective practitioner: Toward a new design for teaching and learning in the professions.* San Francisco: Jossey-Bass.

Tabor, S.W. (2005). Achieving significant learning in e-commerce education through small business consulting project. *Journal of Information Systems Education, 16*(1), 19-26.

Timmons, J.A. (1980). A business plan is more than a financing device. *Harvard Business Review, 58*(2), 28-34.

Turban, E., King, D., Lee, J., & Viehland D. (2006). *Electronic commerce: A managerial perspective*. Upper Saddle River, NJ: Pearson Prentice Hall.

APPENDICES

Appendix A. Sample course outline

The following is an outline of a topical syllabus designed for a 15-week, three-hour-per-week e-commerce course, as adopted by the author for the Spring 2007 semester:

Week 1: Introduction: course rules and requirements; overview of the course and e-commerce
Week 2: Business Planning Basics; How to Write an E-Business Plan
Week 3: E-Business Economics, Models, and Strategies
Week 4: E-Commerce Technologies
Week 5: Customer Interface of E-Business
Week 6: Launching an E-Business Successfully
Week 7: E-Business Operations
Week 8: E-Commerce Marketing
Week 9: Internet Marketing Tools
Week 10: E-Business Analysis and Control
Week 11: Business Plan Presentations and Discussions
Week 12: Web 2.0 Impact on Business
Week 13: Media Transformation
Week 14: E-Commerce Security Issues
Week 15: Macro Environment for E-Commerce: Legal, Ethical, and Social Impacts

Appendix B. Business plan evaluation form

Plan evaluated:

Evaluator:

Your Final Decision (select one):

1. Invest as lead
2. Invest as joint
3. May invest after minor modification
4. May invest after major modification
5. Decline

Detailed Evaluation:

1. Revenue (Feasible? Realistic? Sustainable? Improvement?)
   - Target market
   - Revenue model
   - Size and growth
2. Cost (Feasible? Realistic? Sustainable? Improvement?)
   - Fixed cost
   - Variable cost
   - Cost driver and control
3. Financing (Feasible? Realistic? Sustainable? Improvement?)
   - Investment requirements
   - Cash flow and breakeven
   - Timeframe
4. Comments
   - Overall strengths
   - Overall weaknesses
   - 3 critical success factors (and did they address them?)

Appendix C. Examples of past student business plans

Name	Description	# Students	Semester	Note
AtYourService.com	Errand and concierge service provider	2	Spring 2004	Finalist 2004*
DressCode.com	eCommerce subsidiary for supplier of school and workplace uniforms	3	Summer 2003	Based on a real company
eDinner	Online dining reservation service	3	Spring 2003	
Healthnet.com	Health information network for small and medium health-care providers	2	Spring 2004	
Hospitality Biz Assist	Business service for small and medium companies in hospitality industry	3	Spring 2005	2nd place winner 2005*
KAS Publications	Online information provider of ergonomic policies and news for small restaurants	3	Spring 2003	
Medsafe	Biometric security solutions provider for the health care industry	3	Summer 2003	Based on a real business
MyLunchBreak.com	Health lunch delivery service	2	Spring 2005	
Out n' About Care	Canadian nurse placement service	3	Spring 2004	
PostcardMarketing-Wizard.com	Postcard marketing service for wedding professionals	3	Summer 2003	Business in development
Real Estate IRA	Trust service for real estate investment using IRAs	3	Spring 2005	Finalist 2005*
SJ Computer Systems	Reseller of refurbished computers	2	Spring 2003	
StrategicPoints	Service provider of web session and transaction analysis to online retailers	3	Spring 2004	Winner 2004*
Tattoo Town	Online service provider to the body art industry	2	Summer 2003	
Telecom.net	Internet security solutions provider	3	Spring 2003	Winner 2003* Business in development
Textbooks4Less.com	Textbook exchange service	3	Spring 2003	
Touchplay	Producer of touch screen overlay integrated with interactive websites	3	Spring 2004	Finalist 2004*
Tracker	Point-of-sale information tracking device maker and service provider	3	Summer 2003	
YoUr Interest First	Fee-only financial advisory service	3	Spring 2005	Finalist 2005*

* *Of the business plan competition*

Appendix D. Sample student feedback

"The business plan was a lot of work, but it did ensure that we learned." (Spring 2005)

"The business plan was great. It helped us develop interest in the class." (Spring 2005)

"The business plan was a lot of work, but I learned a lot." (Spring 2005)

"Business plan [is the what] I like most about this course. [We should have] more time spent on the business plan." (Spring 2004)

"I like the business plan and the fact that the class discussions were oriented toward it." (Spring 2004)

"[I like the fact] that we competed with other students in business plan competition, which was a simulation of the real world... I think the class would have been better if we started an e-commerce company or website from [our business plans]." (Spring 2004)

Chapter XIII
Cyber Schools and Special Needs: Making the Connection

Shellie Hipsky
Robert Morris University, USA

Lindsay Adams
PA Cyber Charter School, USA

ABSTRACT

Cyber schools for K-12 students are growing in number. It is vital that appropriate strategies are devised to meet the needs of students with exceptionalities. The PA Cyber Charter School serves 468 students who have individualized education plans. Parent surveys were thematically analyzed and revealed six predominant themes including: communication, interests, focus, less-stigma from the special education label, education differences in comparison to other methods, and cyber school shortcomings. The study also utilized the action research model to determine and present the techniques and strategies that are working in the PA Cyber Charter School for their students with special needs. Teacher-tested documents included in the appendix were based on the study, and a model for special needs strategies in the cyber learning environment has been established through this chapter.

BACKGROUND

Cyber Schools

Twenty-two states thus far have established cyber schools to administer curriculum for students who range from kindergarten through twelfth grade (Borja, 2005). In the state in which this study was conducted, "There are 12 cyber charter schools in Pennsylvania educating about 13,245 students this school year. That's up from about 10,000 students last year." (Duncan, 2006). A cyber charter school is ultimately responsible for demonstrating that the goals for the school, and therefore the students, are met or the school will "cease to exist" (Center for Charter Reform, 2002). The school's charter is revoked if it does not perform.

Miron, Nelson, and Risley (2002) established areas of cyber charter school innovation:

- Providing an innovative way to reach at-risk students who have dropped out of traditional schools.
- Offering a wider range of classes to their students. Students can be offered different (often advanced) instruction compared with courses that may be available in their local district's schools.
- Providing structure and assistance to parents who were previously home schooling their children. Enrolling formerly home-schooled students in cyber schools increases the amount of public oversight and guidance.
- Enabling students with health/medical/social problems that preclude attendance at a traditional school to continue their education from home or from a hospital or rehabilitation center. (p. 116)

Exceptionalities in the Online Learning Environment

Cyber charter schools are required to meet all federal laws and regulations for special education including the individuals with disabilities education act (IDEA), no child left behind (NCLB), the americans with disabilities act (ADA), and Section 504 of the rehabilitation act.

The RPP International (2000) national study suggests that a slightly smaller percentage of students with disabilities are enrolled in charter schools (8 percent) than in public schools (11 percent). In 1998, Cyber Village Academy was started by Neima and Bilyk as the first online charter school to serve Minnesota students who had health impairments. Unfortunately, they struggled with software that was built for college students and corporate training; the lessons could not be edited and they were not engaging (Bilyk, 2005). That same year, the first cyber charter school in Pennsylvania, SusQ-Cyber was formed.

According to Miron and Nelson's (2000) executive summary of research on Pennsylvania's charter schools, the schools received, "high marks from parents who said that their students' special needs were not well served in other schools which suggests that parents with students with special needs are generally satisfied with the progress their children are making in charter schools."

Purpose

Miron, Nelson, and Risley (2002) encouraged research in cyber schools and special needs students:

Some charter schools appear highly successful in serving students with special needs, while others appear unable (or according to some critics, unwilling) to serve such students. It is worth further examination of charter schools' strengths and barriers to serving students with special needs, particularly in schools with radical new formats such as cyber-schools. (p.128)

With cyber schools on the rise, more special needs students will be receiving their educations in this format. Without a map, the new special education teacher in the cyber environment will find herself lost. There is a tremendous gap in research that needs to be filled with practical techniques derived from practitioner-based learned experience and techniques that work for students with special needs.

METHODOLOGY

Role of the Researchers

During Hipsky's work as an educational consultant for the Tri-State Area School Study Council, she was given the task of auditing PA Cyber Charter School's curriculum, Lincoln Interactive. During the process, it became apparent that many of the questions that were being raised were based on how students with exceptionalities were having their needs meet. Hipsky approached Adams, a special education teacher in the online environment, regarding the concept of an action research project based on parent survey results. This chapter includes the results which are both practical and teacher-friendly because they are based on proven techniques from the classroom.

Description of Site and Participants

The PA Cyber Charter School delivers tuition-free, computer-based instruction to students in grades kindergarten through 12th directly to their homes. It is currently serving 6,200 students who live in 460 school districts throughout Pennsylvania. The courses of study offered by PA Cyber are not only accredited, but also diverse and flexible. Both real-time and self-paced classes are available and each student is able to customize his own educational plan through a face-to-face interview upon enrollment. An instructional supervisor is then assigned to each student to ensure that the student maintains an appropriate pace and meets graduation requirements.

Due to the nature of the PA Cyber Charter School as an online learning environment that has some special needs students included in regular education courses, this group was chosen as an information-rich, purposeful sample. "Information-rich cases are those from which one can learn a great deal about issues of central importance to the purpose of the inquiry" (Patton, 2002, p. 230)

Based on a total school population of 6,200, Figure 1 demonstrates the number of students who have individualized plans for their learning. The PA Cyber Charter School's students include: 40 students who are served in the speech and language realm, 74 who are Chapter 15 students, 134 who have gifted IEPs, and 468 students who have IEPs. Figure 2 demonstrates the exceptionalities

Figure 1. Individualized education plans

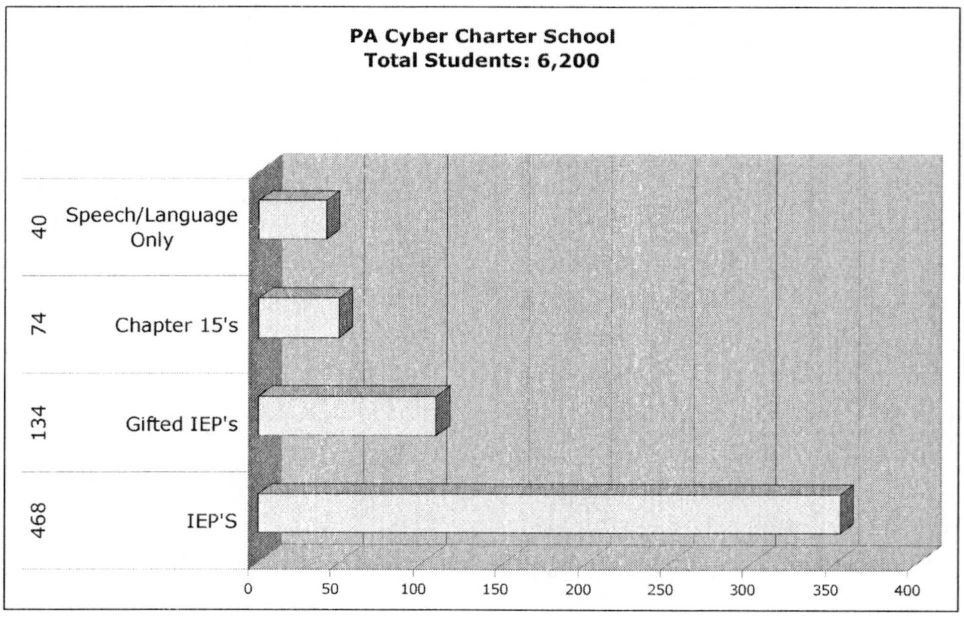

Figure 2. Exceptionality categories served

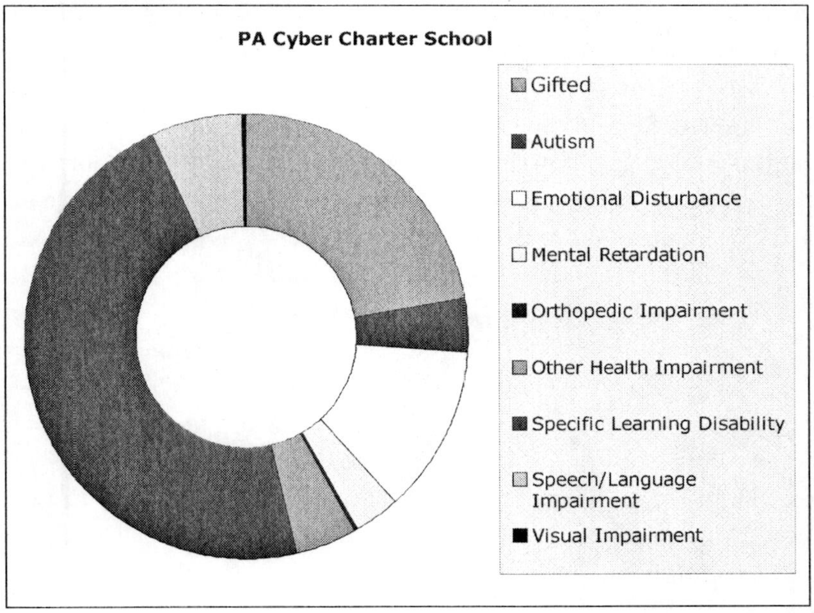

that are served at the school; the conclusion can be drawn that the students with special needs are predominately in the category of specific learning disabilities at 281 students. Only one student is served in the visual impairment category. To accommodate the student who is blind, voice recognition and software for the visually impaired have been utilized.

Data Collection and Data Analysis

The parents of students who receive services for their exceptionalities at the PA Cyber Charter School were contacted via e-mail to respond to the open-ended questions listed in Box 1. to prompt writing on their experience.

The analysis of the qualitative data that derived from the survey responses was examined for thematic connections and reoccurring patterns (Denzin & Lincoln, 1994) based on the issues that were the focus of discussion in the responses. Quotes from the parents begin each subheading in the results section in order to provide insight prior to delving into the specific techniques that the teachers and curricular developers utilize in the specific cyber environment.

Komives (2001) correctly realized that learning from experience does not occur without reflec-

Box 1. Parent survey questions

1. How is the special education program at the cyber school meeting your child's needs?
2. What was your child's previous educational setting (brick and mortar, private school, home school, etc?) and how is the cyber school setting different with regards to special education services?
3. What do you feel are the shortcomings of cyber education for students with special needs?
4. What do you feel are the greatest benefits of cyber education for students with special needs?

tion. After the themes were realized, the special education teacher who is the co-author if this article examined the findings and looked at her own practice in her classroom that supports the parents' sentiments. Action research is "practical inquiry" with practitioner-tested actions (Bernauer, 1999). Adams utilized the action research model to determine the techniques and strategies that are working in the PA Cyber Charter School for students with special needs.

RESULTS

Themes that emerged from the e-mailed parent surveys were: communication, interests, focus, less stigma from the special education label, education differences in comparison to other methods, and cyber school shortcomings.

Communication

It is not different than the public school system as far as requirements, set up, et cetera. Actually having a teacher used to communicating by e-mail and remote methods is far better than the traditional IEP representation from public school where there are other distractions and not used to prioritizing communication with parents. I know that my child is being contacted and I know my child is contacting the school. I am comfortable that my needs are being addressed and I have e-mail access anytime I have a question. Response is (in my opinion) much better. The scope of services is also much bigger.

I credit the personal service and monitoring to her success and the willingness of her counselor to not only help out and advise but to mentor my daughter as well.

Communication progress is monitored and parents are contacted at least weekly for all students. Teachers and students privately communicate during class via text and audio. The resource teacher can provide support in a virtual breakout room (small group session) to facilitate participation.

PA Cyber sticks to the "within 24 hours" guarantee on two important communication techniques. Students receive quick feedback on their questions on the discussion board. Also, individualized feedback is provided to each student through email.

The teachers implement resource logs to keep students with IEPs and parents informed of students' performance during virtual synchronous classes. The teacher-created feedback instrument, Cyber Resource Teacher Communication Log is included in the appendix of this document.

Interests

I feel that my child's needs are being met for the first time in her life since she has been a student of the PA Cyber Charter School for the following reasons: her counselor at her past school never even met with her even after several phone calls to the counselor and the principal. What little that was implemented was never followed up on. As a parent and teacher myself I monitor my child's progress, but what's even better is the personal service she has received from PA Cyber Charter School. She is for the first time happy about school. The counselor has helped her select courses that are right for her and that are courses that meet the requirements but fall within my daughter's interest. I could never imagine my daughter going back to a regular public school, and I teach in one!

The school provides assistance for the students in creating schedules. They allow for some student choice in course selection. Due to the variety of courses that can be provided at the same time in a cyber environment as opposed to a brick and mortar school, they have more flexibility to choose courses based on their own interests.

Figure 3. Cyber classroom snapshot

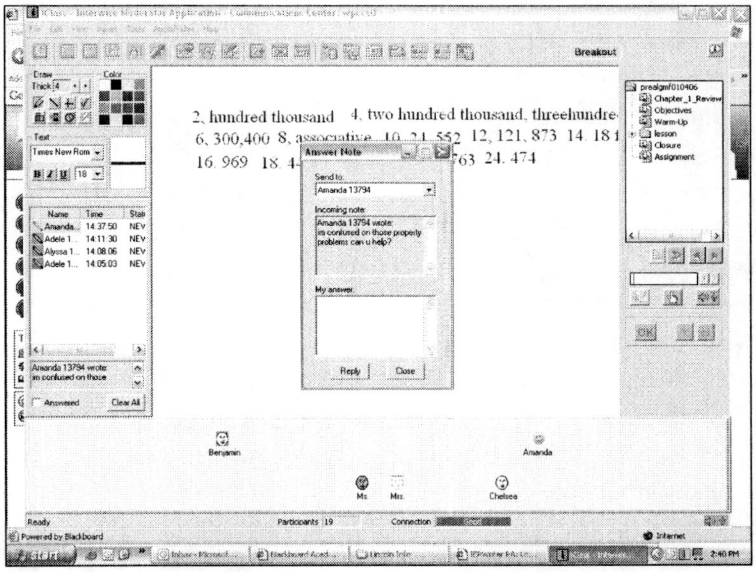

Copyright 2007, Lindsay Adams. Used with permission.

The flexibility of class pacing can be a wonderful resource for a student with a special need. The students can choose a combination of virtual classes and self-paced classes. Within this context they are able to receive more support in areas of need and move ahead in strengths.

Focus

My child can focus on academics and not worry with the social distractions, movement, fear, feeling like a little fish in a big pond. She was lost, unmotivated and frustrated. She now has more direction and takes her work more seriously. I no longer worry if my child can "make it through the day.

There is also less distraction. In a regular classroom there is always someone talking or moving around. This is a great distraction to most special education kids.

For a child who was not succeeding in the traditional classroom, cyber school can eliminate the typical classroom distractions. Many students have reported feeling "comfortable" in this environment to the resource teacher. Transition and other "downtimes" are eliminated for students with behavior or attention problems and therefore instructional time is maximized.

People who are not familiar with the online environment express concern about whether or not the teacher of the whole group can establish the students' understanding of the concepts that are being taught. The teacher often asks comprehension questions to determine that her students are learning. With the new technology, "emoticons" (smiley faces and confused looking faces) can be found on the teacher's desktop so that she can check in with any student who needs more review.

Figure 3 demonstrates the use of emoticons in a resource source room setting. The snapshot is the resource teacher's view during a breakout (small group) session in the virtual classroom. Benjamin is leading the group. He has the ability to control what is shown on the whiteboard and to give other participants permission to speak.

The note in the center of the board is seen only by the teacher. Amanda has privately requested assistance on her assigned math problems. The teacher now has the opportunity to either reply to the message through text or engage Amanda in a private audio conversation.

The students engage in self-assessment of their ability to focus during the lessons and participate. The students fill out behavior self-monitoring sheets. After class they confer with the teacher on their behavior, what they can improve, or that they excelled in that day during class. The useful teacher-created student behavior sheets titled the "Cyber Communication Monitoring Sheet" and the "Cyber Class Participation Monitoring Sheet" are included in the appendix.

Less Stigma from the Special Education Label

As far as the special needs student is concerned I think this is the ideal setting for her. There is so much stigma attached to being a learning support student, this is automatically removed with cyber school. I am not saying this is the best thing for everyone but it has allowed my daughter to pursue a good education and to actually like what she is doing and to look forward to school.

I feel the greatest benefit with this type of schooling is acceptance. My son often felt singled out or experienced ridicule or demeaning comments in the private school. Here he was no different than anyone else. He's always been self conscious about himself. Here no one can see him. He built up his confidence level so much just within the first month. The first week of school he involved himself in class and even 'IM'ed the kids. He loves it. Now he interacts well with his peers.

In a typical school, a student who is twelve years old physically, yet who functions developmentally at a nine year old level, could not walk into the fourth grade class in order to fill in those gaps in his learning. However, cloaked in the anonymity of the cyber school environment this can happen for a student with an IEP without the stigma.

Without his or her peers knowing it, the student can and does receive adjusted "To Do" assignment lists to help the student manage and prioritize individualized assignments. The lists usually communicated to student and parent through e-mail. Modifications of tests and assignments are made. The curriculum is adapted to match the material to student's ability level. Testing accommodations (i.e. reading tests to students in a private conversation in a virtual classroom or taking the test in a resource room, additional study sheets, and additional time) are provided.

Education Differences in Comparison to Other Methods

...it is good for the parents. Teaching is a tough job. If you choose to home school you have to match the regular tough job of teaching with some creativity for the special needs child. It's very exhausting. I did it for three years and I am thrilled to pass the torch off to very capable caring hands. I have the benefit of having him home with him getting a good education.

My son left the private school after completing third grade. When starting the other virtual school in fourth, he actually was put in third grade language arts and math. He never finished it. They moved him onto fourth grade material. He only got through it half way. He just didn't want to work. He improved so much with services and the school wouldn't compensate for the improvement. He was in sixth grade at the time. That didn't help him at all. He was, I think, ashamed of where he was. He started here in seventh grade. He was tested and was on target for language arts and tested between a 5th and 6th grade level in math. He was placed in fifth grade math. Math is difficult for him sometimes. After the first quarter, his grades are "As" and "Bs." He even got a high "A" in Math. This school is top notch in my book.

Too much for too few to handle in public school. Cyber is more individualized and I feel my child is getting the attention she needs. My daughter is responding much better because she feels she is connected to her special education teacher.

My son started in a private school. I thought it would be better with a smaller class. Because it was such a small school, they didn't have a special education system. They had just hired someone for learning support when we left. Also being in a private school I could only get speech services for him. I brought him home using another virtual school. That school started out good. Unfortunately, it became a struggle getting cooperation from the special education department. For instance, I had to beg for two years to have a particular evaluation done for my son. Paperwork was always delayed. WPCCS had everything done in short order. There was very little delay in services from switching schools. As a matter of fact the biggest delay came from the provider and the school playing phone tag. There has been cooperation and understanding from all of his teachers.

The resource room settings are one on one which provides an ideal ratio for a student with a special need. The sessions are created according to student and teacher schedules.

The learning management system, Blackboard, provides automated information about the students' grades. Students can benefit from immediate feedback for all objective assignments, including tests and quizzes.

Teachers are there in person when it is needed. For example, a situation which needs face-to-face participation is during standardized tests. The statewide tests are taken in real time at a testing center. Accommodations can be provided for the student whose IEP requires special services. Special educations teachers also attend school-sponsored field trips and activities in order to help with behavior or other issues that might arise and help guide the students' education beyond the computer.

Cyber School Shortcomings

The shortcomings are for parents who don't know how to be involved because they need to be. Parents need to be an active resource too. If a child needs face-to-face interaction it may be a problem. Also if there isn't an ability to work within a computer environment this could be difficult—however there is support for that too at PA Cyber and instruction is very basic so easily understood.

The Blackboard platform used by the PA Cyber Charter School does require some learning by the parents. Fortunately, when the parent enrolls a child in the program they are included in the face-to-face interview, during which they are taught about the technology and student expectations. They sign a terms sheet to agree to maintain contact and be available when the student is working on the curriculum. Brett Geibel, the Director of Technology, explained the support that is offered to parents:

PA Cyber has a training staff dedicated to student training to assure they have the tools needed to be successful. While there is no training offered directly to the parents, they are encouraged to sit in on student training so that they too can become familiar with our processes and student learning applications. Obviously, if the parent requests specific training, we would oblige, and parents are encouraged to contact their Instructional Supervisor with any questions. Usually, the instructional supervisor is able to answer them in short time, or at a minimum, direct them to the appropriate resources. (personal communication, March 8, 2006)

The teachers are also available to provide technical support, to answer questions, and to listen to concerns via phone or e-mail.

Timmy: A Success Story

I must say that I am in awe of the changes in Timmy. Going from never finishing third grade material (in PAVCS-he completed third grade in a private school), only going half way through the fourth grade material, and dealing with his sensory stuff...This year is unbelievable. I asked him how he felt about it, and he loves it. Not once have I heard him complain about school in general, math at all (he even wrote the teacher an e-mail telling her she is a good teacher), the three-hour run in the afternoon, and not even about getting up so early. He is doing everything on his own now. Every now and then I will check up on him (well, I think it might be once a week). He keeps his own schedule. Hardly ever do I have to remind him to get to class. I've seen him age about three years since he started school. It brings me to tears every time I think about it. And his grades for the first quarter...I can't even put into words what that has done.

He is so much more confident. With school starting before our church's youth group, he went into youth group with a sense of confidence. He is now blending in with his peers like never before. Expressive communication was very difficult for him. He gets frustrated with his speech therapist for working on it with him. This is like speech everyday, and he doesn't even connect it. I was concerned about him being distracted down where we are, so I put him in his room. His sister is in her room also. I was concerned that he would pick things up to play with and not pay attention. Obviously, he is attentive. Now the only thing is for me to change my line of thinking of whom he is and his capabilities. He is totally different. He still struggles in things sometimes, but I believe that will come along too.

I was also impressed with the teachers. Timmy handled it pretty well when his science teacher left. He was Timmy's favorite teacher. I wanted to not give too much detail to the teachers about him so I could see what he could do while not getting treated much different. The few circumstances we had, I emailed the teachers to explain some things, and they were very cooperative. I'm sure they don't understand everything and they were still cooperative. I didn't get that much understanding with the special education teachers at his previous school after the first year. They all said how they love having him in class and how well he is doing. They'd tell me 'no problem, you work with him at home and we'll work with him here and it will be ok.' They are more concerned with his education than anyone I've seen. Timmy apparently is clowning around a little with the kids and the teachers. He's so relaxed. It's awesome!

DISCUSSION

Limitations and Directions for Future Research

The study was run at one specific school. A broader study in the future could include a larger sample size with a greater variety of exceptional education categories. Future studies could examine techniques from multiple schools.

The technology and curricula for cyber schools are new, dynamic, and often changing. This constant updating and modification provides limitation because those techniques that are cutting edge today can be obsolete tomorrow. Further studies can look at the latest strategies for special needs in the cyber school learning environments.

SUMMARY AND CONCLUSION

This chapter examined the literature on cyber schools and special needs students. A parent survey provided insightful quotes and the action research conducted by Adams served to establish strategies for students in special education who learn via cyber school. appendices have been included for practical use in the cyber environment to monitor communication. These teacher-tested documents include: (1) Cyber Resource Teacher Communication Log, (2) Cyber Communication Monitoring Sheet, and (3) Cyber Class Participation Monitoring Sheet. A model for strategies has been established through this chapter.

There are possible shortcomings with the difficulties of the use of technology for the parent who is there to help the student. Cyber school is not right for all students, specifically students with severe special needs who need constant guidance in a face-to-face environment.

The cyber school environment does create anonymity and privacy for the students with an IEP. It is a place where students with special needs can receive modifications and adaptations to the regular curriculum. The cyber environment can open doors to learning for the student in exceptional education when proven strategies are utilized.

REFERENCES

Bernauer, J. A. (1999). Emerging standards: Empowerment with purpose. *Kappa Delta Pi Record, 35*(2), 68-70.

Bilyk, R. (September, 2005). E-learning that goes beyond text and graphics. *T.H.E. Journal, 33*(2), 36-37.

Borja, R.R. (May, 2005) Cyber schools status. *Education Week, 24*(35), 22-23.

Brandao, C. (2002). Teaching online: Harnessing technology's power at Florida virtual school. *T.H.E. Journal, 29*, 1-6.

Center for Charter Reform. (2002). Beyond brick and mortar: Cyber charters revolutionizing education. *CER Action Paper.* Retrieved on December 10, 2005 from http://edreform.com/index.cfm?fuseAction=document&documentID=1001

Denzin, N. K., & Lincoln, Y. S. (1994). *Handbook of qualitative research.* Thousand Oaks, CA: Sage Publications.

Duncan, D. (April, 2006). The cyber school option. *Valley News Dispatch.* Retrieved on December 6, 2006 from http://www.pittsburghlive.com/x/pittsburghtrib/s_449113.html

Komives, S. R. (2001). Learning leadership: As individuals and in communities of practice. *Concepts and Connections, 9*(3),11.

Miron, G., Nelson, C., & Risley, J. (October, 2002). *Strengthening Pennsylvania's charter school reform: Findings from the statewide evaluation and discussion of relevant policy issues.* The Evaluation Center: Western Michigan University. Retrieved on December 10, 2005 from http://www.wmich.edu/evalctr/charter/pa_5year/

Miron, G., & Nelson, C. (October, 2000). *Autonomy in exchange for accountability: An initial study of Pennsylvania charter schools.* Executive Summary. The Evaluation Center: Western Michigan University. Retrieved on December 10, 2005 from http://www.wmich.edu/evalctr/charter/pa_reports/pa_final_rpt.pdf

Patton, M. Q. (2002). *Qualitative research and evaluation methods.* Thousand Oaks, CA: Sage Publications.

RPP International. (2001). *Challenge and opportunity: The impact of charter schools on school districts.* Washington, DC: Office of Educational Research and Improvement, U.S. Department of Education.

APPENDICES

Appendix A. Resource teacher communication log

Student:

VC Teacher:

Resource Teacher:

Class:

Date:

Time:

Number of times student participated verbally:

Number of times student exhibited off-task or inappropriate behavior:

Describe:

Briefly describe your interactions with the student:

Briefly describe your interactions with the teacher (if any):

Areas of concern:

Assignments given:

General Comments (student on time? prepared for class? small group interactions? etc):

Appendix B. Communication monitoring sheet

In my conversations with my classmates this week, was I…

Respectful?____

Truthful?____

Kind?____

Did I…

Talk about things that no one else was talking about?_______

Brag about my own abilities?________________________

Make fun of other students?_________________________

What was good about my communication this week?

____

____

How can I improve for next week?

____

____

Appendix C. Class participation monitoring sheet

Number of times I volunteered in class:_____
Number of times I was called on in class:____
Number of times I answered verbally:_____
Number of times I answered in a note:______
When I felt frustrated, did I…
____ express my feelings?
____ ask for help?

Chapter XIV
Game Mods:
Customizable Learning in a K16 Setting

Elizabeth Fanning
The University of Virginia, USA

ABSTRACT

A game mod describes a modification within an existing commercial, computer-based game that has been created by a user. By game modding, a user can participate in the creative process by taking the setting of their favorite game and customizing it for entertainment purposes or to convey information. For years, commercial computer-based game developers committed considerable resources towards preventing users from "hacking" into or "hijacking" their games. Now several computer-based game developers provide editors with their products to encourage users to create content, and to allow educators, for instance, to take advantage of the benefits and production quality of commercial computer games to create customized instruction. This chapter focuses on mainstream, accessible games with straightforward modding tools that can be easily integrated into a learning environment.

What Do Computer Games Have to Do With Learning?

> *Anyone who thinks there is a difference between education and entertainment doesn't know the first thing about either.*
>
> —Marshall McLuhan, Communications Theorist

INTRODUCTION

Learning theorists from Piaget to Jonassen contend that profound, lasting learning culminates from the participant exploring, discovering, and interacting with their environment and culture to assimilate and create new meaning within their personal schema (Donaldson, 1984; Jonassen, 1992; Satterly, 1987). For a computer-based, constructivist learning environment, the quality of the user's learning experience is vested in the extent to which the computer responds in a way that is consistent with the learner's information processing needs (Jonassen, 1988). The level of the user's interactivity and consequent sense of empowerment and control over their learning experience will affect the extent to which surface or deep learning will occur (Jonassen, 1988). Studies using computer games in learning settings, particularly the classroom, indicate that while student test scores may not improve significantly from using games, students do learn on a more profound level, and are able to describe, for instance, why an answer to a test question is correct or incorrect (Squire, 2002b). While this outcome appears marginal at this point, it is worth exploring what computer games *do* afford a user: empowerment, motivation, insight, and engagement (Gee, 2003; Prensky, 2001).

How one might harness and channel a game's learning opportunities into the classroom in a way that empowers self-directed learning and the development of conceptual tools? Recognizing that emerging and even current learners have most likely grown up with a mouse in hand or at least developed considerable schema shaped by interacting with computer-based technology, computer games have gone beyond satiating the game playing public as a dalliance or source of entertainment and evolved into a meaningful, socially expressive medium, a platform for discussion and reflection that continues after the game session is over and outside the context of the game. However, the resources needed to create a commercial, computer-based game are formidable, in many cases requiring the expertise of game designers, computer artists, and programmers, not to mention robust marketing support. Many have endeavored to create educational games for the classroom and workplace, but most have neither the resources nor expertise to match the production quality and comprehensiveness of content characterized by more mainstream, commercial computer-based games.

Given these requirements and constraints, how might one harness and channel a game's learning opportunities into the classroom? Perhaps game mods could provide a means for educators to use the quality and basic format of commercial games to create customized instruction for enabling students to create meaning in their own learning. A game mod describes a modification within an existing commercial, computer-based game that has been created by a user. To do this, a user works with the game's existing assets to alter a small segment of the game's graphics, text, audio, or interactivity. In effect, a user can participate in the creative process by taking the setting of their favorite game and customizing it for entertainment purposes or to convey information.

MODS: RULES OF THE GAME AND TERMS OF ENGAGEMENT

For years, commercial computer-based game developers committed considerable resources towards preventing users from "hacking" into or "hijacking" their games (Holt, 2004); however, and perhaps in keeping with the spirit of gameplay, many game users consider these prohibitive efforts simply another challenge to master within the game environment (Holt, 2004). Now several computer-based game developers are providing editors with their products to encourage users to create content (Marriott, 2003; Prensky, 2003). It is important to note that these editors do not reveal the entire code, but only enough for the user to create several levels of modification (Holt, 2004; Marriott, 2003; Prensky, 2003).

Why do commercial game developers even offer this much? According to Chaptmann (2004), Holt (2004), and Prensky (2003):

- Within the game cultures, "cool" game companies encourage modding; they are more respected for their responsiveness and their

show of confidence in their users' technical competency
- The game developers are ensured continued play and sales, especially as the user can make the game continue to expand to more levels

In keeping with the gaming culture, a game's modding capability comes with several rules spelled out in the license that comes with the game's software:

- The modder cannot make money off of their mods
- The modder needs to *own* the base games
- The modder cannot *combine* mods from different games into one mod

To what extend does modding serve the modder? According to Hold (2004) and Prensky (2003):

- The mods are free to make
- Modding provides a way for gamers to make their own games ("I can do it better"). The modder creates a new free game, and extends gameplay of a game that may otherwise satisfy for two to three weeks
- The modder can upload their mod to a modders' forum to showcase their work, and participate in a large community of workers, fans, and game players

Note too that modding can be done at different levels, from a simple change to the appearance of a character to a more complex creation of a completely new setting, complete with AI. For the purposes of classroom use, however, what follows is an examination of how to bring modding into the classroom as a creative learning exercise without the encumbrance of complicated prerequisite technical skills.

Tools for the Classroom

Two popular computer games, *Civilization*, created by Sid Meier and published by Firaxis software, and Electronic Arts' *The Sims,* created by Will Wright, provide easy-to-use, easy-to access tools for customizing gameplay. *Civilization* is considered the penultimate commercial educational game (Squire, 2004). Played alone or with several players, *Civilization* is a geographically oriented and fact-driven world history game. Its gameplay requires the user to use maps and resources to manage the development of a civilization, based on the limits and opportunities presented by topography, resources, cultural affectation, and interactions with surrounding cultures (Squire, 2001, 2002a, 2004). By comparison, *The Sims*, created by Electronic Arts, is another popular, but less educational game that allows user customization as well, but its outcomes are shaped less by geographic and political landscape and more by social considerations (Squire, 2001, 2004). *The Sims* is often described of as less of a game and more of a simulation (hence, its name) or a toy, in that it is not designed around a more concrete goal (Fortugno & Zimmerman, 2005). Others argue that the goal of *The Sims* is to succeed in keeping *The Sims* going (Wright, 2003). Regardless of the arguments, the game does suggest a potential for application beyond simply engaging the user in a digital dollhouse. *The Sims* has already been modded for use in language curricula (Squire, 2004). Instructors have gone into the code layer to change the prompting language from English to French, for instance, to encourage incidental language learning via the gameplay experience. Others have taken its modding capabilities even further and into exciting new direction by creating public service advertisements[1] and replications of music videos[2] with *The Sims* video capture tools. Such machinema, as videos that are created using user-friendly editing and storytelling tools like those provided in *The Sims*, are also popular venues in Web spaces like www.youtube.com, which allows users to upload videos for anyone to review and even rate.

BUILDING A PLANTATION

To see how game modding can be used in a classroom setting, we used the popular, more flexible game *The Sims*, created by Electronic Arts. *The Sim's* game engine also allows and encourages the user to create buildings as simple as a modest bungalow, for example, or as complex as an historical landmark, such as Thomas Jefferson's Monticello. We endeavored to create a virtual Monticello (Figure 1) to see how far we could go with *The Sims'* game engine to create a mod based on the structure and its social interaction, and then determine the extent to which the final product might foster motivation, game literacy, or media dialogue.

We spent 20 hours using *The Sims*' tools to create Monticello, based on research of the plantation's blueprints and house plans as well as the historic significance of the time in which it was built. While *The Sims* has an engine that allows for elaborate construction, it did present some limitations:

- Pop art and modern home furnishings that the user can access to "decorate" the structure, thus interfering with the "historical integrity" of the house.
- Timed programming variables that can be difficult to alter. At one point, a school bus drove by and took Patsy Jefferson with it. It was necessary to wait until another gameplay session for Patsy to return (by bus).

To customize a mod further, a user can venture, as we did, into the online community proliferating around *The Sims* to find tools for importing furniture. Some of these tools are created by modders not affiliated with *The Sims* who often expanded on existing *Sims* editor code to enable more elaborate modifications (Figure 2).

Figure 1. Virtual Monticello

Similarly, to create more "real life" Monticello-based characters, Sims Creator (Start/Programs/Maxix/Sims/Sims Creator) was used to create a specific type of person, from personality attributes to physical appearance and gender (Figure 3).

Reactions: The Learner

To begin our study, we opened the mod to a group of four high school students. Two of them were regular *Sims* players. At first, they wanted to sit back and watch, hoping to observe an historical reenactment of what might have taken place in the plantation they had all visited in person at least once.

As is typical of female *Sims* players (Chu, Heeter, Egidio et al., 2004), the girls were more interested in what people in the house were doing and how they got along. Sally Hemmings worked on the first floor. Another cooked in the basement. A butler wandered between the entryway and Thomas Jefferson's bedroom, where Jefferson stood pensively. Eventually, the cook caught on fire and the girls used their *Sims* skills to attempt to save the cook—despite their efforts, death intervened. In time, the focus group also explored the mischief that players typically get into by "fooling" the game and directing characters to behave badly to see what would happen (Perhaps this is the digital age version of teasing the cat or a prank phone call). Mischief accomplished or not, the focus group collectively expressed interest in accessing the blueprints for the plantation's gardens so that they could recreate them within the mod. Hence, despite the shortcomings of our virtual Monticello, our focus group validated Gee and answered the call that beckoned from the computer game: to pursue self-directed learning by solving complex problems (Gee, 2003). Such self-directed exploration to go beyond the boundaries of one's current understanding to create and assimilate meaning is also of course, rich in constructivist learning implications (Jonassen, 1992; Satterly, 1987).

Reactions: The Classroom Teacher

We then presented the Monticello mod to five middle school and high school teachers, for whom we demonstrated the Monticello mod and then asked to explore it on their own. Afterwards, we discussed with them the extent to which they found

Figure 2. Jefferson's revolving bookstand

Figure 3. Peter Fosset description

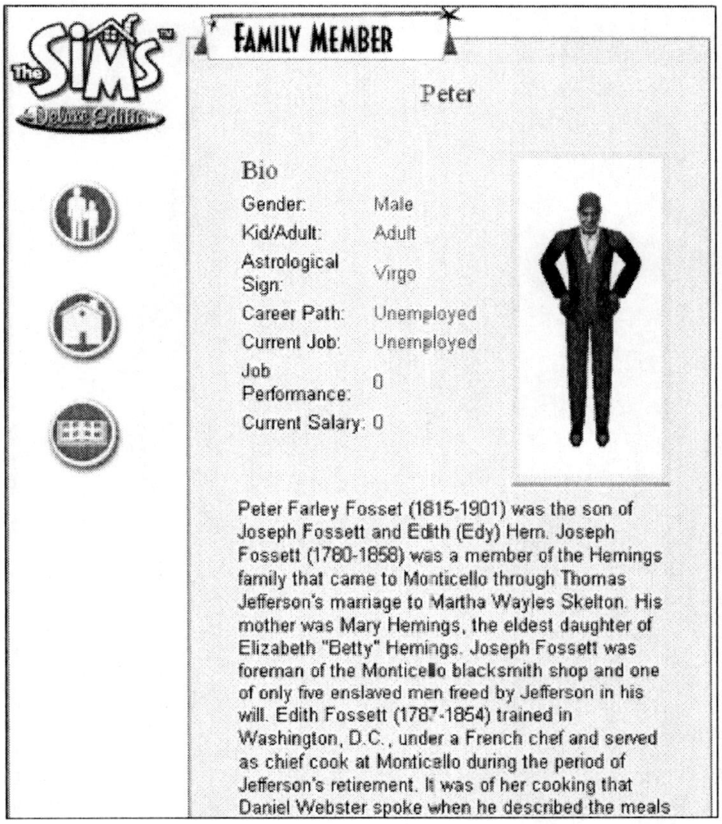

such a tool feasible for classroom use and how they might use it for learning. All of the instructors in the group were familiar with *The Sims*, but none had previous experience with it.

The instructors' first response to the virtual Monticello mod was that it looked like a digital replacement of the traditional, second grade diorama in the shoebox assignment. Initial issues with the mod were that it does not quite accurately represent the plantation physically. One indicated that they would like the ability to:

- Import historically relevant artifacts into the framework
- Change character attributes to accurately represent historic personalities
- Change the rules so that characters could not earn "creativity points" based on artistic accomplishments, but rather earn points based on something that would have been important for that character during a given time period—or remove the point option all together. "History is not a game," as one ardent history teacher explained
- Reshape the "going shopping" metaphor in the game engine that allows users to spend *Sims* dollars to construct and furnish the house. We used "cheats" to access *Sims* dollars to create this mod, following a financing strategy similar to that which Jefferson sometimes used to construct and furnish his own Monticello.

Figure 4. Jefferson in his study

The teachers also expressed concern that a student might be more engaged in the gameplay than the learning. One suggested that in classroom, he would want to use a mod like the virtual Monticello to emphasize its historical significance, toning down the novelty of its gameplay. Another expressed that that they were "excited about the possibility of bringing the past to the present" with mods, citing their value as "an observation tool," and liked the idea of being able to go in and move around, adding, "I can explore the past and explore interactions!" By doing so, he felt that his students could gain insight through exploration and inference by considering the period characteristic values and experiences of historical figures that might shape their choices and behaviors.

Even with its programming limits, most teachers felt that their students could use such mods to experience a "good approximation" of the past, one that would help them to construct their understanding of the time period and issues that shaped decisions. Given this utility, one instructor noted that historically contextual game mods could fortify the development of his students' mental models—and that as a teacher, he would have better insight in his students' "historical reasoning," based on the decisions they made within the framework.

"I could present [a mod] as an alternative assessment, opposed to a test or paper," he explained, adding, "That you can change things is very exciting for the study of history."

Contextualizing the Mod in the Classroom

As teachers explored how they might use the virtual Monticello mod in a classroom setting, their ideas for implementation included:

- **A class inquiry:** The teacher could run the mod on the projector and discuss how to interact with it with the class. Later, the class could work in groups on their own mods to explore and create "shared meanings." The instructor could alternate between using class and group mods to promote inquiry and discussion. As one teacher pointed out, "There's a lot of value just in creating the scenario and having the students do the research for that."

- **An examination of character:** "I want to see Monticello and send in Nat Turner," one explained, to see how a character programmed with Nat's attributes would react in a plantation setting. One instructor suggested replacing Jefferson with Napoleon for the same reasons
- **A yearlong study:** One teacher suggested that the creation of the scenario could be an ongoing project, divided across academic quarters, each dealing with a different part of its construction in a way that parallels the course content delivery

Integrating any type of computer game into a curriculum shifts the culture of the classroom from one that is teacher-driven to a more user-centered learning environment; the teacher becomes a facilitator rather than a proffer of knowledge (King, 2003;Squire, 2004). However, when computer games are introduced into the classroom, the teacher has an amplifying effect on the learning outcome. If the teacher is appropriately prepared to use a game in their curriculum, it augments the learning success. Unfortunately, the inverse is true if the teacher is less aware of how to facilitate learning using computer games (Squire, 2004).

Perhaps we can begin managing this amplifying effect by focusing on what the teacher *can* do to prepare for bringing game-based learning into the classroom, by:

- Identifying the equipment and entry behaviors required for the game
- Determining the time required for the game to make an impact on learning. Typically, a student needs time to understand how to use a game before they can begin learning with it
- Recognizing the types of learning computer-based learning facilitates, which is more knowledge-based and to an extent, subjective
- Determining how computer-based learning can support the learning goals
- Choosing an appropriate game or game editor for creating mods
- Contextualizing the game into the curriculum in a way that encourages exploration, discovery, and the development of "conceptual tools"
- Including appropriate scaffolding as well as offline activities that encourage reflection, dialog and "shared meanings" among the learners beyond the context of the game

It is also worth exploring if an IT specialist should be familiar with the utility of game-based learning and how to use it in a classroom setting as well in order to support the classroom teachers in meeting their learning goals. This same person could be called upon to create simple learning mods or skeletons for the students to work with, as specified by an instructor, or based on a given curriculum's learning goals.

Finally, and perhaps most importantly, the instructor needs to be "game literate," recognizing and understanding that computer games are a means of expression and representation (Green, 2004), and that like reading and writing, game literacy develops when the learner has the tools and ability to create their own games (Clark, Perrone, & Repenning, 2005; King, 2003).

CONCLUSION

The purpose of this discussion is to examine how game mods might provide an affordable, comparable, and customizable alternative to wildly successful commercial games, an alternative that could be used in a classroom setting to facilitate a meaningful learning experience. Current research also suggests that this medium may also have application in therapeutic settings for recreating and responding to a past trauma or to explore and practice different ways in which to respond to stimuli and a range of social interactions. However, the success of the use of game mods in a learning or therapeutic setting will depend not only on the novelty of a good idea, or how comfortable the instructor or facilitator and learner are with using computers and making simple changes to existing applications, but also on how well the modding activities are contextualized within the curriculum. Creating meaning still needs to be forged within an evocative and relevant context.

REFERENCES

Boards. The Creative Edge in commercial Production. (September, 2005). *Public Health Department, Belgium - Teenage Mum.* http://www.boardsmag.com/screeningroom/animation/1146/

Chu, K., Heeter, C., Egidio, R., & Mishra, P. (2004). Girls and games literature review. *Michigan State University Mind Games Collaboratory.* Retrieved March 10, 2005 from http://spacepioneers.msu.edu/girls_and_games_lit_review.htm

Chaptman, Dennis. (2004). Video-games in the classroom. *Wisconsin Technology Network.* Retrieved from http://www.wistechnology.com/article.php?id=513

Clark, D., Perrone, C., & Repenning, A. (2005). Webquest: Using WWW & Interactive Simulation Games in the Classroom. *First Monday.* Retrieved from http://www.firstmonday.dk/issues/issue5/perrone/

Donaldson, M. (1984). *Children's minds.* London: Fontana.

Evers, J. (2004). My daughter killed her brother in "The Sims." *IT World.Com.* http://www.itworld.com/App/4201/041221sibssims/

Fanning, E. 2006. The sims in therapy: An examination of feasibility and potential. Under review for publication by *The Journal of the American Art Therapy Association.*

Fortugno, N. & Zimmerman, E.. (2005). Soapbox: Learning to play to Llearn — lessons in educational game design. *Gamesutra.* Retrieved from http://www.gamasutra.com/features/20050405/zimmerman_01.shtml

Gee, J.P. (2003). *What video games have to teach us about learning and literacy.* New York: Palgrave Macmillan.

Green, H. (2004). Researchers and teachers push for games in schools. *The Institute of Education, University of London.* Retrieved March 10, 2005 from http://ioewebserver.ioe.ac.uk/ioe/cms/get.asp?cid=1397&1397_1=10817

Gorilla Mask. (2005,September). *Trapped in the closet performed by the sims (Part 1).* Retrieved from http://gorillamask.net/rksims1.shtml

Holt, T. (2004). How mods are really built. *Serious Games Summit DC.* Retrieved March 10, 2005 from http://www.cmpevents.com/GDe04/a.asp?option=C&V=11&SessID=3305&Mgt=0&RVid=0

Jonassen, D. (Ed). (1988). *Instructional designs for microcomputer courseware101.* Hillsdale, NJ: Lawrence Erlbaum.

Kim, L.S. (1995). Creative games for the language class. *Forum,* 1(33), 35.

King, B. (2003). Educators turn to games for help. *Wired News.* Retrieved March 10, 2005 from http://www.wired.com/news/games/0,2101,59855,00.html ().

Marriott, M. (2003). Games made for remaking. *The New York Times*. Retrieved March 10, 2005 from http://www.nytimes.com/2003/12/04/technology/circuits/04modd.html?ex=1135227600&en=0fc160f73ad53e42&ei=5070

Prensky, M. (2001). *Digital game-based learning*. New York: McGraw Hill.

Prensky, M. (2003.) "Modding"—the newest authoring tool. *Marc Prensky*. Retrieved from http://www.marcprensky.com/writing/Prensky%20-%20Modding%20-%20The%20Newest%20Authoring%20Tool.pdf

Satterly, D. (1987). Piaget and education. In R. L. Gregory (Ed.). *The oxford companion to the mind*. Oxford: Oxford University Press.

Squire, K. (2001). Games to teach project. *Education Arcade*. Retrieved from http://www.educationarcade.org/gtt/

Squire, K. (2002a). Cultural framing of computer/video games. *Game Studies*, 2(3).Retrieved from the internet September 14, 2007 at http://www.gamestudies.org/0102/squire/

Squire, K. (2002b). Replaying history: Learning world history through playing civilization III. Retrieved from http://website.education.wisc.edu/kdsquire

Squire, K. (2004). What happens when games go into any classroom situation? *Serious Games Summit DC*. Retrieved March 10, 2005 from http://www.cmpevents.com/GDe04/a.asp?option=C&V=11&SessID=3250

Wired. (2003). *Every sims picture tells a story*. Retrieved October 9, 2006 from http://www.bmedia.org/archives/00000330.php

ENDNOTES

1 http://www.boardsmag.com/screeningroom/animation/1146/

2 http://gorillamask.net/rksims1.shtml

Chapter XV
Project Management in Student Information Technology Projects

Maria Delia Rojas
Murdoch University, Australia

Tanya McGill
Murdoch University, Australia

Arnold Depickere
Murdoch University, Australia

ABSTRACT

Universities teach project management to information technology (IT) students. The project management principles that students have previously learned are often put into practice in a project course, intended to give final year students the experience of applying their knowledge to real or simulated projects. This chapter reports on research that investigated the use of, and usefulness of, project management in student IT projects. The results show that there was a wide range in the application of project management practices, with students being more likely to produce the initial documentation associated with some of the project management knowledge areas than to make use of it throughout the project to monitor the project's progress. The results also showed that the number of project management guidelines applied in student projects was not linked with IT project success. However, there was a strong relationship between project management plan quality and obtaining a good software product.

INTRODUCTION

Universities all over the world teach project management to information technology (IT) students (Goold, 2003; Reif & Mitri, 2005; Stein, 2002). The project management principles and system development methodologies students have previously learned are often put into practice in an IT project course, intended to give final year IT students the experience of applying theoretical concepts and practical techniques to real or simulated student projects (Batra & Satzinger, 2006; Ellen & West, 2003). The research reported in this chapter investigates the use of, and usefulness of, project management in student IT projects.

Student projects are usually defined and scoped to run on a one or two semester basis within an academic program and are not as complex as industry projects (Jih, 2003). Within the time limitation placed on these projects, students have to plan, design and implement their systems and create relevant documentation. While student projects are not comparable in size and complexity to industry projects, the rigor expected is the same as for industry projects. Past experience reveals that IT students find it difficult to manage their project for reasons such as lack of understanding of project management tools and techniques (Abernethy & Piegara, 2007; Lowe, 2000; Pournaghshband, 1990).

The Project Management Institute's (PMI) 'project management body of knowledge (PMBOK)' provides a solid base of standards, procedures and practices for managing all types of projects and is used by many organizations to apply project management principles to projects (Freedman, 2002). The goal of project management guidelines is for project managers to achieve better outcomes in projects. IT students can also make use of project management guidelines to try and achieve the same goal.

Project Management and the PMBOK Guide

Project management is defined by the PMI as the application of knowledge, skills, tools, and techniques to project activities to meet project requirements (Project Management Institute, 2004). The PMBOK Guide is a handbook that provides broadly accepted knowledge and practices that are generally applicable to most projects. There has been widespread consensus as to the value and usefulness of these guidelines (Schwalbe, 2004). The PMBOK Guide consists of five project management process groups, and is also divided into nine key sections called the project management knowledge areas. These knowledge areas are further divided into their component project management processes, which describe the activities that need to be fulfilled for each knowledge area. In addition, each of the nine knowledge areas has specific project management tools and techniques which help to carry out the activities in each process. The project methodologies and practices presented in the PMBOK Guide are used to control and manage projects and cover every aspect of project development.

The Role of Project Management in IT Projects

Generally, project management is considered important for three reasons. First, project management can clarify a project's goals because it makes the project manager produce documentation which identifies the project's unique characteristics which have to be addressed throughout the project. Secondly, project management will enable a project manager to identify the required resources, thus assuring the project's stakeholders that resources are being effectively managed. Finally, project management can help to succeed in the achievement of both project and organizational goals.

There has been some research into the value of project management in IT projects. An early study by Pinto and Slevin (1988) tested the importance of factors that are believed to be critical to project success. Each of the critical factors was tested independently against project success and the results showed that having a project schedule and plans was significantly related to project success. They concluded that project managers need to create project schedules and plans and use them on a regular basis.

More recently, the Standish Group's CHAOS project investigated the scope of software project failures and the major factors that cause software projects to fail. The results showed that project success rates have increased since 1994 and that this is partially attributable to better project management, including the availability of better tools to monitor and control progress, better skilled project managers, and better management processes (The Standish Group, 2001). This study also found that 46 percent of successful projects used a formal project management methodology, compared to 30 percent of challenged and failed projects. Hence, having a formal project management methodology appeared to increase the chances of success by about 16 percent (The Standish Group, 2001).

This research is supported by the findings of Aladwani (2002), who studied the mediating effect of project planning between three project uncertainty variables and IT project success, and showed that IT project planning was the most important contributor to IT project success. Gowan and Mathieu (2003) tested a model of the relationship between technical complexity, project size, use of a project management methodology and project performance. The results showed that the use of a formal project management methodology is positively related to project performance, particularly when project size is large.

This research illustrates the value of applying project management practices to IT industry projects and hence it is vital for IT students to learn and apply the standard practices to manage projects successfully. Phillips, Fairholme and Luca (1998) noted that while student project teams address some project management issues, many focus more on the development of the product. IT students need to be aware of the value of project management and they need to be encouraged to use project management principles.

Du, Johnson and Keil (2004) conducted a project to find out what project management topics are being covered in information systems curriculums, and concluded that project management practices have not been fully incorporated into university IT degree programs despite the increased emphasis on project management in the most recent Information Systems model curriculum (Gorgone et al., 2002). They argue that preparing future IT graduates to apply project management guidelines will increase the success rates of industry projects. This view is also reflected by Reif and Mitri (2005).

RESEARCH QUESTIONS

The study reported in this chapter was conducted to explore the use of, and usefulness of, project management in student IT projects. It considered both project management in general, and more specifically, the application of PMBOK guidelines.

Phillips, Fairholme and Luca (1998) argued that it is important for IT students to be aware of project management guidelines. Therefore, the first research question relates to awareness of project management principles:

- Are IT students aware of project management principles that can be applied to student projects?

An objective of the research was to establish how IT students apply PMBOK Guide practices and to identify any relevant project management knowledge areas that are not applied. The second and third research questions therefore relate to actual application of project management principles:

- Do IT students apply PMBOK Guide practices in their IT development projects?
- What are IT students' perceptions of the usefulness of PMBOK Guide practices?

The final objective of the research was to identify whether the application of project management principles increases the likelihood of students completing IT projects successfully. Therefore, the final research question is:

- Does the application of project management principles increase the success of student IT projects?

Given the evidence that use of project management increases the likelihood of industry project success (Aladwani, 2002; Gowan & Mathieu, 2003; Pinto & Slevin, 1988; The Standish Group, 2001) it is likely that this is also the case for student projects. Therefore the following hypotheses were proposed:

H_1: The application of project management practices increases the chances of completing student projects successfully.

H_2: Increasing the quality of project management plans increases the chances of completing student projects successfully.

THE STUDY

The research sample consisted of final year IT students enrolled in an IT project course at an Australian University. Students formed their own project groups to undertake the project. However, groups were subject to approval by the course coordinator to ensure that they were well balanced in terms of the skills of the group members. Each group had approximately five students. The group members were assigned different roles but after a few weeks the roles were rotated.

The students had a range of IT project management backgrounds prior to the project. However, most students had completed at least a systems analysis and design course which includes an introduction to project management.

The data for this study was partly collected by means of a questionnaire, administered during the final weeks of the IT project course. The questionnaires were given to the course coordinator, who distributed questionnaires to all students during project group meetings. It was stressed that the completion of the questionnaire was voluntary, that all information would be kept confidential and that data would be used only for the purpose of the study. Forty one students completed the questionnaire. They represented 14 project groups. Evaluations of project management plan quality and IT project success for the groups were also used in the analysis. These were based on course assessment undertaken by the project group supervisors.

The Questionnaire

The questionnaire included the following sections (see Appendix 1 for a complete set of the questions asked):

- **Background information:** Background information including gender, age and major(s), was collected for each student.
- **PMBOK awareness:** Each participant was asked to rate their awareness of project management and PMBOK guide practices before starting and after completing the

IT development project. The answers were measured on a 5 point scale from 1 'Very little' to 5 'A lot.'

- **PMBOK use:** The section of the questionnaire relating to PMBOK use was divided into seven sections that correspond to seven of the nine project management knowledge areas: project integration management, project scope management, project time management, project quality management, human resource management, project communication management, and project risk management. Project cost management and project procurement management were not included in the questionnaire because these projects were not given a budget and did not obtain people or sources from an outside organization. Use of the guidelines that correspond to the seven knowledge areas were measured with a series of question that were answered 'Yes' or 'No.'
- **Perceived usefulness of PMBOK guide practices:** Students' perceptions of the usefulness of the individual PMBOK practices were measured on 5 point scales from 1 'Not useful' to 5 'Very useful.'

Additional Study Variables

The following variables were measured separately from the questionnaire after completion of the projects.

- **Quality of project management plan:** The IT students worked in groups and hence produced only one project management plan per group. Project management plan quality was measured based on the mark that each group of students received from their supervisor for their project plan.
- **Project management practices applied:** The number of project management practices that were applied by each group was used as a measure of overall application of project management principles. In each student group there were up to five students. In some groups all students were involved in all parts of project management, but in some groups students were involved in only some. Therefore, if any member of the group had carried out a particular practice it was counted towards the total for the project group and this was used as a measure of the number of project management practices carried out.
- **IT project success:** Traditionally project success has been measured in terms of the project objectives of time, cost, and scope known as the 'triple constraint' (Brock, Hendricks, Linnell et al., 2003; Schwalbe, 2004). The triple constraint makes it relatively easy to evaluate project success by comparing the actual time, cost and performance, with the planned time, cost and performance objectives. However, in these student projects only one of the three criteria was applicable. There was an absolute deadline for completion set, and the projects were not given a budget. Therefore, this study used only one of the triple constraint criteria. This was the scope and software quality of the software product that was created for the client. As was mentioned before, the participants worked in groups and they developed only one software product per group. This software product was marked by the team's supervisor. Consequently, each group had only one IT project success score.

RESULTS AND DISCUSSION

IT Students' Awareness of Project Management Guidelines

The first research question related to student awareness of project management principles. Table 1 shows the levels of awareness of project

Table 1. IT students' awareness of project management and PMBOK practices

		Mean	Max	Min	Std. Dev.	Diff.	Sign.
General PM	Before Project	3.33	5	1	0.971		
	After Project	4.34	5	3	0.656	1.01	0.000
PMBOK	Before Project	2.17	5	1	1.138		
	After Project	2.93	5	1	1.233	0.76	0.000

management in general, and of PMBOK both before, and after, completing the IT project. These were compared using t tests. Students' awareness of general project management practices was significantly higher after completing the IT project course ($t=-6.58$, $df=39$, $p<0.001$). Likewise, students' awareness of the PMBOK was also significantly higher after completing the IT project course ($t=-4.868$, $df=40$, $p<0.001$). Therefore, doing an IT project raised the awareness of both project management in general and PMBOK Guide practices.

As discussed earlier, it has been argued that it is important for IT students to learn project management theory and to then practice it in student projects. This study confirms the role of IT student projects in reinforcing project management knowledge obtained from earlier courses.

However, awareness of general project management was higher than awareness of PMBOK Guide practices both before, and after, completing the IT project. That is, IT students know more about project management in general than they do about the specifics of the PMBOK practices. Before the IT project, the average level of awareness of project management in general was 3.33 and the average level of awareness of PMBOK Guide practices was 2.17. The means for awareness after the IT project increased for both general project management (mean=4.34) and PMBOK Guide practices (mean=2.93). However, even though there was an increase in the average there were still 14 percent of all students who had very little awareness of the PMBOK Guide practices after the IT project. On the other hand, no students had low levels of awareness of general project management after completing the IT project as the lowest level reported was 3.

Project management is important, however knowing just about project management in general is not sufficient. IT students need to be aware of the PMBOK Guide as it is one of the main project management frameworks that provide standards to manage software development projects in industry (Abernethy & Piegara, 2007; Freedman, 2002), and practical experience in how to use and apply the PMBOK framework will be useful when they seek and gain employment. The gap between general project management awareness and specific awareness of the PMBOK practices found in this study may, however, be a result of the way in which project management was taught to these students. Universities should consider this, and Du et al.'s (2004) finding that project management practices have not been fully incorporated into university IT degree programs, when planning their project management teaching.

Do IT Students Apply PMBOK Guide Practices?

Awareness of project management guidelines is not enough, project management guidelines have to be applied to obtain benefits, and the study

Table 2. Project management practices that IT students applied

PM Practices	PM Knowledge Areas	Yes		No	
		Count	%	Count	%
Use Microsoft Project to develop schedule	Project Time Management	40	97.6	1	2.4
Use template to create project management plan	Project Integration Management	38	92.7	3	7.3
Created scope statement	Project Scope Management	38	92.7	3	7.3
Regularly distributed project information	Project Communication Management	34	82.9	7	17.1
Discuss project's progress	Project Communication Management	33	80.5	8	19.5
Created scope management plan	Project Scope Management	26	63.4	15	36.6
Monitored project results	Project Quality Management	26	63.4	15	36.6
Created risk management plan	Project Risk Management	24	58.5	17	41.5
Created communications plan	Project Communication Management	21	51.2	20	48.8
Use skills matrix	Human Resource Management	20	50.0	20	50.0
Created quality plan	Project Quality Management	20	48.8	21	51.2
Follow project plan (WBS, Gantt Chart)	Project Time Management	19	46.3	22	53.7
Monitor risks	Project Risk Management	18	46.2	21	53.8

investigated whether IT students are applying PMBOK Guide practices. The investigation of the application of project management guidelines showed that not all IT students followed the PM-BOK Guide practices. Table 2 shows the number of students that applied each of the PMBOK Guide practices. The project management principles are ranked from most used to least used.

Table 2 shows that there was a wide range in the application of project management practices. The most frequently used project management practice was the use of Microsoft Project to develop project management schedules and this was done by 97 percent of the participants. The second most used project management practices were the use of templates and the creation of scope statements, both done by 92 percent of the participants. The third most frequently used project management practice was regularly distributing project information to other group members and this was done by 82 percent of the students. The least used project management practices were following the project plan (46.3 percent), and monitoring risks, which was done by only 46 percent of the participants.

Table 2 also shows which project management knowledge area each practice is from. Several of the project management knowledge areas were represented by several practices in the questionnaire and there were differences in the number of participants who used the various project management practices within the areas. For example, whilst 97 percent of the participants used Microsoft Project to develop the project management schedule only 46 percent followed that schedule. This project management practice is part of project time management. Similarly, whilst 58.5 percent of participants said that they created a risk management plan, only 46.2 percent said they monitored risks. In addition, most students used project communication management practices to distribute project information (82.9 peprcent) and discuss the project's progress (80.5 percent) but only half of the participants (51.2 percent) created communication plans.

These results also suggest that IT students are more likely to produce the initial documentation associated with some of the project management knowledge areas than to make use of it throughout the project to monitor the project's progress. Students were much more likely to create both project management schedules and risk management schedules than to follow them. One reason for this might be that the project management plans are too hard for students to follow because they may not have a full grasp of how to use them. Another reason might be that following project management plans was not part of the assessment for the project course and students might be more encouraged to use project management plans if they were awarded points or grades for following them. Therefore, if instructors want IT students to apply more of these guidelines then the course's content and assessments should be aligned with PMBOK Guide practices.

What are IT Students' Perceptions of the Usefulness of PMBOK Guide Practices?

The third research question relates to students' perceptions of the usefulness of PMBOK Guide practices. Table 3 presents ratings of the perceived usefulness of each of the project management practices ranked from most useful to least useful. In general, IT students did find project management practices useful as they rated the usefulness of all project management practices above the middle of the scale. Even the project management practices that were ranked in the bottom half were still perceived to be relatively useful. Therefore, students do believe that project management practices help them in their projects.

The results showed that communication plans were perceived as the most useful practice with a mean usefulness rating of 4.10. By using these plans, IT students learned to plan how they could distribute project information among group members and report their project performance to verify how well the project was moving towards meeting its goals.

The project management practice that students perceived as second most useful was the use of templates which had an average usefulness rating of 3.85. IT students used templates as a starting point to create a project management plan for their project. They had to identify the sections of the template that were appropriate for their project as not all the sections are relevant to a given project. The outcome of following templates was that students produced project management plans to co-ordinate project activities, guide project execution and control. Participants commented how valuable it was to use a template to create the project management plan for their project.

Table 3. Usefulness of the project management (PM) knowledge areas to student projects

Project Management Practices	Mean	Std. Dev.	Max	Min
Usefulness of communications plans	4.10	1.081	5	2
Are templates useful?	3.85	0.910	5	2
Usefulness of risk management plans	3.56	1.026	5	1
Usefulness of project time management	3.54	0.869	5	1
Usefulness of scope statements	3.18	1.048	5	1
Usefulness of quality plan	3.15	0.988	5	1
Usefulness of skills matrixes	3.10	1.215	5	1

The project management practices that were perceived as least useful were scope statements (mean=3.18), quality plans (mean=3.15), and skills matrixes (mean=3.10). These project management practices were not part of the course assessment and as a result it appears that only a few students applied them. Another possible reason for these results is that not all IT students within a student group were involved in completing a particular task. Therefore, some participants who answered that skills matrices are not useful may not have been involved in that project activity and hence may not have realized their value.

Does the Application of Project Management Principles Increase the Chances of Completing Student IT Projects Successfully?

Previous research (Gowan & Mathieu, 2003; The Standish Group, 2001) has provided evidence that the use of project management increases the likelihood of project success in industry projects, and this study investigated whether this is also the case for student projects. The fourth research question was considered in two ways. To address the first hypothesis, the number of project management practices used was correlated with the IT project success score. The results of this indicate that there was no relationship between IT project success score and number of project management practices that were applied (r=0.090, p=0.761). Thus the first hypothesis was not accepted. It appears that the key to project success in IT student projects is not based in how many project management principles student groups use.

To test the second hypothesis, the quality score for each group's project management plan was correlated with the IT project success score. The results indicated that there is a strong relationship between the quality of the project plan and IT project success score (r=0.710, p=0.004), thus the second hypothesis was accepted. This result indicates that having a good project plan is associated with obtaining a good software product from the student project. This is consistent with the results of research into the success of industry projects (Aladwani, 2002; Gowan & Mathieu, 2003; Pinto & Slevin, 1988; The Standish Group, 2001).

These results imply that student IT project success is not based on how many project management principles are applied but how well they are used. Therefore, the quality of project management practices plays a big part in how successful the project is going to be.

CONCLUSION

The main focus of this study was to investigate the role of project management in student projects. The major limitations of this research are the small sample size, and the fact that the study was only conducted in one university. Future research should further explore the issues raised in this study in other universities and with larger groups.

The research revealed that students' awareness of general project management and PMBOK Guide practices was higher after completing an IT project course. However, IT students were more aware of general project management than the specifics of the PMBOK Guide. As the PMBOK Guide is so widely used in industry (Abernethy & Piegara, 2007; Freedman, 2002), the level of awareness and use of the PMBOK practices should be raised.

The study also showed that there was a wide range in the project management guidelines that IT students applied in their projects. The study identified the project management practices that were used the most (e.g., use of Microsoft Project to develop a project schedule, using templates to create project management plans, creation of a scope statement) and those that were the least used (e.g., creation of a quality plan, follow the work breakdown structure and monitoring of risks). The results also suggest that creating documentation appears to be easier for students than using it to monitor project progress. It is therefore recommended that assessment is aligned with PMBOK Guide practices to encourage students to apply more project management guidelines in their student projects.

The study also investigated whether the application of project management guidelines increases the chances of completing IT projects successfully. The results showed that the number of project management guidelines applied was not associated with IT project success. However there was a strong relationship between having a good project management plan and obtaining a good software product.

Grundy (1997) suggested that IT students need to be immersed in a project situation to put into practice what they have learned about project management in previous courses, and many authors have argued that the project activities have to mirror the real world for IT students to learn what needs to be done in industry projects (Abernethy & Piegara, 2007; Ellen & West, 2003; Lowe, 2000; Nance, 1998). The findings of this study support the importance of enabling students to practice their project management skills in a software development project and suggest that if maximum value is to be obtained, skill use should be linked to assessment.

REFERENCES

Abernethy, K., & Piegara, G. (2007). Teaching project management: An experiential approach. *Journal of Computing Sciences in Colleges, 22*(3), 198-205.

Aladwani, A. M. (2002). IT project uncertainty planning and success: An empirical investigation from Kuwait. *Information Technology & People, 15*(3), 210-226.

Batra, D., & Satzinger, J. W. (2006). Contemporary approaches and techniques for the systems analyst. *Journal of Information Systems Education, 17*(3), 257-265.

Brock, S., Hendricks, D., Linnell, S., & Smith, D. (2003). A balanced approach to IT project management. *Proceedings of the 2003 Annual Research Conference of the South African Institute of Computer Scientists and Information Technologists on Enablement through Technology* (pp. 2-10).

Du, S. M., Johnson, R. D., & Keil, M. (2004). Project management courses in IS graduate programs: What is being taught? *Journal of Information Systems Education, 15*(2), 181-187.

Ellen, N., & West, J. (2003). Classroom management of project management: A review of approaches to managing a student's information system project development. *Journal of American Academy of Business, 3*(1/2), 93-97.

Freedman, R. (2002). Adopting PMI standards in an IT environment. Retrieved October 28, 2005, from http://techrepublic.com.com/5100-6330-1038859.html

Goold, A. (2003). Providing process for projects in capstone courses. *ACM SIGCSE Bulletin, 35*(3), 26-29.

Gorgone, J. T., Davis, G. B., Valacich, J. S., Topi, H., Feinstein, D. L., & Longenecker, H. E. (2002). IS 2002 model curriculum and guidelines for undergaduate degree programs in information systems. Retrieved February1, 2007, from www.acm.org/education/is2002.pdf

Gowan, J. A., & Mathieu, R. G. (2003). The role of project methodology in large-scale IS project management. In *Proceedings of the Ninth Americas Conference on Information Systems* (pp. 1328-1332).

Grundy, J. C. (1997). A comparative analysis of design principles for project-based IT courses. *Proceedings of the 2nd Australasian Conference on Computer Science Education ACSE '97* (pp. 170-177).

Jih, W. (2003). Simulating real world experience using accumulative system development projects. *Journal of Information Systems Education, 14*(2), 181-191.

Lowe, G. S. (2000). Preparing students for the workforce. *Proceedings of the Australasian Conference on Computing Education ACSE '00* (pp. 163-169).

Nance, W. D. (1998). Experiences with an innovative approach for improving information systems students' teamwork and project management capabilities. *Proceedings of the 1998 ACM SIGCPR Conference on Computer Personnel Research* (pp. 145-151).

Phillips, R., Fairholme, E., & Luca, J. (1998). Using e-mail to improve the experience of students doing a group-based project unit. *Showing Best Practice at ECU. The Proceedings of the 1998 Best Practice Showcase.* Retrieved September 30, 2004, 2004, from http://www.cowan.edu.au/eddev/98case/phillips.html

Pinto, J. K., & Slevin, D. P. (1988). Critical success factors across the project life cycle. *Project Management Journal, 19*(3), 67-75.

Pournaghshband, H. (1990). The students' problems in courses with team projects. *Proceedings of the 21st SIGCSE Technical Symposium on Computer Science Education* (pp. 44-47).

Project Management Institute. (2004). *A guide to the project management body of knowledge (PMBOK Guide)* (Third ed.). PA: Project Management Institute Inc.

Reif, H. L., & Mitri, M. (2005). How university professors teach project management for information systems. *Communications of the ACM, 48*(8), 134-136.

Schwalbe, K. (2004). Introduction to project management. In J. Locke & A. Poirier (Eds.), *Information technology project management* (Third ed.) (pp. 1-29). Cambridge, MA: Course Technology.

Stein, M. V. (2002). Using large vs. small group projects in capstone and software engineering courses. *The Journal of Computing in Small Colleges, 17*(4), 1-6.

The Standish Group. (2001). Extreme CHAOS. Retrieved October 28, 2005, 2005, from http://www.standishgroup.com/sample_research/index.php

APPENDIX

Appendix A. Questionnaire used in the study

What is your gender?
- ☐ Female
- ☐ Male

How old are you?
___ yrs old

What is your major(s)? Tick all that apply.
- ☐ Computer Science
- ☐ Applied Information Technology
- ☐ Multimedia Information Systems
- ☐ Business Information Systems
- ☐ Information Systems Development
- ☐ Internet Computing
- ☐ Games Technology
- ☐ Internetworking and Security

What is your project group's name?________________________________
____

How aware were you of project management practices before ICT333?

Very Little				A lot
1	2	3	4	5

How aware are you of project management practices now?

Very Little				A lot
1	2	3	4	5

How aware were you of Project Management Body of Knowledge (PMBOK) practices before ICT333?

Very Little				A lot
1	2	3	4	5

How aware are you of Project Management Body of Knowledge (PMBOK) practices now?

Very Little				A lot
1	2	3	4	5

Where did you learn project management?____________________________
____
____

How did you learn project management?_____________________________
____
____

Continued on following page

Appendix A. continued

Project Integration Management

Did your group use the template provided by the course coordinator (or any other template) to create the project management plan?

☐ Yes ☐ No

Was it easy to follow the template(s)?

Difficult to use			Very easy to use	
1	2	3	4	5

Why?____

How useful do you see templates as being when applying them to IT projects?

Not useful			Very useful	
1	2	3	4	5

Would it be better for a group to create their own documents rather than following existing generic templates?

☐ Yes ☐ No

Why?____

Project Scope Management

Did your group create a scope statement to define the scope of your project?

☐ Yes ☐ No

Did your group create a scope management plan to manage the scope and integrate changes to it?

☐ Yes ☐ No

How useful are scope statements and scope management plans to manage your project's scope?

Not useful			Very useful	
1	2	3	4	5

Project Time Management

Did your group use Microsoft Project to assist the development of the project's schedule?

☐ Yes ☐ No

Did your group follow the project plan closely (Work Break Structure, Gantt Chart) to check the deadline of tasks?

☐ Yes ☐ No

How useful are project time management tools/techniques?

Not useful			Very useful	
1	2	3	4	5

Continued on following page

Appendix A. continued

Project Quality Management

Did your group create a quality plan to identify which project management quality standards are relevant to your project?
☐ Yes ☐ No

Did your group continually monitor project result(s) to determine if they comply with the relevant quality standards?
☐ Yes ☐ No

How useful is a quality plan to manage the project's quality?

Not useful Very useful

1 2 3 4 5

Human Resource Management

Did your group use the skills matrix provided by the Course Coordinator to assign task(s) to the correct person?
☐ Yes ☐ No

How useful are the skills matrixes to allocate people to the right task?

Not useful Very useful

1 2 3 4 5

Project Communication Management

Did your group create a communications plan to verify how the team would communicate with each other?
☐ Yes ☐ No

Did your group regularly distribute project information to team members e.g. agendas, minutes, etc.
☐ Yes ☐ No

Did your group frequently discuss your project's progress (written progress reports) with your supervisor?
☐ Yes ☐ No

How useful is it to have communication plans to distribute information to other team members?

Not useful Very useful

1 2 3 4 5

Continued on following page

Appendix A. continued

Project Risk Management

Did your group create a risk management plan to decide how to approach and plan the risk management activities of the project?
☐ Yes ☐ No

Did your group monitor and control the risks that might occur in the project?
☐ Yes ☐ No

Are risk management plans useful?

Not useful			Very useful	
1	2	3	4	5

In your opinion, has it been beneficial to work on a real life project with real clients in this course?
☐ Yes ☐ No

Why do you feel this way?____

____

____

Chapter XVI
Teaching TCP/IP Networking Using Practical Laboratory Exercises

Nurul I. Sarkar
AUT University, New Zealand

ABSTRACT

Motivating students to learn TCP/IP network fundamentals is often difficult because students find the subject rather technical when it is presented using a lecture format. To overcome this problem we have prepared some hands-on exercises (practicals) that give students a practical learning experience in TCP/IP networking. The practicals are designed around a multi-user, multi-tasking operating system and are suitable for classroom use in undergraduate TCP/IP networking courses. The effectiveness of these practicals has been evaluated both formally by students and informally in discussion within the teaching team. The implementation of the practicals was judged to be successful because of the positive student feedback and that students improved their test results. This chapter describes the practicals and their impact on student learning and comprehension, based on the author's experiences in undergraduate computer networking courses.

INTRODUCTION

Almost any computer science, engineering, and business curriculum includes some basic courses on transmission control protocol/Internet protocol (TCI/IP) networking. Unfortunately, motivating students to learn TCP/IP network fundamentals can be difficult not only because students find the subject rather abstract when it is presented using a lecture format, but also because of very limited resources designed to supplement the teaching of TCP/IP networking is publicly available. Research has shown that students learn TCP/IP networking better, and feel more engaged with their courses, if they are given practicals that illustrate theoretical TCP/IP networking concepts (Midkiff, 2005; Richards & Waisbrot, 2002; Sarkar, 2006; Sarkar & Craig, 2006). Students gain first-hand experiences in TCP/IP networking by hands-on practical work, for example, setting up a TCP/IP network, installing and configuring a server, IP sub-netting, TCP/IP connectivity, Telnet, and anonymous file transfer protocol (FTP). Therefore, we have prepared some practicals that facilitate an interactive, hands-on learning experience in TCP/IP networking. These practicals are designed around Linux, and can be used either in the classroom as a demonstration to enhance the lecture environments, or in the computer laboratory to provide hands-on learning experience at an introductory level (either first year or second year undergraduate courses).

The basic concept of TCP/IP networking is described in many textbooks (Forouzan, 2003; Kurose & Ross, 2005; Mansfield, 2003; Raymond, 2005; Stallings, 2007), and TCP/IP network performance is discussed extensively in the networking literature (Gotsis, Goudos, & Sahalos, 2005; Hassan & Jain, 2004; Man, Hasegawa, & Murata, 2006; Xin & Jamalipour, 2006;. Zeng & Trajkovic, 2005). Sloan (2004) describes TCP/IP lab materials which may be suitable for practical work in the computer laboratory. However, these lab materials are still in work-in-progress. A number of sophisticated network simulators exist for building a variety of TCP/IP network models (Fall & Varadhan, 2007; *OPNET Technologies*, 2007; Zeng, Bagrodia, & Gerla, 1998). Nevertheless, by setting up and configuring actual TCP/IP networks the students gain first-hand experience that cannot be gained through computer simulation and modeling, and which plays a crucial role in motivating them to learn about TCP/IP networking. To date, we have focused on developing practicals to support teaching and learning traditional TCP/IP networking courses at undergraduate level (second year). The hands-on learning approach to teaching TCP/IP networking has been successfully applied for two years now in the networking and telecommunications course (undergraduate computer science/IT curriculum) at AUT University. The course covers the main aspects of TCP/IP networking, including client-server networking, local-area network (LAN) administration and management. The practicals and other materials are revised annually based on student feedback, and the use of the latest version is discussed in this chapter.

The main contribution and strength of this chapter is the emphasis on practicals as a means to achieve effective student learning at undergraduate level. The most innovative aspect of this chapter is the structuring of the laboratories and crafting of the specific exercises to be effective in complementing the lecture content of the course. The course context in which the practicals are presented is described next.

COURSE CONTEXT AND LEARNING OUTCOMES

The networking and telecommunications in which the aforementioned practicals are introduced has a total of 56 contact hours assigned. These courses are at Level 6, or second-year degree level at AUT University (Sarkar & Clear, 2000; Sarkar & Petrova, 2001). The Level 5 'hardware and software infrastructure' is the prerequisite for this course. The course goal is to produce graduates who will be able to work for networking and telecommunication companies worldwide.

The first half of the course covers networking fundamentals, while the second half focuses on TCP/IP networking. On completion of this course students will be able to:

- Discuss the relative advantages and disadvantages of various data transmission media, topology, data encoding, and protocols employed in telecommunication networks,

- Draw a detailed network diagram for the solution of a business case scenario, involving designing a network spanning multiple floors in two or more buildings,
- Gain practical experiences with server hardware and software, including installation and configuration of a customized TCP/IP network.
- Configure IP sub-netting and explore TCP/IP connectivity and interoperability using telnet and FTP, and
- Gain practical knowledge in network administration and management.

The last three learning outcomes are met by practicals described in this chapter which is about 50 percent of the course total.

DETAILS OF PRACTICALS

Table 1 lists the six practicals that have been developed to date and have trialed over two years in the networking and telecommunications undergraduate course. Each practical is conducted as a two-hour session, held weekly in a network laboratory at AUT University.

Table 1. Practicals and related networking concepts

Practicals	Networking concepts
Linux installation	Network installation and administration
Network administration	Network mask, IP address, broadcast address, MAC address, configuration table
Linux configuration	Host configuration, administration and management
IP subnet addressing	Classes of IP networks; subnet mask and network sub-netting.
TCP/IP connectivity 1	TCP/IP connectivity and practical exercises
TCP/IP connectivity 2	Client-server architecture; telnet; FTP; domain name server; name resolution

Laboratory 1: Linux Installation

This Linux practical shows students how to install a Linux host. Students (in pairs) install the Redhat Linux (from a CD) onto a removable hard disk (removable hard disk is discussed later in the chapter). Students are asked to keep a journal of the major steps of Linux host installation, including problems and possible solutions. By installing the Linux host, students achieve the following learning outcomes: (1) prepare Linux installation boot disk; (2) partition the removable hard disk for Linux installation; (3) set up the root password; (4) add a user account; (5) configure password authentication; and (6) install various software packages.

After Linux installation, students must reboot the machine to verify that the host is up and running as expected. In completing this practical, students develop a better understanding in Linux host installation and configuration.

Laboratory 2: Network Administration

This practical shows students how to login to the newly installed Linux host (described in Laboratory 1) as a root-user and do some practical activities related to network administration. In completing this practical, students achieve the following learning outcomes: (1) use various commands such as *ifconfig, arp, route*, and *hostname* to investigate the network status of the host machine; (2) check the network connectivity of the machine using the loop-back interface; (3) change the network configuration of the host; (4) check the network connectivity of the host machine within and outside the classroom; (5) record the entries in the network interface configuration table; and (6) use appropriate commands to find network mask, IP address, broadcast address, and MAC address for each interface.

Laboratory 3: Linux Host Configuration

In this practical, students further extend their knowledge of Linux host configuration gained in Laboratory 2. Students are encouraged to explore the Linux host configuration by completing the following tasks: (1) experiment with network configuration files; (2) identify the network interface cards; (3) assign super-user rights to a user; (4) use Linux commands to display the contents of files; and (5) check and interpret the content of various files; and (6) experiment with various network configurations of the host machine.

In completing this practical, students develop a better understanding in Linux host configuration and network management.

Laboratory 4: IP Subnet Addressing

The various classes of IP addresses, subnet mask, and network sub-netting are introduced during lectures where students are also given a handout on TCP/IP networking. This Linux practical shows students how to configure (with IP addresses) three sub-networks shown in Figure 1. Students gain a sound knowledge and understanding of IP subnet addressing by working through a problem based on a small business case scenario as follows.

"An organization has been assigned a class B address: 156.62.0.0. You are given the task to configure (with IP addresses) the three sub-networks, each has 10 hosts and a router connects them. The network mask used by the organization is 255.255.255.0. Assign a valid IP address to each network interface and show the broadcast address and the subnet ID for each sub-network."

In completing this practical, students gain experience in TCP/IP network design and IP sub-netting.

Laboratory 5: TCP/IP Connectivity 1

This practical provides a walk-through in exploring TCP/IP connectivity. Students are given the following exercises to explore TCP/IP connectivity:

1. Determine the appropriate bandwidth required for connecting AUT University to the Internet.

Figure1. A LAN comprising three sub-networks linked by a router

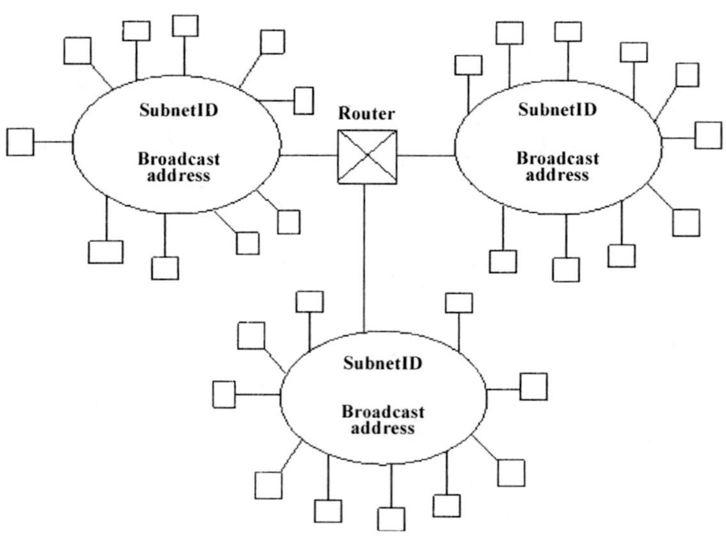

2. If a router (R) can connect up to K networks, how many routers are required to connect to N networks? Write down an equation that gives R in terms of N and K.
3. Suppose five million new computers are added to the Internet in a nine month period, if computers are added at a uniform rate, how much time elapses between two additions.
4. Draw a diagram of a TCP/IP network that consists of two sub-networks connected by a router.

In completing this practical, students develop a better understanding of TCP/IP network connectivity.

Laboratory 6: TCP/IP Connectivity 2

This Linux practical shows students how to create TCP/IP connectivity and interoperability using telnet and FTP. The basic concepts of FTP, telnet, address resolution protocol (ARP) and domain name systems (DNS) are introduced during laboratory sessions. Students are asked to create a text file and then transfer it to a Linux host using FTP. Students then establish a telnet session between a MS Windows client and the Linux host, and locate the text file on the Linux host. Students are encouraged to explore TCP/IP connectivity and interoperability using various telnet and anonymous FTP sessions.

Additional Projects

The following practicals are being considered:

- **Routing:** The learning activities include configuring static routing, routing table information, configuring routing information protocol (RIP) and interior gateway routing protocol (IGRP) routings.
- **TCP/IP security:** The learning activities include wireless access to campus network, email server, and the Internet. Students can experiment with a wireless access point and observe communication traffic as the access point is notified and user data is transmitted.
- **Open:** A category for student-suggested projects.

BENEFITS OF PRACTICALS

The TCP/IP practicals provide the following main benefits:

- **Hands-on:** The practicals are hands-on and reinforce learning.
- **Usefulness:** The practicals are easy to set-up and can be used either in the classroom, or in the computer laboratory.
- **Challenging:** The practicals provide the students with a challenging yet friendly environment where they can test their knowledge of TCP/IP networking.
- **Economical:** Having students set up Linux host on removable hard disks has the advantage of leaving laboratory PC configurations unaltered for other users.
- **Reusability:** Students can reuse removable hard disks for other courses.

TCP/IP PRACTICALS IN PRACTICE

There are 25 computers (networked PC; MS Windows XP client) in a typical network laboratory at AUT University, which allows us to accommodate up to 24 students in a laboratory (one PC is used for lab tutor's demonstration) for practical work. These computers can use removable hard disk drives so that students can install different operating systems and to modify various system settings without changing the standard classroom disk image for other users. At the beginning of the lab sessions, we provide each student with a removable hard disk (hard disk cost is included in the course fee, and the students are allowed to

Figure 2. A customized TCP/IP network in a typical network laboratory at AUT University

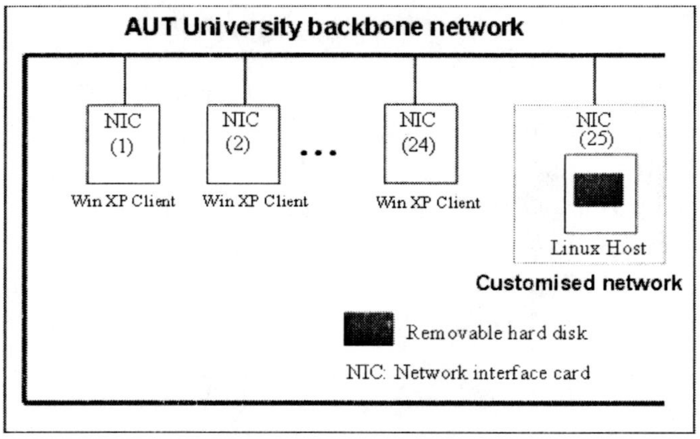

take away their disks at the end of the course). Students can use the same disk for other courses, for example, Web development.

Figure 2 shows a customized TCP/IP network set up in a typical network laboratory at AUT University. At the beginning of each lab session, each student gets a PC in the lab and using it by inserting a removable hard disk into the drive bay.

Students are well disciplined in the laboratory and quickly get used to follow some basic rules and regulations. For example, at the end of each lab session, students must return their removable hard disks to tutor. They also need to ensure that all PCs are reconnected back to the AUT University network and have their log sheets signed. Students can further explore the TCP/IP practicals outside the scheduled lab times by booking a time slot.

The author has tested the practicals during 2003-2004, in the networking and telecommunications undergraduate course (second year) and the author's experience with practicals has been favorable overall. The practicals are easy to use and follow, and by participating in the learning activities, students became increasingly motivated to learn more about TCP/IP networking.

Evaluation by Student Feedback

To assess the educational value, the practicals were evaluated extensively by students. The formal evaluation of the practicals was conducted in the classroom by a lecturer, and the anonymity of the respondents was assured. As part of the formal evaluation process students were asked to complete a short, six-question questionnaire as follows:

- **Prior knowledge:** How well did you understand TCP/IP networking concepts before entering this course?
- **Easy to follow:** How easy (overall) did you find the TCP/IP practicals to use and follow?
- **Measure of success:** How effective were the practicals in helping you to improve your understanding of TCP/IP networking?
- **Hands-on:** Would you like to have more practicals of this kind as part of your course?
- **Time to explore:** Would you like to have an extra time (apart from schedule lab time) to further explore the practicals?
- Learning a new OS: Would you prefer to learn TCP/IP networking with another operating system (OS) such as MS Windows?

Figure 3. Student response: Graphs illustrating the number of respondents in each category for each of the six questions in the questionnaire.

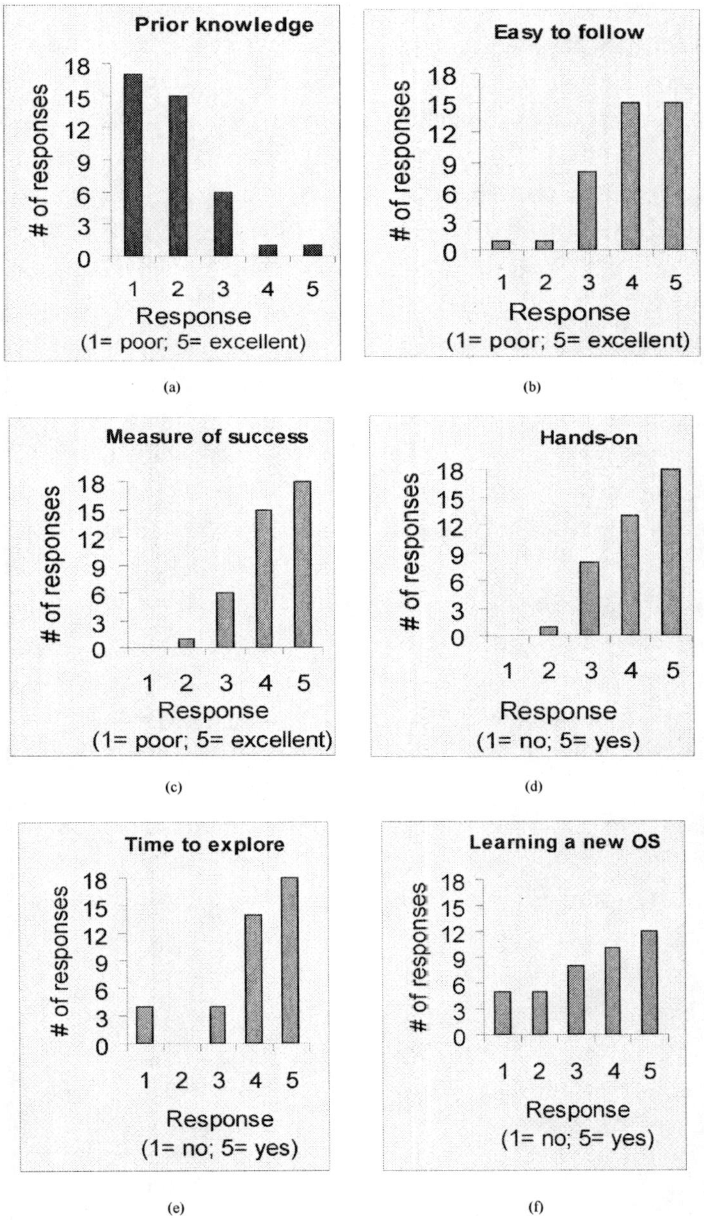

A five-point ordinal evaluation scale was used in the questionnaire. For Questions one to three, one on the scale was Poor, and five was Excellent; for Questions four to six, one was No and five was Yes. Over two semesters, 40 undergraduate students (about 60 percent male and 40 percent female) from networking and telecommunications completed the questionnaire and their responses are plotted in Figure 3. The responses were interpreted as follows.

Thirty students have indicated that they had no prior knowledge of TCP/IP networking before entering this course. Three students indicated that they had basic knowledge of TCP/IP networking; whereas, the remaining seven students were neutral (Figure 3a).

The practicals were found to be reasonably easy to use and follow. Twenty-eight students were satisfied with the current version of the practicals. Two students expressed some concerned, and the remaining 10 students were neutral (Figure 3b).

Thirty-three students indicated that the practicals had clearly assisted them in developing a better understanding of TCP/IP networking. However, one student was not totally satisfied with the current practicals; and the remaining six students were neutral (Figure 3c).

Thirty-one students indicated that they would like to have more practicals in the course. One student was not very interested in trying more practicals, and the rest eight students were neutral (Figure 3d).

Thirty-two students indicated that they need extra times outside the scheduled lab times to explore the practicals. Four students indicated that the scheduled lab time was adequate to explore practicals, and the rest, four students were neutral (Figure 3e).

Twenty-two students indicated that they would like to try another network operating system to learn TCP/IP networking. Ten students indicated that they were not very interested in trying another network operating system; and the remaining eight students were neutral (Figure 3f).

IMPACT OF PRACTICALS ON STUDENT PERFORMANCE

As mentioned earlier, the practicals have been used in networking and telecommunications undergraduate course and the experience of both the lectures and students is very positive. However, we have restructured the networking and telecommunications and have created several new 15-credit courses, including Networking 2 and Advanced network technologies. Some of the practicals have been used in Networking 2 during 2006.

Students are highly motivated in learning TCP/IP networking by doing hands-on practical activities, a fact also confirmed by student feedback (Figure 3c). In the past (e.g., during 2001-2002), only occasional computer simulation and animation were used to reinforce the theoretical concepts. However, a deeper knowledge of a customized TCP/IP networking has been achieved by students who completed practicals. Particular topics in which students' learning and comprehension have been improved were as follows:

- TCP/IP network design and configuration
- IP subnet addressing
- TCP/IP connectivity
- Network administration and management
- Installation of a multi-user operating system

To estimate (quantitatively) the impact of practicals on student performance, an analysis of the student grades in the written exam for the last five years (2001-2004 and 2006) has been conducted. Students' performance in the final exam with and without practicals is shown in Table 2.

As seen in Table 2, the overall student average pass rate in the final exam in 2003, 2004 and 2006

Table 2. Comparison of student performance in the final exam with and without practicals

Semester	Year of study	With practicals	Without practicals	Average pass rate (%)
2	2001		√	74
1	2002		√	76
2	2002		√	70
1	2003	√		94
2	2003	√		95
1	2004	√		96
1	2006	√		96
2	2006	√		100

is higher compared to students in 2001 and 2002. Since the final exams were of comparable style and levels of difficulty, the higher pass rate can be accounted for by the students in 2003, 2004, and 2006 who used practicals; whereas, the students in 2001 and 2002 did not get an opportunity to use the practicals at all. One could argue whether the results of the written final exam precisely reflect laboratory work. It should be noted that the final exam (closed book) was comprehensive and covered the material taught throughout the semester (i.e., 14 weeks). Theory presented in lectures is enhanced via laboratory demonstrations, and practical lab exercises.

By inspecting Table 2, one can observed that the students have done slightly better in 2006 compared to 2004. This is mainly due to the lab materials being refined to eliminate points of confusion for the students and also the lab tutor building up a knowledge base of common laboratory problems.

CONCLUSION

We have developed a series practical exercises that can be used either in the classroom for class demonstrations to enhance the lecture environments, or in the computer laboratory for hands-on learning experience in TCP/IP networking. With practicals, the teaching and learning of TCP/IP networking were made more interesting and applied. Fundamental theoretical concepts were illustrated with the use of interactive practical exercises. Students were given the opportunity to identify their misconceptions and reconstruct their knowledge, which helps them internalize, rather then absorb, the basic concepts of TCP/IP networking (Chen, 2003).

Objectifying computer networking concepts is one of the aspects of the constructivist approach towards the teaching and learning of TCP/IP networking. The practicals supported constructivist pedagogical approaches and improved students' participation in the learning activities. Students evaluated the practicals, and their responses to the questionnaire were highly favorable. They indicated that they had found the practicals easy to use and that these helped them gain a deeper knowledge and understanding of TCP/IP networking. The practicals also had a positive impact on student learning and comprehension. Results show that the students with practical experience scored better in the final exam than those without practical experience.

More practicals, such as routing and network security, are under development. The practicals described in this chapter can be applied to courses in other fields, such as computer engineering. The author's ongoing effort is to develop more networking courses following this practical learning approach. More information about practicals can be obtained by contacting the author.

ACKNOWLEDGMENT

The author would like to thank the anonymous reviewers for their constructive comments to improve the overall quality of the chapter. The author also wishes to thank K. Petrova for her contribution to the TCP/IP practicals.

REFERENCES

Chen, C. (2003). A constructivist approach to teaching: Implications in teaching computer Networking. Information Technology. *Learning and performance Journal, 21*(2), 17-27.

Fall, K., & Varadhan, K. (2007). *The Ns Manual. The VINT Project.* Retrieved March 5, 2007, from http://www.isi.edu/nsnam/ns/

Forouzan, B. A. (2003). *TCP/IP Protocol Suite* (2nd ed.). McGraw-Hill.

Gotsis, K. A., Goudos, S. K., & Sahalos, J. N. (2005). A test lab for the performance analysis of TCP over ethernet LAN on windows operating system. *IEEE Transactions on Education, 48*(2), 318-328.

Hassan, M., & Jain, R. (Eds.). (2004). *High performance TCP/IP networking — concepts, issues, and solutions*. New York: Pearson Prentice Hall.

Kurose, J. F., & Ross, K. W. (2005). *Computer networking: A top-down approach featuring the internet* (3rd ed.). New York: Addison Wesley.

Man, C. L. T., Hasegawa, G., & Murata, M. (2006). ImTCP: TCP with an inline measurement mechanism for available bandwidth. *Computer Communications, 29*(10), 1614-1626.

Mansfield, N. (2003). *Practical TCP/IP — Designing, using, and troubleshooting TCP/IP networks on Linux and Windows*. Addison-Wesley.

Midkiff, S. F. (2005). An experiential course in wireless networks and mobile systems. *IEEE Pervasive Computing, 4*(1), 9-13.

OPNET Technologies. (2007). Retrieved February 20, 2007, from www.opnet.com

Raymond, R. P. (2005). *Business data networks and telecommunications* (5th ed.). Englewood Cliffs, NJ: Prentice Hall.

Richards, B., & Waisbrot, N. (2002, June 24-28). *Illustrating networking concepts with wireless handheld devices.* Paper presented at the 7th Annual SIGCSE Conference on Innovation and Technology in Computer Science Education (IT-iCSE'02), (pp. 29-33). Aarhus, Denmark.

Sarkar, N., & Clear, T. (2000, June 30 - July 3). *Developing a new course for the software development pathway on the AUT Bachelor of applied science programme.* Paper presented at the 13th Annual Conference of the National Advisory Committee on Computing Qualifications (NACCQ), (pp. 303-309). Wellington, New Zealand.

Sarkar, N., & Petrova, K. (2001, July 2-5). *Teaching computer networking & telecommunications: a network analysis and software development approach.* Paper presented at the 14th annual conference of the national advisory committee on computing qualifications (NACCQ) (pp. 379-384).Napier, New Zealand.

Sarkar, N. I. (Ed.). (2006). *Tools for teaching computer networking and hardware concepts* (Vol. 1). Hershey, PA: Information Science Publishing.

Sarkar, N. I., & Craig, T. M. (2006). Teaching wireless communication and networking fundamentals using Wi-Fi projects. *IEEE Transactions on Education, 49*(1), 98-104.

Sloan, J. D. (2004). *A remotely accessible network laboratory — TCP/IP Laboratories. http://webs.wofford.edu/sloanjd/netlab/IPLabs/labs.htm.* Retrieved December 27, 2004, from http://webs.wofford.edu/sloanjd/netlab/IPLabs/labs.htm

Stallings, W. (2007). *Data and coputer communications* (8th ed.). NJ: Prentice Hall.

Xin, F., & Jamalipour, A. (2006). TCP performance in wireless networks with delay spike and different initial congestion window sizes. *Computer Communications, 29*(8), 926-933.

Zeng, W. G., & Trajkovic, L. (2005, August 22-24). *TCP packet control for wireless networks.* Paper presented at the IEEE International Conference on Wireless And Mobile Computing, Networking And Communications (WiMob'2005) (pp. 196-203).

Zeng, X., Bagrodia, R., & Gerla, M. (1998). *GloMoSim: a library for parallel simulation of large-scale wireless networks.* Paper presented at the Twelfth Workshop on Parallel and Distributed Simulation (pp. 154-161).

Section III
Assessment

Chapter XVII
Assessment of ICT Status in Universities in Southern Nigeria

Sam E. O. Aduwa-Ogiegbaen
University of Benin, Nigeria

Raymond Uwameiye
University of Benin, Nigeria

ABSTRACT

The aim of this study was to investigate the influence of faculty affiliation and teaching experience on the use of the Internet by faculty members in six first-generation universities in Southern Nigeria. A total of 476 faculty members from nine faculties across the six universities participated in the study. The data for the study was collected by means of a questionnaire survey and this was deemed appropriate as it allowed the views of all the participants to be sought on a Likert-type scale options. The results of this study provide a number of insights: (a) the faculties of engineering, science and arts in that order were the foremost users of the Internet for instructional purposes; (b) the faculties of education and agriculture were the least experiences in the use of the Internet; and, (c) faculty members with less than five years teaching experience use the Internet more than older faculty members. The recommencation was made that universities in Nigeria should invest more in ICT facilities.

INTRODUCTION

In recent years, information and communication technology (ICT) has impacted on all aspects of society. The glaring potential of the computer for education is enormous and experts in the field have long recognized this (Stonier & Conlin, 1985). In the developed countries of the world, technology has revolutionized the way work is done and the Internet has reduced the world considerably to the much talked about global village. Also, the use of ICT is increasing in schools though its application is still very varied. It has also been ascertained that the use of ICT in the developed countries in teaching subjects across the curriculum is increasing but good practices are not common (Tearle, 2003). There is no doubt that with its introduction into educational practices, ICT has widened the scope of opportunities in secondary and tertiary levels of education. The changes in teaching and learning which ICT has brought into the classroom pose a special challenge to all categories of teachers and students and it is their prerogative to seek ways and means of maximizing its potential for the benefit of all. The high expectation of the role of ICT in teaching and learning placed a particular responsibility on the present day teacher to evolve good practice in utilizing ICT in schools.

In recent years, many researchers (Andrews, 1997; Hoffman, 1996; Tearle, 2003) studied several aspects of ICT implementation in schools and their works are of particular relevance to this study. These studies have emphasized the strong feeling that individual staff in educational institutions in the developed countries wanted to keep striving for excellence in applying ICT in teaching and learning. The process of ICT implementation has important characteristics which include paying specific attention to preparation and planning, access to resources, shared responsibility, training and support.

In the developed countries of the world, there are good examples of schools where ICT is regularly used by almost all staff working in all curriculum areas to enhance teaching and learning. However, in the developing countries, such as Nigeria, the banking sector, oil and gas exploration and processing, communication (the global system for mobile communication), and many other private sector driven aspects of the economy have all embraced ICT and integrated it into their workplaces.

Higher education in Nigeria faces great challenges in response to changes in its internal and external environment and this include the issue of widening access to educational technology, particularly how to integrate ICT into educational practices. The lack of access to information technology and its requisite skills contribute to an inability to compete in the mainstream economy and mobility to participate meaningfully to civil society. Lack of access to information technology impedes success in academic pursuit, the skills necessary to work in knowledge driven society, and ability to prosper in modern society.

Already the Manufacturers Association of Nigeria (MAN) at a meeting held at Ibadan and reported in the Guardian Newspapers Nigeria Ltd of July 23, 2002, indicated that there is a manifest mismatch between formal education system, especially university education in Nigeria, and the needs of employers especially in the manufacturing sector. This according to the association portends grave danger for the growth of value–added industry in the country. The association further stressed that "a proper mix between vocational education and industry requirements has not been achieved" (Guardian Newspapers, Nigeria Limited, 2003). It has also been observed that the current floppy manpower training and development scheme has greatly reduced the possibility of industry's utilization of graduates of tertiary institutions over the last few years. Consequently, the Nigerian universities were challenged to provide the leadership to transform the Nigerian society and the economy through the relevance of its knowledge and skills.

Need For ICT in Higher Education in Nigeria

The Association of African Universities (AAU) at its tenth general conference held in Nairobi, Kenya, on February 5–9, 2001 determined general policies and core priorities for African universities in the next four years. The major areas of focus at the conference include leadership and management, quality of training and research, information and communication technology (ICT), and women in African higher education. At the conference, there was general consensus among the over 250 participants (which included vice-chancellors, rectors and senior academics from

163 member universities) that African universities must enhance their information and communication technology in order to participate effectively in the global information age (Whitaker, 2001). The conference, according to Whitaker (2001) charged African universities to use ICT as a tool in higher education management—to track students, faculty, budgets, and so forth—and in the classroom, to facilitate teaching and learning. The conference asserts that ICT will allow African academics to participate actively in global research using the Internet, publish articles, and exchange ideas with colleagues. It is not in the best interest of African universities to continue to graduate students who are not fully prepared for the information age. According to Sept (2004)

It is ironic that in the so-called Information Age we are still graduating passive, solitary learners poorly equipped to cope with the explosion of information resources competing for attention. (p. 49)

Nigeria's universities and other higher institutions need ICT for the following reasons:

- The integration of modern information and communication technologies into the administrative, teaching and learning process to enhance the tools and environment for learning.
- To allow materials for teaching and learning to be presented in multiple media for multi–channel learning (Haddad, 2003)
- Videos, television, and computer software are excellent instructional media that motivate learning process. Research shows that when learning activities are authentic, challenging, multidisciplinary and multisensory, students are quickly motivated to engage in learning activities.
- In an information age, the students today need to learn to manage information. They need to ability to sort, retrieve, store, and organize information. The computer creates such learning environment
- Nigeria students need to have access to word processing programmes with which they can write term papers and other assignments. The computer can help the student to create multimedia term papers, integrating such media such as graphics, sound, and motion for a more complete presentation (Smaldino, Russell, Heinich et al., 2005)
- In a knowledge-based society and a world that has turned to a global village, the sharing and expansion of knowledge and information are the key elements for economic, political and social growth. As a result of this the speed at which information and knowledge move around the world has been greatly accelerated by the rapid expansion of the Internet and communication technology (Johari, 2006).
- To develop affection and intellectual capacity, the students need to imbibe the spirit of inquiry and exploration which create a sense of adventure for the learners. This spirit of adventure does not present itself in the traditional classroom setting "where questions and answers are established a prior and are unrelated to students' interest and where research is reduced to a word in the textbook" (Haddad, 2003, p. 5) ICT provides the students with the potential to engage in inquiry learning and to explorer knowledge with minimum guidance from the teacher.
- Nigerian universities need ICT to provide unlimited access to world wide information sources. In the past few decades, educators have seen the enormous increase in data communications and network expands in size, sophistication, and the ability to communicate over distances (Picciano, 2006). The need to install local area networks (LAN) and wide area network (WAN) has become imperative in higher education in Nigeria as these will offer the university

administrators, teachers and students a platform through which they can communicate with colleagues at distant and local places. It will enhance their work, develop their research ability and support their functions in such a way as if geographical boundaries are no more.

- Nigeria universities can not afford to be on the wrong side of the digital divide. If computer facilities and Internet facilities are not provided for staff and students a discrepancy in access to technology resources will place university teachers and students at a great disadvantage. To go beyond the traditional approach of offering courses, the university in Nigeria must acquire and use technology in three interrelated components (1) educational technology courses, (2) integration of technology into university courses, and (3) introduction of technology into student field experience.

There is no doubt that the conventional modes of delivering knowledge in higher institutions in Nigeria consist of lectures, tutorials, and seminars. Concomitant with the rapid expansion in the number of students entering the universities in Nigeria, there is an increase in the student body. Many would be undergraduates come to the universities with different range of educational experience, ability, qualifications, age and motivation. As a result, the different levels of knowledge, ability and motivation pose a great challenge to faculty members to deliver materials at a pace and level that would suit the capabilities and personal circumstances of the learners. This is where ICT can play a significant role in achieving better results.

Since functional education, that is the development of relevant skills, science and technology, hold the key to rapid and sustainable economic growth, the transmission of knowledge and relevant skills becomes the major challenges of Nigerian universities. The universities in Nigeria can not provide the much desired leadership training without appropriate resort to information and communication technology (ICT). Information and communication technology can extend what people can do and it is an effective learning tools. Because ICT can make learning easier, more attractive and provides immediate feedback, it extends significantly man's capabilities as well as alters his thinking process positively. For the teacher, ICT alters his vision, potentials, priorities and his work methods and techniques. For the learner, ICT is learner-centered as it supports open, independent and learning materials online as international networks are upgraded in the World Wide Web. Bailey and Cotler (1994) ascertained that the Internet is a cost-effective method for providing improved learning for students. When properly utilized, the Internet can improve student's opportunities for better rapport and interaction with faculty members. According to Abdelraheem and Al Musawi (2003), faculty members and their students have greater flexibility in terms of time, pace and place in asking questions and getting answers. Kent (1996) said that students' professional skills can be improved in tertiary institutions when the Internet is properly utilized.

Barriers to ICT Use in Nigeria's Educational System

The growth and use of information and communication technologies (ICTs) in higher education is a global phenomenon. The possibility of using ICT has major implications for teaching and learning in the universities in the developed and developing countries of the world. It constitutes a global challenge to higher institutions to change their mode of operation and organization. However, in integrating ICT into teaching and learning, there are problems and obstacles. Finley and Hartman (2004) identify some of these obstacles to include; lack of common vision and strong leadership, few incentives to change teaching practices and shortage of hardware and software. A review of the literature shows that there are many barriers to technology use in education. Fisher, Dwyer, and Yocam (1996), Sandhottz, Ringstaff and Dwyer (1997), Schofield (1995), Starr (1996) identify infrastructure problems, lack of adequate training, cultural issues, inequality for minorities and low-income students, limited funds and resources, limited access and large number of technical problems as some of the major barriers to integration of ICT in education. Aduwa-Ogiegbaen and Iyamu (2005) identify the following barriers to ICT; (a) higher cost of computer hardware and software (b) weak infrastructure (c) lack of human skills and knowledge to fully integrate technology into teaching, (d) lack of relevant software, and (e) limited access to the Internet

Generally in Africa, particularly in Nigeria, the cost of access to the Internet is high. Lelliot (2000) ascertained that five hours per month Internet access costs approximately $500, which is about 10 times that of the United States, while per capita income is at least 10 times less. The educational technology of a country rests squarely on top of the national telecommunications and information facilities. Computer equipment are made to function with electricity but there is no part of Nigeria today that can boast of electricity supply for 24 hours as the supply of power is not stable and constant. Nigeria's telecommunication company (NITEL) has just been sold to a transnational company after decades of non-performance and decay; though wireless technologies such as VSAT (very small aperture terminal) have been introduced thereby giving option to Internet service providers. This has even made the cost of ICT higher in the country.

A basic requirement for ICT use in education is access to computers in schools communities, and households. Most educational institutions in Nigeria lack computer facilities and many teachers and students have no access to the technology. Even where there are few available computers many teachers do not use them, either because they don't know how, are satisfied with their current approach to teaching, feel that using ICT is too fraught with difficulties, or that they have no sufficient time to devote to the type of learning best supported by technology (Lelliot, 2000). In Nigeria there is an acute shortage of trained personnel in the application of software, operating systems, network administration and local technicians to service and repair computer facilities (Aduwa-Ogiegbaen & Iyamu, 2005). Okebukola (1997) indicated that those who are designated to use computers in Nigeria do not receive adequate training, at worst, do not receive any training at all.

The greatest challenge to technology use in Nigeria is how to establish reliable, cost effective Internet connectivity. The few Internet providers in Nigeria, many of which are in partnership with foreign information and communication companies, charge high fees for their services and they provide unsatisfactory services to their small clientele. Again, the few Internet providers and their major customers who establish cybercafés are mostly found in the major cities. Connectively beyond major capital cities poses a potential problem in creating national ICT strategy (Darkwa & Mazibuko, 2000).

The obstacles to the use of ICT facilities in African universities have been highlighted by Association of African Universities (2000). These obstacles include: lack of coherent institutional plan for introducing ICTS in universities, poor and unreliable maintenance of ICT facilities, insufficient computer facilities for staff and students, low-level awareness and utilization capacity among faculty and staff of institutions, incomplete local area networks, inadequate training programs for critical skills to manage and support ICT functions, absence of systematized skills for integrating technology into teaching and learning, and lack of recognition for staff efforts and innovation in ICT development and application as past of promotion criteria. The Association of African Universities recognized that many universities are poorly funded with the exception of few, and that many of them lacked the wherewithal to meet the basic requirements for academic growth. The association also ascertained that higher institutions without ICT facilities, human and financial resources, as well as utilization of ICT in the year 2003 will not be able to discharge the functions for which they are established and will play little or no role in the advancement of knowledge.

Status of ICT in Nigerian Universities

As a result of the need for African universities to deploy ICT facilities for the advancement of knowledge, it is pertinent to establish how universities in Nigeria have been using information and communication technologies in the discharge of their educational functions. In 2003, the National University Commission (NUC), the regulatory body for the universities in Nigeria, organized a two-day workshop on information and communication technology (ICT). The workshop, which drew participants from ICT managers from the universities across the country, was addressed by many computer experts in the country. The National Universities Commission's executive secretary stressed the need that Nigeria's education system must embrace ICT to make them competitive in the international global system. The Universities were told that they have no alternative than to embrace ICT revolution in order to assert themselves and maintain global competiveness. At the workshop, the universities were urged to embark on massive implementation and commissioning of campus-wide local area network, digitalization of library materials, computerization of management information system, aggressive e-learning programme and enable Internet connectivity in order to match their counterparts in developed countries (The Guardian Newspaper, 2003)

The workshop served as a spring board for many universities in Nigeria to begin to install ICT facilities in their campuses. The system received massive boost as government injected funds into higher education which enabled 60 percent of the universities to have Web presence and Internet facilities for staff and students. The funding level has helped to address some of the ICT development issues in the universities. The National Universities Commission (NUC) and the Federal Ministry of Education have initiated several ICT projects. These include, school Net, National Open University of Nigeria, and the National Virtual Library. The NUC has encouraged Universities to use at least 20 percent of their teaching and research grants for ICT development. (Guardian Newspaper, 2004, p.39).

As a result of this initiative by the Federal Ministry of Education and the National Universities Commission, many universities in the country began to embrace the ICT culture. In Southern Nigeria, the Obafemi Awolowo University (Ile-Ife), the University of Nigeria (Nsukka), the University of Benin, and others have embraced ICT to a level where teachers and their methods are being positively influenced. This range form online access via the Internet (NUC), or intranet/learning centres (University of Jos). The Carnegie Corporation and MacAthur Foundation have made grants available to some University in Nigeria to help establish their ICT programme (Guardian Newspapers, 2004, p. 39).

The Obafemi Awolowo University (Ile-Ife) ICT laboratory has been equipped with computers and other accessories—software engineering and simulation, modeling laboratory, process control laboratory, as well as instrumentation and telecommunication laboratory. The department of computer science and engineering has been providing training support for its staff and students as well as personnel from other universities. In 2005, the University of Nigeria (Nsukka) launched the first phase of e-learning facilities. The e-learning facilities in the University were supported by AfriHub Inc., of the United States of America, Zinox Technologies, of Nigeria, and SKY Terra Communications. AfriHub will provide e-learning facilities in 40 universities in Nigeria, with the University of Nigeria, in Nsukka, as the first beneficiary. The ICT parks in the two campuses (Nsukka and Enugu) have 100 personal computers (Guardian Newspapers, 2005, p. 41).

At the University of Benin (Benin City) over 350 computer systems with peripherals have been acquired and installed in various academic departments since 1999 at a cost of 28 million naira (about $200,000). In the same period, over one and half million naira ($21,000) was expanded to upgrade the university's e-email and Internet services and with the VSAT, in 2002, the university community now has unlimited access to the Internet (Anao, 2004). The University of Benin also expended 23.6 million naira (about $170,000) for LAN cabling of nearly all units in the University. This ICT project provides approximately 2,500 network nodes across the campus to provide Internet access to almost all offices, laboratory and lecture halls in the university. The university also set up a central records processing unit (CRPU) in 2001 at a cost of about 3.5 million naira (about $25,000) to handle the computerization of all students and staff records. "The university has also spent about three million naira (about $21,000) on the design, testing and application of both the students and personnel records software" (Anao, 2004, p. 192). The computerization of all the operations of the main library of the university, which commenced in 2002, cost 10 million naira (about $140,000) using a specially designed Oracle-based application. At the same time, the bursary section of the university was computerized with the sum of five million naira (about $36,000).

In 2001, the university in collaboration with Broad Technologies Limited established UB technologies, a joint-venture company with the university holding 49 percent equity and the partner holding 51 percent equity. UB Technologies provides Internet service primarily to the university community. It commenced operation with a VSAT facility and several cyber cafes have been established on campus. This project cost 13.43 million naira (about $100,000). With the massive investment in ICT facilities, the University of Benin now admit, register and provide students with their accommodation via the Internet. Little wonder that the University of Benin was recently rated second among the 945 universities in the 54 commonwealth countries for massively deploying ICT facilities for educational services to staff and students, and community relevant research. The university is the first in sub-Saharan Africa to be so rated, and it received presidential commendation for the high rating. The university scaled the high profile educational assessment through the deployment of ICT facilities for educational service. The University of Benin ICT facilities have enabled it to provide powerful tools and a new paradigm for the execution of the functions and roles of higher education in Nigeria. Other universities in the country which have acquired the requisite ICT facilities are also bracing up to the challenges of educating contemporary students who are using or should be using technology to play, shop, bank, conduct research and converse (Howard, 2006).

In Nigeria, there has been legislature calls to educate students with ICT resources to enable them cope with the demand of information society. Howard (2006) has also indicated the dilemma of "employers seeking university graduates who are conversant and adept with the current technologies, particularly computer-based ones" (p.198). Furthermore many employers require workers to have the skills necessary to collaborate, work with teams and share information across local and global networks and interact with others across cultural boundaries.

REVIEW OF LITERATURE

In order to know the attitude of lecturers toward ICT and to ascertain the influence of faculty affiliation and teaching experience on faculty members' use of the Internet in universities in southern Nigeria, it is pertinent to reflect on existing literature on the matter. When considering how ICT can be used to support higher education, some teachers are uncomfortable and a major factor that influences technology use is teachers' attitude toward ICT. Higher education teachers hold a variety of conceptions of teaching and these in turn influence their attitude and approach. It seems too often that teachers' attitudes create barriers that impede innovations in the use of ICT to enhance students' learning. Many teachers do not have positive attitudes toward the use of technology in education due to the fear that technology may replace or displace them in the classroom. Others are afraid of technology because they have not learned to use it, thereby feeling that technology is too fraught with technical difficulties and complications. Many teachers who have access to technology may not use it because they feel satisfied with their current approach or traditional approach to teaching.

However, most teachers are excited by information technology and are willing to integrate it into their classroom practices. In a study to ascertain the attitudes of teachers toward the use of technology in teaching, Falba, Strudler, and Boone (1999), confirmed that virtually all faculty believe that technology integration in teacher education is important. Strudler et al. (2003) reported that preservice teachers have positive perceptions of technology and its usefulness in educational settings. Knight, Knight, & Teghe (2006) documented that international research into the attitude and skills of educators indicates that they have difficulties embracing rapid changes that they face in utilizing ICT. Liu, Maddox and Johnson (2004) have ascertained that attitude variables such as degree of enjoyment in using technology, motivation to learn to use technology and importance of technology to students' learning are significantly related to the use of computer technologies by educators.

In an evaluation of the use of electronic learning environments at Erasmus University, in Rotterdam, Wieland (2005) asserts that majority of the teachers uses the electronic learning environment in 75-100 percent of their courses and that most lecturers spend an average of 0-2 hours per week on working with the electronic learning environment using personal computers at the University. Becker (2000) found that teachers in their first few years of teaching are the more likely than older teachers to consider ICT to be essential classroom tool. Naquin (2000) also found that teachers with four to five years of teaching experience were more likely to use ICT than their older counterparts. Findings by Vodanovich and Piotrowski (1999) indicated that academic psychologists have more positive attitude toward ICT for instruction, while Cohen (2000) found that the faculties of sciences, social sciences, and liberal arts, in that order, use computer mediated instruction often than others. Dennis and Espinoza (1997) however found that business and technology faculties were the most experienced with ICT while the faculty of education was the least experienced in ICT use. Abdelraheen and AL Musawi (2003) found that differences in ICT use in terms of college affiliation favoured the science faculty.

In an era when teachers who are experts in the subject areas and masters of their craft in the classroom find themselves in an uncomfortable situation to relearn how to teach in a new environment with little or no support, it is interesting to investigate faculty members attitude toward ICT and how faculty affiliation and teaching experience influenced ICT use in Nigerian universities. Since the use of the Internet for delivering instruction is growing in popularity teachers should "recognize that technology is part of the curriculum and should be viewed as one of the curriculum components" (Okojie & Olinzock, 2006, p.34). To be able to integrate technology into the curriculum, teachers need to develop positive attitude toward ICT and be able to learn how to utilize it for various learning activities. According to Okojie and Olinzock (2006), a teacher who has developed positive mind-set toward the use of technology is more likely to have favourable attitude to the use of technology for instruction, and also able to create ICT networks, guide and mentor students on how to use those networks.

The following research questions and hypothesis have been raised to guide the study:

1. What is the faculty members' attitude toward ICT?
2. Is there a difference in Internet usage in regard to faculty members' responses according to teaching experience?
3. Is there a difference in Internet usage among faculties?

HYPOTHESIS

Method of Study

There is no significant difference in ICT usage among faculty members with reference to teaching experience.

This study investigates teachers' attitude toward information and communication technology and the influence of faculty affiliation and teaching experience on Internet usage in Nigerian universities. The study, which utilized the survey research design, drew a sample of 520 academic staff from nine faculties from six first-generation universities in southern Nigeria. The sampling technique used was the proportionate stratified random sampling procedure. This ensures greater representativeness of the sample relative to the population. The six first-generation universities were used for the study to the exclusion of others mainly due to the fact that these universities are over thirty years old and have had most of their programmes fully accredited by the National Universities Commission (NUC), the supervisory body for universities in Nigeria. Also, these older universities have in recent years been involved in acquiring ICT facilities for teaching and learning.

Research Instruments

Two research instruments were used in the study. Items in the questionnaires were devised by the researchers after extensive review of literature. The first instrument measured faculty attitude toward ICT with Likert-type scale items which have four-point levels of strongly Agree (4), Agree (3), Disagree (2) and strongly disagree (1). The negative items were scored in reverse order. The second questionnaire also has a Likert-type scale items which were designed to elicit information on the frequency participants used the Internet for instructional purposes. The items in the questionnaire were given a 5-level Likert-scale options ranging from "Never" (level 1) to very frequently (level 5) to rate the frequency of Internet usage for various academic activities. The two questionnaires were tested for their reliability by using the Cronbach Alpha to ascertain their internal consistency. The reliability coefficients were 0.70 and 0.75 respectively for both questionnaires.

Though a total of 520 copies of each of the questionnaires were distributed to the selected sample, however, 476 copies of them were duly completed and returned.

Data Analysis

The data were collected and analyzed using percentages, mean, standard deviation, rank order and one way analysis of variance. Post hoc analysis was carried out where there were significant differences in Internet usage. Data were tested at the .05 level of significance.

RESULTS AND DISCUSSIONS

Attitude of Faculty Member Toward Computers

The attitude survey consisted of 18 items which were worded positively and negatively. Responses to the items are shown on Table 1. The attitude survey revealed that faculty members have positive attitude to computer as shown in the table.

Influence of Faculty Affiliation on ICT Usage

Data relating to the issue of the influence of faculty affiliation on Internet usage among faculty members were analyzed using mean, standard deviation and rank order. Table 2 shows the results.

The results in Table 2 showed that academic staff in the faculty of engineering were the foremost users of the Internet in the universities, as the faculty were ranked number one among the faculties. The faculty of science followed having ranked second while the faculty of arts came in third. This finding partially supports that of Cohen (2000) and Abdelraheem and AL Musawi (2003) who found that differences in ICT use in terms of college affiliation favoured the science faculty. The faculty of engineering which came first is closely related to the science faculty as both utilize scientific and laboratory-based practical approach to solving educational problems.

At the middle of the pack, the law, social science and medicine faculties were fourth, fifth and sixth in that order. One would have expected the faculty of medicine to be one of the foremost

Table 1. Faculty member's responses to attitude scale

	Statement	Mean	Std. Dev	Decision
1.	I enjoy using the computer	2.59	0.49	Agree
2.	I am tired of using the computer	1.59	0.49	Disagree
3.	Computers give me opportunity to learn many new things	3.60	0.49	Agree
4.	I feel very comfortable working with computer	3.56	0.50	Agree
5.	I will do less work with computer	1.71	0.79	Disagree
6.	I am not scared of computers	3.59	0.49	Agree
7.	Working with computers makes me uneasy	1.59	0.65	Disagree
8.	I can learn more from books and reference materials	3.60	0.49	Agree
9.	It is important for teachers to learn how to use the computer	3.79	0.41	Agree
10.	I enjoy browsing the World Wide Web	3.59	0.45	Agree
11.	The more often I use the computer the more I enjoy it	3.60	0.49	Agree
12.	Using a computer can be quite frustrating	1.53	0.65	Disagree
13.	I will attend workshops training programmes, conferences to enhance my computer usage	3.56	0.52	Agree
14.	Computer would stimulate creative endeavour in students	3.31	0.98	Agree
15.	Computer prevents the normal social interaction between the teacher and the students	1.64	0.86	Disagree
16.	Computers are indispensable tools in educational institutions	3.63	0.48.	Agree
17.	The use of e-mail helps make a course more interesting	3.19	0.85	Agree
18.	Computers are not easy to use	2.08	1.11	Disagree

Table 2. Mean, standard deviation, and rank order of influence of faculty affiliation on internet usage

Faculty	N	Mean	Standard Deviation	Rank
Agriculture	52	53.1154	10.0365	9th
Arts	38	59.1053	13.0959	3rd
Education	60	55.7000	10.2846	8th
Engineering	56	63.9464	7.2499	1st
Law	49	58.4286	6.2816	4th
Medicine	57	56.8246	15.9297	6th
Pharmacy	44	56.5909	11.0038	7th
Science	60	61.7000	11.0212	2nd
Social Science	60	57.1833	11.3921	5th

users of the Internet to gain access to current development in the field especially in this era of HIV/AIDS upsurge and the quest to develop a cure for it. But, when one takes a good look at the working conditions of Nigerian doctors, especially those in teaching hospitals, one would discover that they are very busy trying to teach and at the same time treat patients. This may be responsible for their placing in Internet usage.

The faculties of pharmacy, education, and agriculture occupy the last three positions respectively. The finding on the Faculty of education is supported by that of Espinoza (1997) who ascertained that the faculty of education was the least experienced in ICT usage.

However, a major finding in this study which should be of great interest is that each faculty has more than 50 percent of its members using ICT for instructional purposes. This is revealed in Table 3 showing percentage ICT usage.

Table 4. ANOVA for experience variable

Source	Sum of Squares	df	Mean Square	F	Sig
Between Groups	1000.96	2	500.458		
Within Groups	60364.244	473	127.620	3.921	.020
Total	61365.160	475			

Table 3. Showing percentage of internet usage among faculties

Faculty	% of use
Engineering	71.05
Science	68.56
Arts	65.67
Law	64.92
Social science	63.54
Medicine	63.14
Pharmacy	62.88
Education	61.89
Agriculture	59.02

If more than 50 percent of faculty members in Nigerian universities are using Internet for instructional purposes that should be a right step in the right direction. Considering the fact that the Internet is a recent phenomenon in Nigerian educational setting, the result obtained here showed that the sky is the limit for Internet usage in Nigerian universities.

Influence of Teaching Experience on Internet Usage

To investigate the research question on the influence of teaching experience on Internet usage, one-way analysis of variance (ANOVA) was used

Table 5. Multiple comparisons turkey HSD for teaching experience

(I) teaching experience	(J) teaching experience	Mean Difference (I-J)	Sig
Less than 5 years	Between 5 and 9 years	3.6094*	.024
	10 years and Above	3.3650*	.039
Between 5 and 9 years	Less than 5 years	-3.6094*	.024
	10 years and Above	.2444	.976
10 years and Above	Less than 5 years	-3.3650*	.039
	Between 5 and 9 years	.2444	.976

**The mean difference is significant at the .05 level.*

to analyze the data obtained. Table 4 shows the result.

From Table 4, it is clear that F calculated value of 3.921 is statistically significant at level 0.05. This shows that there are differences in Internet use based on teaching experience. To find the sources of differences, multiple comparison Turkey HSD was used. The findings are in Table 5.

Table 5 shows that the mean for the teachers with less than five years teaching experience is greater than the means for the teachers with five to nine years and 10 years and above teaching experience. This difference is large enough to be significant at the .05 level of confidence. This finding indicates that teachers with less than five years teaching experience use the Internet more than others. This in a way supports the findings by Becker 2000, and Naquin (2000) that teachers in their first few years of teaching are more likely to use ICT more that their older counterparts.

The reason for the findings here may not be far fetched. The Internet is a recent phenomenon in the educational system in Nigeria and as a result older faculty members may need more time to adjust their old pedagogical practice. According to Thomas (1987), teachers' unwillingness to use technological media is their relative satisfaction with current teaching techniques. Again, most of the faculty members with less than five years teaching experience are in the categories of graduate assistant (with good bachelor's degrees), assistant lecturers (with master's degree) and those who have just completed their doctorate degrees. For the graduate assistants and assistant lecturers, they are still training to obtain doctorate degrees in their various fields. This may make them hungrier for current information in their disciplines not only to update their knowledge, but also to satisfy supervisors demand for robust research procedures. In fact, it is a common practice for older faculty members to request their supervisees to browse the Internet for recent literature, especially in a situation where current books and journals are hardly available in most Nigerian universities' libraries.

CONCLUSION AND RECOMMENDATIONS

As access to higher education widens and the traditional patterns of study become less attractive because of their numerous shortcomings, ICT is likely to figure increasingly as a delivery mode in higher education in Nigeria. Information and communication technology (ICT) has considerable benefits than the traditional methods where there are large students to be taught. With the increase in students enrolment in universities in Nigeria, and indeed elsewhere, ICT is likely to play more significant role in instructional delivery, especially now that many Nigerian universities are equipping departments and faculties with ICT facilities and linking them with the World Wide Web.

This research is one of the pioneering research efforts on Internet in Nigeria and it will add to the few existing literature in the field in the country. As an emerging educational phenomenon, ICT use will continue to become popular not only amongst faculty members but also university students. It is noteworthy to observe than in spite of its recent introduction to tertiary education in Nigeria; ICT is fast gaining ground in the universities as this study has revealed that over 50 percent of faculty members in each faculty are already using the Internet for instructional purposes. Though, the result obtained in this study revealed that the engineering, science and arts faculties are in the forefront in Internet usage, the rest faculties are not doing badly too.

The study also revealed that faculty members with less than five years teaching experience use the Internet more than others. There may be several reasons for thus. The older faculty members may not be using them because they are afraid of the new phenomenon or are reluctant to change classical teaching methods as they may not see much advantage in using new technology (Sorgo, 2003).

However, older faculty members and new faculty members alike need to learn how to integrate ICT into the curriculum. "When ICT is pedagogically integrated into the course design and adapted for current environment, it can enable and support enhance forms' of learning" (Kirkwood & Price, 2006, p. 8). Changes brought about by ICT in the academic environment require educational and training institutions to operate in different ways. ICT integration in the curriculum requires concerted efforts and continuous teamwork among faculty members and through individual and collective performance institutional transformation which allows participants to explore technologies unhindered can be engendered.

In order for the older faculty members to use ICT more in the teaching activities, there is need for universities in Nigerian to provide adequate professional development activities to faculty members. Kirkwood and Price (2006) recognize that the main problem with current professional development in relation to ICT use is the concentration of effort on the technological aspects. As a result of this, they ascertained that "professional development for staff has largely focused on developing teacher's technical skills" (Kirkwood & Price, 2006, p. 9). They stressed that this has led to the tendency of using ICT tactically, mainly to support existing teaching practice rather than having any transformative influence in response to the changing environment. For ICT to have any strong technological influence, teachers should be given opportunities to reflect on existing practices and their own teaching beliefs. Because faculty members hold a variety of conceptions of teaching and these in turn shape their approach, they should be made to understand that professional development is the scholarship of teaching, and therefore, holds the key to effective use of technologies. For professional development policies and practices to be successful, Kirkwood and Price (2006) suggest two ways to accomplish this:

First, professional development activities need to go beyond the individual and adopt a holistic approach to development that includes departmental, faculty, and senior university managers. Second, the focus needs to be on underlying pedagogical theories and practices and their effects in particular educational contexts. (11).

In line with the findings of this study, the researchers recommend:

1. That universities in Nigeria should invest more money in procuring ICT facilities for both teachers' and students' usage,
2. The universities should conduct workshops and seminars on the use of computers, software, and Internet specifically addressing the needs of the education, agriculture and pharmacy faculties and also the needs of the older faculty members. This will assist to allay the fears of faculty members who may be reluctant to use new technologies,
3. The older faculty members seem to use technology less than the new ones. This finding is no surprise as researches have often revealed that most teachers, especially those with long years of teaching experience, are intimidated by technology and are comfortable with their own teaching styles. Professional development of teachers should be at the forefront of any successful integration of technology in the school curriculum. University teachers in Nigeria need formal training and sustained and constant support from colleagues with technology know how to help others learn how best to integrate technology into teaching,
4. Faculty members in universities in Nigeria need to go beyond the use of the Internet to access resources send e-mail to friends or download course material as is the current practice. They need to transform from information consumers into information

producers and share resources with peers in other countries. The exchange of ideas and good practices with other teachers in their subject areas helps in many ways to master the use of technology, and

5. Since the new learning provided by information and communication technology is now quite different from what majority of faculty is familiar with, teachers have to cope with some uncertainties. Technology as a tool is now being used to transform the classroom into interactive and inquisitive learning environment. Faculty members need to be able to transform the learning environment from a static one-way flow of information from the teacher to an academic arena which is dynamic and where student-centred learning activities are promoted. Since the new learning environment provided by ICT is different from the old order, teachers will need competences and skills like creativity, flexibility, logistic skills, administrative and organizational skills and collaborative skills.

The Future

There is no doubt that the use of ICT in teaching and learning is now a new policy issue in most of the universities in Nigeria. The future looks bright. However, a major stumbling block that might limit technology integration into the curriculum is the high cost of ICT facilities. Though funding of the federal universities has improved considerably in the past four to five years, money available for technology integration may not be sufficient to meet the need of staff and students. Again, most state universities established by state governments were founded for political reasons but the funding in that sector leaves much to be desired. There is a technological literacy challenge for universities in Nigeria. Such challenge requires that all students and their teachers should be technologically literate in order to foster an enriched and more flexible learning environment. Stakeholders in university education need to develop positive orientation to technology integration to the future. Faculty members can and will embrace ICT if they are provided with the requisite professional development and the necessary support.

REFERENCES

Abdelraheem, A., & Al musawi, A.S. (2003). Instructional uses of internet services by Sultan Qaboo's University faculty members. *International Journal of Instruction Media, 30*(1), 45-59.

Aduwa-Ogiegbaen, S.E., & Iyamu, E.O.S. (2005). Using information and communication technology in secondary schools in Nigeria: Problems and prospects. *Educational Technology and Society, 8*(1), 104-112.

Anao, A.R. (2004). *Stewardship report of professor A.R. Anao, Vice-chancellor; University of Benin: February 10, 1999 — February 9, 2004.* Benin City: Ambik press, Ltd.

Andrews, P. (1997). Secondary teacher's perception of computer availability: A qualification study. *Journal of Information technology for Teacher Education, 6*(3), 321-335.

Association of African Universities. (2000). Technical experts meeting on the use and application of information and communication technologies in higher education in institutions in Africa. Retrieved January 22, 2007, from http://www.aau.org/english./documents.aau-ktreport-p2.htm

Bailey, E., & Cotler, M. (1994). Teaching via internet. *Communication Education, 43*(2), 184-193.

Becker, H. (2000). Internet use by teachers. Retrieved February 25, 2005, from http://www.crito.uci.edu/TLC/findings/internet-use/stArt-spage.htm

Cohen, J. (1995). *Computer mediated communication and publication productivity among faculty association of Jesuit colleges and universities (AJCU) Institution.* PhD Dissertation, SUNY, Buffalo.

Darkwa, O., & Mazibiko, F. (2000). Creating virtual learning communities in Africa: challenges and prospects. *First Monday*, 5(5).

Dennis, G., & Espinosa, S. (1997). Models of internet use: Approaches from various disciplines. Retrieved February 14, 2005, from http:/www.coe.uh.edu/insite/elec.pub/HTML1997/tg.denn.htm.

Falba, C.J., Strudler, N.B., & Boone, R. (1999). College of education faculty use of technology: A snapshot. In Price, J., Willis J., Willis, D., Jost M. & Boger-Mehall, S. (Eds.) *Proceedings of Society for Information Technology and Teacher Education 10th Annual Conference.* Charlottesville, VA: Association for Advancement of Computing in Education.

Finley, L., & Hartman, D. (2004). Institutional charge and resistance. Teacher preparatory faculty and technology integration. *Journal of Technology and Teacher Education, 12*(3), 319-337.

Fisher, C., Dwyer, D.C., & Yocam, K. (Eds.). (1996). *Educational and technology: reflections on computing in classrooms.* San Francisco. Apple Computer, Inc., Jossey-Bass Publishers.

Guardian Newspapers Nigerian Limited. (2002). Nigeria's floppy skill development scheme worries manufacturers, July 23.

Guardian Newspaper (2003) Tuesday, October 12 2003 p. 39.

Guardian Newspapers (2003) Tuesday, December 2003, page 45.

Guardian Newspapers (2005) Tuesday, January 18 p. 41.

Haddad, W.D. (2003). Is Instructional technology a must for learning? *Techknowlogia.* Jan-March, Oakton: Knowledge Enterprise Inc.

Hoffman, B. (1996). What drives successful technology planning? *Journal of Information Technology for Teacher Education, 5*(1&2), 43-55.

Howard, W.G. (2006). Socrates and technology: A new millennium conversation. *International Journal of Instructional Media, 33*(2), 197-204.

Johari, A. (2006). Creating a knowledge-based society through e-learning in Korea. *Educational Technology Research and Development, 54*(5), 529-540.

Kent, A. (1996). Technology: Moving faster. *Australian Accountant, 66*(1), 26.

Kirkwood, A., & Price, L. (2006). Adaptation for changing environment: Developing learning and teaching with information and communication technologies. Retrieved May 10, 2006, from http://www.irrodl.org/index.php/irrodl/article/viewArticle/294/624

Knight, C., Knight, B.A., & Teghe, D. (2006). Releasing the pedagogical power of information and communication technology for learners: A case study. Retrieved February 5, 2007, from http://ijedick.dec.uwi.edu/viewarticle/php?id=16T&layout=html

Lelliot, A. (2000). Online education in africa: Promises and pitfalls. Retrieved January 22, 2007, from http://generalupdate.rau.ac.za/infosci/conf/Wednesday/lelliot.htm

Liu, L., Maddux, C., &Johnson, L. (2004). Computer attitude and achievement; is time an intermediate variable? Retrieved February 5, 2007 from http://www.accessmylibrary.com/coms2/summary_0286-18279752_ITM

Naquin, D. (2000). The integration of technology at Loudon Campus, Northern Virginal community college; A faculty survey. *Inquiry, 5*(1), 58-65.

Okebukola, P. (1997). Old, new and current technologies in education. *UNESCO Africa, 14*(15), 7-18.

Okojie, M., & Olinzock, A. (2006). Developing a positive mind-set toward the use of technology for classroom instruction. *International Journal of Instructional Media, 33*(1), 33-41.

Picciano, A. G. (2006). *Educational leadership and planning for technology.* NJ: Pearson Prentice Hall.

Sandholtz, J.H., Ringstaff, C., & Dwyer, D.C. (1997). *Teaching with technology: Creating student-centred classrooms.* New York: Teachers College Press.

Schofield, J.W. (1995). *Computers and classroom culture.* Cambridge NY: Cambridge University Press.

Sept, J. (2004). The stone age in information age. In W.E. Becker & M.L. Andrews (Eds.), *The scholarship of teaching and learning in higher education,* (pp. 47-80). Bloomington, IN: Indiana University Press.

Smaldino, S.E., Russell, J. D., Heinich, R., & Molenda, M. (2005). *Instructional technology and media for learning.* NJ: Merrill Prentice Hall.

Sorgo, A. (2003). Searching for information on the internet—What if your students cannot speak English? *International Journal of Instructional Media, 30*(3), 315-319.

Starr, P. (1996). Computing our way to educational reform. *The American Prospect,* 27 (July-August), 50-60.

Stonier, T., & Conlin, C. (1985). *The Cs: Children computers and communication.* London: Wiley.

Strudler, N., Archambault, L., Bendixen, L., Anderson D., & Weiss, R. (2003). Project THREAD: Technology helping restructure educational areas and delivery. *Educational Technology Research and Development, 51*(1), 41-56.

Teale, P. (2003). ICT implementation; what makes the difference? *British Journal of Educational technology, 34*(5), 567-583.

Thomas, R.M. (1985). Educational radio and television. In R.M. Thomas and V.N. Kobayashi (Eds), *Educational technology—its creation, development and gross cultural transfer* (pp. 105-123). Oxford: pergamon press.

Vodanovich, S., & Piotrowski, C. (1999). Views of academic 1-0 psychologists toward internet-based instruction. Retrieved from http:/www.siop.org/tip/backissues/TIP July 99/7 vodanovich.htm

Whitaker, B.E. (2001). Association of African Universities charts goals for 21st century. *International Higher Education, 23(spring 2001).*

Wieland, A. (2005). Evaluation of the use of electronic learning environments: Perceptions of lecturers and students. *Recent research developments in learning technologies.* Rotterdam: FORMATEX.

Chapter XVIII
Using Indices of Student Satisfaction to Assess an MIS Program

Earl Chrysler
Black Hills State University, USA

Stuart Van Auken
Florida Gulf Coast University, USA

ABSTRACT

The purpose of this chapter is to demonstrate a methodology by which management information systems (MIS) alumni evaluate the content of courses and their satisfaction with an entire MIS program. The approach can be used to assess the relevancy of an MIS curriculum. By way of clarification, an MIS program prepares its graduates to be effective in the tasks necessary to design, program, and implement systems that will provide management with timely, accurate and useful information for decision making. This is in contrast to computer science (CS) programs that prepare their graduates to be knowledgeable in the technical aspects of computer hardware and operating systems software. This study first determines if there are any differences in the evaluations of the content of required MIS courses by alumni based upon whether the graduate was using their first year on the job or one's current position as a frame of reference. Next, a factor analysis is performed, using the scores earned by specific courses, to reduce the content value of specific courses into specific factors, thus simplifying understanding of the type of learning that is taking place. A factor analysis is performed both for course content scores during one's first year on the job and, again, in one's current position. Using a global measure of satisfaction with the entire MIS program, the course content factor scores are then regressed against a student's satisfaction with the entire MIS program. This regression analysis is performed, once again, for both one's first year on the job and in one's current position. The implications for evaluating the effectiveness of an MIS curriculum are presented and discussed.

INTRODUCTION

Over the past few years there has been increasing pressure for higher education to demonstrate that it is effective in delivering persons who have the required skills and knowledge of MIS. The Association to Advance Collegiate Schools of Business-International has also incorporated this philosophy into its accreditation procedures. Instead of measuring "inputs" in terms of the qualifications and scholarly performance of faculty, it is requiring schools of business to provide evidence that graduates are being provided with the experiences necessary to develop the skills and knowledge promised in the educational objectives that flowed from a school's conceptual objective or mission statement.

There are several methods of obtaining opinions regarding what an MIS curriculum should contain. Cougar et al., (1995) offered a set of 'guidelines' for an undergraduate curriculum. Lee, Trauth, and Farwell (1995) published their findings of an investigation that involved both academicians and industry representatives. Nelson (1991) published the perceptions of information systems (IS) and end-user personnel. Another method of determining whether students have been given the knowledge, skills and experiences that firms' value would be to survey firms recruiting graduates of an IS program. This is one approach suggested by Van Auken (1991). The survey form could inquire as to how well graduates are meeting the expectations or needs of the employer. This approach also has shortcomings. An employer may not wish to alienate an IS area by being critical of its graduates. Also, the authors have observed that on some occasions employers are concerned with relatively short-term goals. They wish to hire graduates who are productive from the first day on the job, that is, they were trained rather than educated, and are less concerned with the contributions that may be made later in the employee's tenure. In a related study, Banerjee and Lin (2006) surveyed information technology practitioners in an attempt to determine the skills graduates should possess in order to be effective systems analysts. Similarly, Harris, Lang, Oates et al., (2006) evaluate and recommend alternative approaches to the material related to the performance of the tasks associated with the position of systems analyst. These approaches, however, focused on only one type of position, rather that what skills and knowledge one should possess upon entering the MIS profession and how the required skills and knowledge change over time.

Also, if the person is evaluated some time after joining the organization, the assessment may be based upon factors other than job skills. These factors may be loyalty, interpersonal skills, political astuteness and being a team player, rather than the ability to focus their IS education on specific organizational problems.

An interesting approach was that used by Liu, Liu, Lu et al., (2003). These researchers examined job descriptions of positions firms were attempting to fill. By examining these position descriptions they were able to discern that information technology skills in demand were subtly changing over time and recommended that those responsible for designing curricula should assure their programs were current, perhaps by using advertised position description requirements as guidelines.

In order to continually attempt to keep their programs relevant, many IT/MIS/CIS faculty revise their curricula as new hardware, software and methodologies appear in industry. Based upon their survey of curricula, published in 2006, Kung, Yang and Zhang (2006) concluded that significant changes had occurred in the ten years since the curriculum findings of Maier and Gambill (1996) were made known.

The dynamic nature of the requirements for positions in information technology was also noted in the study by Petrova and Claxton (2005). They comment that, "the issue of developing student skills and capabilities adequate to the demands of the (information technology) industry becomes a moving target." It is worthy of note, however, that CIS/MIS/IT programs that focus on concepts, techniques and methodologies, rather than in-depth skill development such as advanced programming, develop a graduate who has greater flexibility and is less open to obsolescence.

Still another method of assessing the effectiveness of an educational program is to have the program evaluated by the graduate, another approach mentioned by Van Auken (1991). The graduate is the primary "consumer" of the educational experience, having invested the equivalent of four years of one's life and a considerable sum of money in his or her program of choice. It is the graduate, it seems, that knows best if he or she was appropriately prepared for the functional area chosen as a career field. Another approach is that used by Gasen et al. (1992) which asked students to evaluate various aspects of the IS program. Gasen's study, however, was primarily focused on the satisfaction students perceived with the entire IS program and did not address the content value of individual courses. As a consequence, the results of administering this survey form to graduates would not be very useful in determining whether specific courses should be continued. Finally, a questionnaire proposed by Hanchey (1995) asked graduates to indicate the value of specific IS topics in addition to the quality of preparation in several areas such as analytical skills, values and ethics, and so forth. Once again, the graduate is not able to evaluate specific course content as to applicability.

PURPOSE OF THE STUDY

The purpose of this chapter is to present and discuss the findings of a survey of graduates of an MIS program offered by an AACSB-I accredited business school on the west coast regarding the value of the content of the required courses in the program both during their first year on the job and in their current position in the MIS field. The chapter builds on the earlier work of Chrysler and Van Auken (2002) both conceptually and analytically and serves as a prototype as to what is possible. Specifically, the objectives of this research were to determine:

- The comparative perceived content value of each of a set of required courses in an MIS program during the first year on the job;
- The comparative perceived content value of each of a set of required courses in an MIS program in one's current position;
- Whether a factor analysis of course content evaluations by alumni using different time frame orientations can identify factors related to the type of learning taking place;
- The contribution toward satisfaction with an entire MIS program by specific course content factors; and,
- Whether the contribution toward satisfaction with an entire MIS program by specific course content factors differs depending upon the point in time that is being used by the alumni as a frame of reference.

METHODOLOGY

A primary concern is the questioning to be asked. Directly related to that is what type of scoring technique to use. The questions must be phrased such that the content of the course being evaluated by the person is viewed without regard to the differing presentation methods of faculty or, indeed, the personality of the faculty member offering the course. This issue should be addressed in the material that introduces the person to a series of courses to be evaluated. By repeatedly focusing the person's attention on the course content, a psychological set would be in place. This should result in the person performing the evaluation by reviewing the course from the frame of reference of the extent to which the course content has been of value in the person's performance.

The scoring issue is always of importance in this type of research. The use of a continuum with numeric values placed at equal intervals on the continuum does not insure ratio scale, or even equal interval, data. Nonetheless, this type of scaling has been used in much research and it appears to be the best available method in this type of situation.

In using this type of scoring, an odd number of values may reveal a tendency for the values selected to cluster about the middle value. The implication, therefore, is that by using an even number of values a person selecting a score is forced to lean toward one direction or the other direction, which may result in subtle differences being identified.

The words used at each end of the continuum, once more, should be selected so they reinforce the fact that what is of importance is the value of the course content. An example of how a course to be evaluated would appear on a questionnaire is shown in Figure 1.

A significant problem is not only what questions need to be asked, but also the framing of the questions regarding the value of the components of a curriculum at a specific time in one's employ and the type of position in which one finds oneself.

It is desirable not only to obtain an objective assessment of the topics of which each course consists, but to link the value of the course topics to a specific environment in which the applicability of the course topics are being evaluated. In the situation of those graduating with a degree in IS, they may find themselves hired to work exclusively in the IS field. However, in some instances one with a degree in information systems is employed in an area within an organization that requires the person to be only performing IS tasks part of the time or, indeed, the person may be selected because of the IS knowledge possessed, but not perform IS tasks at all, such as a computer systems auditor. The responses of only those alumni who indicated their current position was fully in the IS area are used in this study.

Figure 1. An exemplar of the measurement scale utilized to measure course content value

Cobol Language Programming							
Extremely Weak							Extremely Strong
Content Value	1	2	3	4	5	6	Content Value

Since the university did not have the ability to provide a list of all students who had completed the MIS major, different versions of a questionnaire that addressed the issues discussed were sent to all persons who had completed several MIS courses and the graduates were instructed as to which version of the questionnaire to complete based upon whether the student had been an MIS major, MIS minor, or neither. Many students completed MIS courses since some students selected MIS courses as business electives and some students selected a minor in management information systems. However, at the time this study was conducted the MIS program had not graduated a large number of students. When the questionnaires were returned only those who had completed the MIS major version were used.

FINDINGS

Required Course Content Evaluations

The required courses being compared, with brief course descriptions, are found in the appendix. The program of study included the basic, core IS courses. Such courses as network management or Web management or languages such as Java were not required at the time this study was conducted. They were, however, available as electives since they were offered by the computer science department in the college of engineering, computer science and technology. Aside from these required core courses students were required to select a 'pattern' of nine units in the accounting, production/operation or software area. The software pattern contained courses such as C++ programming, visual basic programming, Java language programming, and so forth, again available from the computer science department of the college of engineering, computer science and technology.

First Year on the Job

The mean, relative rank and number of respondents for each of the required courses are shown in Table 1.

As can be seen, the group of courses ranked one through five have significantly higher average scores that the remaining two courses. The accounting information systems course was a junior-level core course designed to provide the MIS student with in-depth knowledge of the design and application of accounting systems, whereas the information center administration was designed to be the management-oriented, capstone course for the MIS program. By way of contrast, the system development practicum course was designed to be the applications-oriented capstone course for the MIS program.

In One's Current Position

The next phase of the study involves analyzing the responses of the same graduates when they evaluated the same set of required courses, but now were required to use their current position as the frame of reference. The mean, relative rank and number of respondents for each of the required courses are shown in Table 2. As can be seen, the courses ranked one through four have average scores significantly above the remaining three courses. The rankings for the courses, along with their average scores have, however, changed significantly due to the change in the frame of reference used by the graduate. The elapsed time since graduation for the alumni ranged from 33 months to 117 months. While one might believe that the time since graduation would have an effect on one's perceived value of the content of a course, there was no significant difference between the length of time since graduation of the alumni and the change in the perceived value of any of the courses. An apparent explanation of this finding is that when one changes his or her position, regardless of the length of time in one's first position, the requirements of the new position cause one to realize which skills and knowledge are now most valuable and relate that change when being asked to evaluate the content of various courses.

To determine the change in the perceived value of a course based upon one's frame of reference, the average score for a course in one's current position was subtracted from the average score during one's first year on the job. The implica-

Table 1. Mean scores for required courses during first year on the job

Course Title	Mean	Rank	N
COBOL Language Programming	4.56	4	30
Accounting Information Systems	3.20	7	30
Structured Systems Analysis	4.67	3	31
Software Project Management	4.89	2	31
Data Base Concepts	4.92	1	31
Information Center Administration	3.65	6	30
Systems Development Practicum	4.47	5	31

Table 2. Mean scores for required courses in one's current position

Course Title	Mean	Rank	N
COBOL Language Programming	3.13	6	30
Accounting Information Systems	3.09	7	30
Structured Systems Analysis	4.89	3	31
Software Project Management	5.37	1	31
Data Base Concepts	5.22	2	31
Information Center Administration	3.74	5	30
Systems Development Practicum	4.85	4	31

tions of the signs of the resulting values are as follows.

If the sign of the difference is positive, the implication is that the value of the course was greater in one's first year on the job than in one's current position.

If the sign of the difference is negative, the implication is that the value of the course is greater in one's current position than it was in one's first year on the job.

If the difference is zero, the implication is that the value of the course in one's current position is approximately the same as it was in one's first year on the job. If both the score for the value during one's first year on the job and in one's current position are high, this implies a course whose content is very valuable on a continuing basis. If both the score for the value during one's first year on the job and in one's current position are low, this implies a course whose content value is questionable and should lead to a critical review by the faculty.

The direction and magnitude of the difference values, shown in Table 3, indicates that the COBOL language programming course and the accounting information systems class appear to have more value in one's entry-level position, whereas the content of the remaining courses has from minimal to significantly more value later in one's career.

Table 3. Differences in mean scores for required courses: First year on the job minus in one's current position

Course Title	Difference	N
COBOL Language Programming	1.43	30
Accounting Information Systems	.11	30
Structured Systems Analysis	-.22	31
Software Project Management	-.48	31
Data Base Concepts	-.30	31
Information Center Administration	-.09	30
Systems Development Practicum	-.38	31

Inter-Relationships between the Changes in Value of the Required MIS Courses

In order to determine what relationships, if any, existed between the changes in perceived value of the required MIS courses, a factor analysis was performed to compute the correlation coefficients between the perceived changes in value of the individual courses to determine how the changes in the value scores formed clusters. The factor analysis resulted in two factors being identified. The two factors explained 64.6 percent of the variance in the differences between the average scores for the required courses.

The factor loadings are as shown in Table 4. The bold values indicate the courses that are the members of a given factor solution. The two factors are consistent with the earlier findings regarding entry-level versus continuing or long-term career value of the content of the courses.

Curriculum Implications of the Direction and Magnitude of the Changes Found in the Perceived Value of the Courses

When faculty design an MIS curriculum they typically attempt to provide students with an education that consists of (1) entry-level skills that

Table 4. Factor loadings of courses

Course Title	Factor 1	Factor 2
COBOL Language Programming	.216	.817
Accounting Information Systems	.006	.808
Structured Systems Analysis	.822	.130
Software Project Management	.754	.138
Data Base Concepts	.622	.283
Information Center Administration	.809	.230
Systems Development Practicum	.836	- .003

assure the graduate will be a productive employee in a minimum amount of time, (2) knowledge of concepts, methodologies and techniques that will be applicable as the graduate moves from an entry-level position to various positions of increasing responsibility and (3) knowledge of the application of managerial theories that will be of value as the graduate moves into beginning then higher supervisory level positions. If the curriculum were achieving the faculty's objectives, one would expect to observe that courses that contain topics that relate primarily to a position that has entry level responsibilities would be perceived as having a high value in one's first year on the job, but be evaluated as having less value in one's later positions. Similarly, one would expect to see some courses that contain topics that are applicable not only to the responsibilities of one in an entry level position, but continue to have applicability as one progresses in one's MIS career. Such courses would be perceived as having a medium to high value in one's initial position and a continued medium to high value in one's current, non-entry level position. Management-oriented courses could be expected to have a relatively low perceived value in one's entry level position, but if one has been in the field long enough to have achieved a role with significant managerial responsibilities, the content of these types of courses could be expected to be assigned a medium to high value at this point in one's career.

Analysis of Course Score Differences in Direction and Magnitude

The large positive difference between the average value score for COBOL language programming indicates the course had a very high value during one's first year on the job, but was of significantly less value as one moved upward in one's MIS career. This course, then, apparently is meeting the goal of providing a graduate with strong value in an entry-level position.

The very small but positive difference between the average value score for accounting information systems, added to the fact that both the value in one's first year on the job and in one's current position were low, indicates the value of this course as required for the MIS major should be questioned or the course content should be modified.

The small but negative difference between the average value for structured systems analysis, added to the consideration that that both the value in one's first year on the job and in one's current position were high, indicates this course appears to be meeting the objective of the faculty in terms of having a continuing high professional value. As a consequence, being required for the MIS major seems quite justifiable.

The fact that software project management initially had a high average score and when viewed in one's current position has an even higher value indicates that this course also is meeting the goals of the faculty and there is ample justification for its required presence in the MIS curriculum.

Similarly, the data base concepts course had both a high initial value score and an even higher value in one's current position, thus indicating a desirable required course for the MIS curriculum.

The information center administration course had both an initial value score and a current value score only slightly higher that the accounting information systems course and one would be inclined, therefore, to question its continued presence as a required course in the MIS curriculum. It is worthy of note, however, that this course focused on the skills and knowledge requirements for managing an entire MIS organization. It is highly unlikely, therefore, that any of the alumni surveyed had an opportunity to utilize the knowledge gained in this course and thus realize the value of the course content.

The systems development practicum evidenced a difference between the mean scores greater than that of the data base concepts course but less than that of the software project management course,

Figure 2. Exemplars of MIS program satisfaction measures

Good Experience	1	2	3	4	5	6	Bad Experience
Good Use of My Time	1	2	3	4	5	6	Bad Use of My Time
Valuable	1	2	3	4	5	6	Valueless
Useful	1	2	3	4	5	6	Useless

with both average scores approaching that of the structured systems analysis course. This implies a high value in one's first year on the job and an even higher value as one moves to positions of more responsibility in the MIS field. Once again, a continuing high evaluation of the course by those moving forward in their MIS career implies the course deserves to be required for those preparing for a career in MIS.

Index of Satisfaction

The MIS program satisfaction variables were measured using six-point semantic differential scales. These appear in Figure 2.

These global measures have been successfully used in attitudinal assessments in the marketing field and more recently they have surfaced in MIS outcome assessments.

FACTOR ANALYSIS

Given a review of assessment literature, one finds rather limited usage of factor analysis. One study has employed it in assessing teaching methods that influence MBA retention (Bailey, 1997), yet no previous applications have been performed with respect to course content value.

Factor analysis is one of the more popular multivariate research techniques that helps reveal the underlying structure of a data matrix. It is particularly useful in analyzing the structural relationships among a large number of variables, for example, those found in outcomes assessment measurements. Two of the main uses of factor analysis are in data reduction and summarization, both of which are germane in this research.

In an effort to explore this data, a principal components factor analysis was employed. This reveals the best linear combination of variables in the sense that the variable combination accounts for more variance in the data than any other linear combination of variables. The first factor extraction is therefore the best summary of the linear relationships exhibited in the data, while the second factor is the second best linear combination, yet orthogonal to the first. The latter is important in analyses of data with a criterion variable and predictors that might be multi-collinear, that is, strongly inter-correlated. By reducing the data to orthogonal factors, the multi-collinearity problem is substantially reduced.

Course Content Value: First Year on the Job

The application of the principal components factor analysis to the six MIS courses that were evaluated as to course content value for one's first year in an MIS position revealed the presence of two varimax-rotated factors. The varimax rotation was followed because it produces factors with both high and low loadings, thus simplifying factor interpretation. These factors possessed eigen values greater than 1.0, which is a common cutoff criterion for the number of factors interpretations. These two factors explained 66.6 percent of the variance in the data. The clustering of the factors is shown in Table 5.

Table 5. Factor analysis results of assessing the Varimax-rotated content value of six MIS courses to one's first MIS position

Courses	Factor 1	Factor 2
COBOL Language Programming	.61	−.26
Structured Systems Analysis	.87	.30
Software Project Management	.75	.31
Database Concepts	.63	.44
Information Center Administration	.02	.89
Systems Development Practicum	.27	.78
Eigenvalues	2.86	1.14

Table 6. Factor analysis results of assessing varimax-rotated content value of six MIS courses to one's current MIS position

Courses	Factor 1	Factor 2
COBOL Language Programming	.12	.60
Structured Systems Analysis	.70	.53
Software Project Management	.84	.40
Database Concepts	.92	.06
Information Center Administration	.03	.84
Systems Development Practicum	.39	.71
Eigenvalues	3.13	1.08

As can be observed, the variables with loadings of .7 or higher have been boxed as a guide to factor interpretation. The .7 criterion was used based on the study's sample size, which was less than 100.

An assessment of the latent dimension holding the courses in Factor 1 in common again evidences the presence of continuing career value as all courses are other than time dependent. In essence, they continue to provide utility or value over one's career. With respect to the second factor, the "glue" that is holding the courses together again denotes an overall managerial perspective. Basically, a foundational grounding (Factor 1) is complemented by a "managerial focus" (Factor 2). This latter factor is also marginally supported by the COBOL programming language course, which included topics in programming management, programming standards and test documentation that may be viewed as skills that are integral to the managerial orientation this program possessed. The revelation of the two factor constructs is significant because they may be related to program satisfaction to determine which one has greater explanatory power, if any.

Course Content Value: In One's Current Position

A continuation of the principal components factor analysis as applied to course content value to one's current MIS position revealed the presence of two varimax-rotated factors. These factors explained 70.1 percent of the variance in data. The revealed factor loadings appear in Table 6.

As can be seen, the factor structures are similar between the two scenarios, yet vary in practical significance with a juxtaposition seen in the COBOL language programming course. Such a shift helps to confirm the usefulness of data captured from different job perspectives. Commonly, only a single-perspective is used and valuable insights can be lost.

Overall Satisfaction Variable

To establish this variable, all of the exemplar indicants of the satisfaction variables that appear in Figure 2 were selected for factor analysis scrutiny. The results reveal the extraction of a single factor that explains 79.8 percent of the variance in the data. The factor loadings appear in Table 7.

Multiple Regression

With respect to outcomes assessment, the issue of variability is a critical one. For example, if all alumni were highly satisfied with a program and their responses to all satisfaction measures denoted a value of six on six-point semantic differential scales, then there would be no variation to explain. Conversely, if all courses had the same

Table 7. Factor analysis results for overall satisfaction variables

Variables	Factor 1
Bad Experience—Good Experience	.85
Bad Use of My Time—Good Use of My Time	.95
Valueless—Valuable	.89
Useless—Useful	.88
Eigenvalue	3.19

course content value, there would be no variation and nothing to explain. However, as observed in the tables showing the evaluation of the content of the required courses, these data do evidence variation, along with overall program satisfaction, thus permitting assessments of independent variable explanatory power along with relative independent variable strength in explaining the observed variation. Since the factor analysis of the semantic differentials yielded a single factor, each respondent's factor score is to be used as an overall satisfaction index.

First Year on the Job

The relationship of the two course content factors, using each respondent's factor scores, to the overall program satisfaction factor score for each respondent produced an R^2 of .212, thus 21.2 percent of the variation in program satisfaction is explained by the two factors. Attesting to the strength of the relationship is a statistically significant F value of 7.66 (2, 57), p = .00. Overall, the two course content factors are explaining MIS program satisfaction. Additional insights into the regression results appear in Table 8.

As can be observed from the presented beta coefficients, both course content factors vary directly (versus inversely) with program "satisfaction." Their relative contribution to the criterion measure is seen in their beta weights. The results reveal that the factors have similar explanatory power. Also, the tolerance values for each course content factor are equal to unity. They explain the amount of variance in an independent variable that is not explained by the other independent variable. Since the two-course content factors are orthogonal (i.e., uncorrelated) and the tolerance values are equal to unity, there is no evidence of multi-collinearity. Basically, both course content factors are about equal in their impact upon satisfaction. Also, the shift from the first job scenario to one's current MIS position may be revealing.

In One's Current Position

The relationship of the two factors (i.e., "knowledge" and "managerial") to overall program satisfaction evidenced an R^2 value of .251. Thus, 25.1 percent of the satisfaction factor's variance is explained by the two factors. Attesting to the relationship is an F value of 9.57 (2, 57), p = .00. While one may question the efficacy of only about 25 percent of the variance being explained by the two course content factors, this is still an important

Table 8. Results of regressing the two course content factors against overall MIS program satisfaction: First year on the job

Variable	Beta Coefficient	Beta Weight*	Tolerance	t-value	P
Foundation Theory and Knowledge	.34	.34	1.00	2.89	.01
Managerial Focus	.31	.31	1.00	2.63	.01

**Beta coefficients and beta weights are equal because the factor score output was standardized to a mean of zero and unit variance.*

Table 9. Results of regressing the two course content value factors against overall MIS program satisfaction: Current position

Variable	Beta Coefficient	Beta Weight	Tolerance	t-value	P
Foundation Theory and Knowledge	.28	.28	1.00	2.44	.02
Managerial Focus	.42	.42	1.00	3.63	.00

finding as other considerations also explain program satisfaction (e.g., teaching styles, placement services, counseling, and the responsiveness of recruiters to the program's graduates). The fact that approximately 25 percent of the revealed variation is explained is indeed useful and indicates that course content value is a definitive part of program satisfaction. To reveal insights into the explanatory power of the two course content factors, Table 9 shows the directionality and the relative power of each factor in explaining program satisfaction.

As can be seen, the beta coefficients have positive signs thus suggesting that both course content factors co-vary directly with the overall satisfaction metric. This would also follow expectations. Additionally, each course content factor bears a statistically significant association with the dependent variable. Still, the "managerial focus" factor manifests a much higher beta weight, thus revealing more relative explanatory power of one's satisfaction with the MIS program than the "foundation theory and knowledge" factor. Also, the tolerance values for each course content factor are again equal to unity. The pattern of results for one's current position is thus noticeably different than that for one's first year on the job.

SUMMARY AND CONCLUSION

The methodology presented and demonstrated in this study reveals insights not only into data patterns but also insights into how the revealed patterns relate to overall program satisfaction. Uniquely, each content factor for each scenario (in one's first year on the job and in one's current position) contributed to the explanation of program "satisfaction." For one's first year on the job, both "foundation theory and knowledge" and a "managerial focus" were near equal contributors. For one's current MIS position, however, a "managerial focus" evidenced more explanatory power than "foundation theory and knowledge" and may be reflective of increasing job responsibilities in the MIS field. These findings also point to the utility of both knowledge and skills within MIS education and training. Thus, arguments in favor of one over the other may be open to question. Overall, slightly more of a knowledge versus skill discrimination existed in the factor analysis for one's current position, yet each factor was not exhibiting an importance in the regression results that would suggest a curriculum shift. Indeed, alumni value the basics and the overall focus of the MIS curriculum, and this includes both knowledge and skills.

Finally, IS students are products of a program and need to manifest desired characteristics to be successful. They are also customers and this consideration is important in the assessment of program relevancy. If alumni are not satisfied, one surmises that one's IS program may not have met the needs of the profession. This study has thus taken a unique look at IS outcomes assessment using time-dependent measures and multivariate analyses to reveal useful insights into IS program relevancy. This methodology is therefore recommended for inclusion in the arsenal of IS program assessment techniques.

REFERENCES

Bailey, J. R., Langdana, F. K., Rotunda, P. D., & Ryan, J.C. (1997). A factor analytic study of teaching methods that influence retention among MBA alumni. *Journal of Education for Business, 72*(5), 297-302.

Banerjee, S., & Lin, W. (2006). Essential entry-level skills for systems analysts. *Journal of Education for Business, 81*(5), 282-286.

Chrysler, E., & Van Auken, S. (2002). Entry level value versus career value of MIS courses: Faculty expectations versus alumni perceptions. *Journal of Computer Information Systems, 42*(3), 38-43.

Cougar, J.D., G. B. Davis, D. G., Dologite, D. L. Feinstein, J. T. Gorgone, A. M. Jenkins, et al. (1995). IS '95: guidelines for undergraduate IS curriculum. *MIS Quarterly, 9,* 341-359

Gasen, J. B., Weistroffer, H. R., & Haynes, K. (1991-92). Development of an Instrument for Assessing MIS Majors. *Journal of Computer Information Systems*, Winter, 20-22.

Hanchey, Cindy M. An assessment model for surveying graduates (1995-96). *Journal of Computer Information Systems*, Winter, 67-75.

Harris, A. L., Lang, M., Oates, B., & Siau, K. (2006). Systems analysis and design: An essential part of IS education. *Journal of Information Systems Education, 17*(3), 241-248.

Kung, M., Yang, S. C., & Zhang, Y. (2006). The changing information systems (IS) curriculum: A survey of undergraduate programs in the United States. *Journal of Education for Business, 81*(6), 291-299.

Lee, D.M.S., Trauth, E. M. & Farwell, D. (1995). Critical skills and knowledge requirements of IS professionals: A joint academic/industry investigation. *MIS Quarterly, 19*, 313-340.

Liu, X., Liu, L.C., Lu, J., & Koong, K. S. (2003). An examination of job skills posted on internet databases: Implications for information systems degree programs. *Journal of Education for Business, 78*(4), 191.

Maier, J.L., & Gambill, S. (1996). CIS/MIS curriculums in AACSB-accredited colleges of business. *Journal of Education for Business, 71*(6), 329-333.

Nelson, R.R. (1991). Educational needs as perceived by IS and end user personnel: A survey of knowledge and skill requirements. *MIS Quarterly, 15,* 503-525.

Petrova, K., & Claxton, G. (2005). Building student skills and capabilities in information technology and e-business: A moving target. *Journal of Information Systems Education, 16*(1), 27-41.

Van Auken, S. (1991, April). Outcomes assessment: Implications for AACSB accredited business schools and marketing departments. *Western Marketing Educators' Association Conference Proceedings*, (pp. 34-37).

APPENDIX

- **Accounting Information Systems:** Analysis of the creation, flow and uses of accounting data in a business organization.
- **COBOL Language Programming:** Application of the syntax rules of the COBOL language, programming logic and the design and debugging of business-oriented programs.
- **Data Base Concepts:** Introduction to database concepts and development of a database application.
- **Information Center Administration:** The management view of the entire information systems area.
- **Software Project Management:** The design, coding and testing of a small-scale business system as a team using COBOL for programming.
- **Structured Systems Analysis:** Examination of, and practice in, the tasks associated with the analysis and design of computerized business information systems.
- **Systems Development Practicum:** Performing an information systems consulting project for a campus functional area as a team.

Chapter XIX
How Students Learned in Creating Electronic Portfolios

Shuyan Wang
University of Southern Mississippi, USA

Sandra Turner
Ohio University, USA

ABSTRACT

This case study investigated the learning experiences that occurred during students' development of culminating electronic portfolios for a master of education in computer education and technology program. The meaning that students gave to their learning experiences and the problems they encountered were also investigated in order to understand how students learn in a technology-enriched learning environment. Data were collected through in-depth interviews, participant observations, and document analyses from seven M.Ed. students before, during, and after developing electronic portfolios. Findings indicate that creating electronic portfolios supports students' mastery of technology-related knowledge and promotes critical thinking and problem-solving skills. Students reported that they learned not only "by doing," but also from peers through collaboration, from reflection on their artifacts, and from synthesizing their electronic portfolios.

INTRODUCTION

Electronic portfolios have been widely adopted as an assessment method in American education, especially as an effective means for representing and developing teacher knowledge (Barrett, 2005; Strudler & Wetzel, 2005; Wetzel & Strudler, 2005). In addition to the advantages of the reduced storage demands, ease of back-up, portability, ability to create links, and development of students' technology skills, creating electronic portfolios provides students with the responsibility of reflecting on their learning and structuring their knowledge and skills (Porter & Cleland, 1995).

As emerging technologies rapidly become commonplace in education, are teachers ready to prepare their students for full participation in a technology rich society? According to McKinney (1998), teachers who demonstrate their competence in technology through the development of an electronic portfolio are more likely to incorporate technology into their own classrooms. As teacher education programs move toward promoting greater integration of technology in the curriculum, electronic portfolios are a means of not only demonstrating content but pedagogical knowledge and technology expertise as well (Franklin, 2005). Thus, if teacher candidates recognize the advantages of developing electronic portfolios, experience the problems encountered in the process, and understand their implications and possible solutions, it is expected that they will be more confident in using technologies in their future classrooms.

PURPOSE OF THE STUDY

The purpose of this case study was to investigate and understand the learning experiences that occurred in the development of electronic portfolios by graduate students. The meaning that students gave to their learning experiences and the problems they encountered were also investigated in order to understand how students learn in a technology-enriched learning environment. The following research questions were addressed:

1. What are the learning experiences of students in developing their electronic portfolios?
2. What meanings do students give to these experiences?
3. What are the problems encountered by students when developing electronic portfolios?

Significance of the Study

As educational multimedia, hypermedia, and telecommunications become more easily accessible, the use of electronic portfolios as a means of authentic assessment has become increasingly popular in undergraduate as well as graduate programs in teacher education. Some educators might question the meaning and value of electronic portfolios versus other forms of assessment in constructing knowledge. Unfortunately, empirical evidenCE to document the effects of portfolios is limited (Barrett, 2005). With in-depth interviews, observations, and document analysis, this research intended to provide first-hand, detailed data to analyze what and how students learned in the process of creating electronic portfolios.

METHODOLOGY

A qualitative case study formed the methodological framework of this study. This method was appropriate because the researcher studied a particular phenomenon in its natural setting (Punch, 2000), and attempted to make sense of or interpret the phenomenon in terms of the meanings people brought to it (Guba & Lincoln, 1994). This research studied a group of unique students, master of education students majoring in computer education and technology, to investigate their learning experiences during the development of electronic portfolios. Therefore, a case study is appropriate for understanding and interpreting their uniqueness.

Research Setting and Participants

The research setting was a large university in a small Midwestern college town. Students in the college of education were provided with the latest instructional technology tools through the curriculum and technology center. They had access to productivity software, educational software, digital cameras, camcorders, scanners, and other technology-related equipment.

Beginning in the spring of 2002, master of education students majoring in computer education and technology in the college of education were given the option of either (a) assembling an electronic portfolio as their graduation assessment, or (b) writing a seminar research thesis. From then on, nearly all students chose to create a portfolio for their graduation projects. In their electronic portfolios, students were expected to synthesize what they had learned in the Master's program, show their growth over time, and document that they had mastered the National Educational Technology Standards for Teachers (NETS·T). NETS·T are performance-based standards developed by the International Society for Technology in Education (ISTE) for improving technology competence in teacher education.

The requirements for the electronic portfolio include a title page that expresses the student's personality, a table of contents to help navigation in the portfolio, and an introduction including a statement of the student's professional goals, philosophy of education, and a résumé. The main part of the portfolio is the evidence that the student has met the NETS·T standards with reflections and future learning goals. Under each standard students list sample artifacts that provide documentation that they have met this standard and a link to the actual artifact. Along with this artifact, students write a reflective paragraph discussing how the artifact they have selected illustrates their growth in meeting this standard and state their goals for future learning related to this standard.

The participants in this study were seven M.Ed. students in computer education and technology. These students were selected because they were willing to participate in this study among the students who chose to create electronic portfolios as their cumulating projects in spring 2003. Four of the participants were full-time teachers while studying part-time in this program. The other three were full-time students. On average, students took two years to complete the Master's program. All students who chose the portfolio option had taken several educational technology classes where they learned technical skills before they assembled the final products such as: Website design, digital imaging, desktop publishing, multimedia authoring, and video and audio editing. At the time they produced their portfolios, they were familiar with most of the software used in portfolio creation such as: FrontPage, Dreamweaver, Photoshop, QuarkExpress, HyperStudio, KidPix, and iMovie. They were skillful in using Microsoft Office Suite and familiar with PDF. All the mentioned skills were useful in developing electronic portfolios.

Theoretical Framework and Literature Review

The theoretical framework that guided this research study was constructivist learning theory. Constructivists regard learning as an active process and believe that students must be given opportunities to construct knowledge through their own experiences in a meaningful context because learners learn best when they actively construct their own understanding (Piaget, 1977). In their view, learning should be whole, authentic, and real. Vygotsky (1978) further claims that it is the collaboration between people that causes learning to occur, not just a rich, interesting environment. Therefore, cooperative and collaborative learning skills should be engaged in class activities. Pedagogically, the assemblage of an electronic portfolio is a classic example of a constructivist activity because the construction of an electronic portfolio enables students to continuously construct and revisit their knowledge, beliefs, and biases about the profession (Foti, 2002).

Constructivists believe that teaching is an active and learner-centered process. This philosophy recognizes that students build upon their own understanding of the world by using what they already know to interpret new ideas and experiences. Constructivists emphasize not only what students know, but also what they do (Geier, 2002). Brown (2002) found in his study that developing a portfolio supported not only identification of prior learning but also led to new learning outcomes. He pointed out that the portfolio process helped students understand how their learning took place due to the need to critically analyze, organize, evaluate, and write about their learning from experiences. Fiedler and Baumbach (2005) conducted a study in which, instead of taking comprehensive examinations, a doctoral student designed her comprehensive portfolio to document competence in all program standards with relevant artifacts and reflections. They concluded that a traditional written comprehensive examination did not offer the same high-fidelity picture of a student's mastery of program standards because "developing an electronic portfolio requires students to offer written reflections and to provide compelling evidence that can apply their knowledge, skills, and abilities in the artifacts they choose to include" (Fiedler & Baumback, 2005, p. 32).

When students create an electronic portfolio, they have to define the portfolio's context and goals, collect and select artifacts, reflect upon the artifacts, design the layout of the portfolio, and present their final products (Barrett, 2001). Artifacts are such items as written assignments, multimedia projects, lesson plans, or other evidence that students have created during their studies. Creating electronic portfolios increases students' hands-on technology skills while documenting student progress. In addition, the actual creation of the portfolio encourages personal improvement and motivation to become active members in their learning process.

The most important benefit from creating an electronic portfolio is the reflection. Porter and Cleland (1995) stated that the power of reflection helped students and teachers move beyond seeing the portfolio as a mere alternative to traditional assessment to appreciating its value as a learning strategy. In this capacity, Porter and Cleland thought that portfolios can become vehicles for reflection in which learners examine where they have been, where they are now, how they got there, and where they need to go next.

This study used the phenomenological approach to guide in the interpretation of meaning that students gave to their experiences. The researchers were mainly interested in understanding the structure and essence of the students' learning experiences in developing electronic portfolios from their own perspective. Therefore, phenomenology helped the participants interpret their experiences and the researchers to interpret the data. The process sought to uncover the meanings within experience and translate felt understandings into words (Creswell, 1998).

Data Collection and Data Analysis

Data collection in this study included semi-structured, in-depth, face-to-face individual interviews, observations, and analysis of documents such as syllabi and students' electronic portfolios.

Interviews

Two rounds of semi-structured individual interviews (about 40 minutes each) were conducted and tape-recorded with students' permission. The first round interviews were conducted at the beginning of the semester before students started creating their electronic portfolios. The purpose of the first round of interviews was to gather information about the students' background information such as their personal preferences and learning styles, their technology skills and the timeline for developing their electronic portfolios, as well as their expectations for the project. Information from the first round interviews helped the researchers plan the research schedule and question protocol for further interviews with each interviewee.

The second round of interviews took place at the end of the term right after students presented their electronic portfolios in a public showcase. The purpose of the second round of interviews was to gather information about students' experiences in creating electronic portfolios such as (a) the learning processes they went through, (b) reflection on their work, (c) the problems they encountered, and (d) the strategies they used to deal with these problems. This interview allowed the researcher to probe deeply, to uncover new clues, to open up new dimensions of a problem, and to secure vivid, accurate, inclusive accounts from informants that were based upon personal experiences (Holstein & Gubrium, 1995).

All the interviews were transcribed, analyzed, and coded right after the interviews took place in order to form the next round of interview questions as follow-ups for further clarification. Notes were taken and questions were either re-worded, added, or taken out during the process. This allowed the researcher to revisit and clarify issues for data collection and analysis. The researcher maintained the participants' words by transcribing the interviews verbatim without editing for grammar or sentence structure in order to display their original ideas.

Observations

Observations help give the researcher a chance to know what her subjects are doing and how they are behaving. She could determine whether their behaviors reflected what they told the researcher. The researcher observed participants in two settings. One setting was in the computer lab where students worked on their electronic portfolios. In this setting, the researcher observed the activities and experiences that the participants went through. She identified elements such as what learning procedures occurred, what kind of problems the students encountered, and how they solved the problems. The second setting for observations happened during students' electronic portfolio presentation showcase. The researcher observed the entire presentation showcase, focusing on the participants' explanation of what kind of artifacts they chose, how they selected the artifacts, and why they thought the artifacts met the standards. The data gained through the observations also served as a ground for the formulation of final interview questions. In other words, some of the questions asked during the second round of interviews were based upon information gathered during the observations of the participants.

Analysis of Documents

The relevant documents, such as the Master's portfolio syllabus, the National Educational Technology Standards for Teachers (NETS·T), and students' electronic portfolios on CDs, were collected and analyzed. Reviewing the syllabus and standards gave the researchers a clear idea about the project requirements. Analyzing students' electronic portfolios helped the researchers understand what students learned during this process, whether the artifacts selected met the standards, and whether participants truly reflected on their achievements and future goals or cursorily described the artifacts selected for the portfolio.

Data Analysis

Data analysis procedures for this study included organizing the data, generating categories, themes and patterns, and writing the report (Marshall & Rossman, 1995). The researcher put together each participant's reflections for each standard, the interview transcripts, a description of the participant's presentation, and researchers' observation notes in order to organize what she heard, saw, and read. Then the researcher generated categories, themes, and patterns. She read all the interview transcripts, observation notes, reflections of each participant, and viewed their electronic portfolios on CD-ROMs for the first time to get a general sense of the data. The data was reviewed a second and third time to generate themes and patterns from each data segment and combine the topics with overlapping meanings. The aim of this stage was to integrate what had been done into a meaningful and coherent picture of the data. Finally, the researcher classified the topics into major categories and themes that directly related to the purpose of the study.

FINDINGS AND INTERPRETATION

Students' Learning Experiences

Students reported that their learning experiences in developing electronic portfolios included learning by doing, learning from peers, learning from reflection, and learning through synthesis.

Learning by Doing

Regardless of whether the participant was interviewed or observed, the phrase participants used most was "learning by doing." Based on the data analysis, learning by doing in this study covers learning from experiences, learning from mistakes, and learning from trial and error. Students stated that creating electronic portfolios not only gave them a chance to refresh their memory about the skills they had learned, but also provided them with opportunities to learn new skills.

Students reported that the first thing they learned in creating electronic portfolios was to be organized. Nearly all the participants mentioned that they had to start organizing files from the very beginning of the program; otherwise, the process could be really messy and frustrating because they had a lot of files from each course they took.

Participants reported that it took them considerable time to select suitable artifacts for each standard. They revised their previous projects, digitized their documents or projects if they could not find the electronic versions, and taught themselves new programs in order to create a visually appealing layout. The majority of students mentioned that they learned what worked and what did not in converting different projects into a compatible format. They learned more in this process than in class because they had to apply what they learned in class to a new situation.

Students mentioned that their learning was reinforced by designing the final product. If a student wanted to design a visually appealing and sophisticated portfolio with many multimedia artifacts, more time and energy were required. If a student was not interested in design aspects, the basic features of the software could be used to design a simple and plain portfolio with more paper artifacts than multimedia projects. As with any subject, the more time spent on developing a project, the more likely that students would increase their knowledge.

Through creating and presenting their electronic portfolios, all participants realized that "anytime you bring a non-human factor into a process, such as computers, you are going to have problems. It was just all the tiny things that caused you a lot of trouble." Therefore, students realized, "We had to get used to those things and be relaxed about things that don't work because we can't change them." At the beginning, students felt panic when things did not work, but then they learned to relax. As Jane said, "If there is a problem that we can't solve, we have to learn to relax about that. Portfolio learners have to learn to reduce anxiety as much as possible."

Nearly all of the participants stated that it would not be hard for them to learn a new function of software or learn an entirely new program after finishing their electronic portfolios because they were confident with technology. As Tom said, "With a good foundation in technology now, we will be able to pick up new technology much faster." This finding is consistent with Hein's

(1991) point of view. He stated that it was not possible to assimilate new knowledge without having some previously structured knowledge to build on. Therefore, the more one knows, the more one can learn.

According to Piaget (1977), students need a rich environment for exploration to assimilate and accommodate new knowledge. This means that learners do not immediately understand and use information they are given. Instead, they learn by fitting new information together with what they already know. Findings from this study indicated that developing electronic portfolios created a rich learning environment for students to explore by having opportunities to assimilate and accommodate new knowledge. They reviewed what they had learned from each class and selected the suitable artifacts for each standard. Then, they reflected on them and designed the layout of their portfolios. The whole procedure required students to fit new information or knowledge together with what they already knew.

Learning from Peers

The learning strategies that students used most were communication, interaction, and collaboration with peers. For example, students discussed design ideas and solutions to technical problems with peers, attended electronic portfolio presentation showcases in previous terms, and viewed former students' electronic portfolios. In the latter instance, students learned by interacting with the projects rather than interacting with a person. Students indicated that most of the time it was peer critiques that helped them improve their projects. Among peers, participants discussed what artifacts they could use to demonstrate the standards and how to solve the technical problems they encountered.

Vygotsky (1978) claimed that it was the collaboration between people that caused learning to occur, not just a rich, interesting environment. Similarly, Dewey (1963) believed that learning required some outside guidance from teachers, peers, or social institutions. The findings from this study imply that nearly all of the students collaborated with peers or asked advice from instructors at each stage of creating their portfolios as described above.

Learning from Reflection

Participants indicated that reflection was the most important part of their learning experiences in developing electronic portfolios. Through reflection, students recognized how the knowledge gained from the Master's program changed them and motivated them to satisfy their individual inquiries. Reflections highlighted not only what had been done, but also what had not been done so that students were aware of their growth as well as their need for further improvement.

Students mentioned that reflecting involved looking at the standards and the artifacts again so that they would understand what their artifacts were about, and how and why their artifacts met the standards. Through these activities, students often realized that they viewed the artifact in a very different way than what they had originally. Students also found at times that the artifact was not as fully developed or complete as previously thought. Students commented on the fact that "this artifact is not my best work." Students used such statements to identify what would have made the artifact better and what procedure they needed to learn to improve their final projects.

All participants mentioned that reflection was a very important tool to see their growth over time because it gave them a chance to see where they had started and where they had ended. Half of the students mentioned how the reflection helped them self-assess their learning acquisition. Self-assessment for them meant analyzing the strong and weak points of what they had done, and looking at the good and the bad parts of a particular learning activity. Through reflection, students understood what they knew and what they needed to know to perform better. This self-assessment helped them set future learning goals.

Dewey (1963) stressed that students did not learn from experience but learned from reflecting on experience. Similarly, Bruner (1996) said, "Thinking about thinking has to be a principal ingredient of any empowering practice of education" (p. 19). Findings from this study imply that reflection helped students' thinking and thinking encouraged their learning. In addition, findings from this study imply that the portfolio did not just substitute for a research paper, but instead, provided opportunities for students to connect professional and classroom experiences and to reflect on interpretations and judgments which most assessments did not allow.

Learning from Synthesis

Students reported that the process of synthesis was one of the most important learning experiences. According to them, synthesis referred to the process when they put all the parts together to show what they had learned. During the synthesis process, students had to review their artifacts again and again to decide how to put them together in a meaningful way so that their final products would showcase what they had learned throughout the program and how they could use this knowledge in their teaching careers.

Findings from this study indicate that creating electronic portfolios was a meaningful task because it provided an opportunity for students to synthesize what they had learned in the Master's program, show their growth over time, and demonstrate to faculty that they have mastered the technology standards. This is consistent with the findings by Jonassen et al. (1999), who stated that "students learn from thinking—thinking about what they are doing or what they did, thinking about what they believe, thinking about what others have done and believe, thinking about the thinking processes they use—just thinking." They concluded, "Thinking mediates learning. Learning results from thinking" (p. 2).

Meaning of Students' Learning Experiences

The findings of this study confirm that students had a variety of learning experiences despite the learning strategies they used. They gained confidence and a sense of ownership of learning through these experiences. Students expressed that these learning experiences were valuable for them because they were related to real life.

In this study, all of the students showed responsibility and initiative toward their learning. The students made deliberate decisions to create an electronic portfolio instead of writing a research paper. They indicated that creating an electronic portfolio allowed them to be active, independent, and motivated in their learning that allowed them to focus on their learning goals and control their own learning. The findings of this study imply that this kind of learning environment could help students control their own learning, and set their learning goals by building upon prior knowledge.

Hein (1991) noted that students must learn to learn as well as accrue knowledge. Students' responsibility and initiative that were required in developing electronic portfolios allowed students to become their own teachers. By being their own teachers, students could concentrate on what they think is important and what they want to learn. Analyzing the problems they encountered, trying out different solutions, and consulting with peers were essential for the students in their process of learning by doing. The findings imply that creating an electronic portfolio can enhance students' ownership of their learning.

The flexibility and self-directed nature of creating electronic portfolios were important for the students and allowed them to be more active and reflective of their learning. The students indicated that creating electronic portfolios gave them the opportunities to think and reflect on their thinking and learning. Students mentioned that reflection helped them to ponder about what they learn and how they learn. They thought that creating the portfolios independently would help them retain the knowledge they experienced.

In this study, the students indicated that they needed to constantly remind themselves about the project they needed to work on. Motivating themselves to be organized with hundreds of files from different classes throughout the whole program was difficult; trying to follow their timeline on each stage in completing their electronic portfolios was even more difficult. This was challenging because it required disciplining themselves, managing their time, and trying to be on task. Students mentioned that when they realized the importance of organizing files it was already too late. Nearly half of the students finished their final projects just before the due date so that they did not have time to check every detail. As a result, a couple of students ran into trouble in displaying their portfolios during presentation. This implies that creating electronic portfolios required students to be self-disciplined, self-motivated, and organized. It also implies that creating electronic portfolios can enhance students' skills for managing time and self.

The purpose of rich environments for constructivist learning is to help students construct their individual knowledge in authentic contexts and develop skills for critical thinking, problem solving, and lifelong learning. In this study, the students reported that electronic portfolios documented their progress over time and were directly related to their learning and teaching. Specifically, students indicated that by reviewing and converting artifacts and designing the layout of their portfolios through hands-on activities and trial and error methods, they would be able to retain the knowledge that they gained in this program. Students felt that they would be able to use the knowledge and skills they learned in developing electronic portfolios in their professional development and teaching in the classroom.

Problems Encountered in Developing Electronic Portfolios

Students reported that they encountered problems during each stage of the process from the time they started to organize their files until they presented their products in the public showcase.

The problems they encountered in the stage of collecting and selecting artifacts were disorganized files and lost files, which caused students trouble in finding the files they needed, or in replacing lost ones. When they converted their multimedia artifacts to compatible formats, they encountered problems caused by different versions of the software or different operating platforms. When they started to design the final layout, they found it difficult to get started. It was also hard for them to decide what software and what format to use for designing the portfolio, how many multimedia and paper artifacts to use, because they could not use them all, and where and how to use them to make the portfolio meaningful and attractive. During the presentation, some encountered long download times, broken links, and changes in the layout. However, it was these problems that provided students with more opportunities to review what they had learned in the program. Solving these problems was part of the process of learning, as discussed in the section of learning by doing.

In addition to these problems, students mentioned the problem of how to use their time properly. Although all participants started collecting artifacts from the beginning of the program, they differed in the steps that they took to finish their final products. Some students started working on the project very early and finished their portfolios weeks before the presentation date. On the other hand, one student finished her portfolio two hours before her presentation. As a result, she was the only presenter who encountered major problems in displaying her portfolio. Therefore, students indicated that because of the unpredictability of technology, you should start early enough to leave some time for solving the inevitable problems.

RECOMMENDATIONS

The following recommendations are provided to help teachers and students maximize the learning experiences in developing electronic portfolios.

Although students received information on developing electronic portfolios when they entered the program, they needed more detailed information when beginning to plan the final products. It would be helpful to have a workshop or meeting where students could view some samples from former students and discuss the design requirements before their final term. It is also recommended that students be trained to organize their files in the first technology class in the program.

This study revealed that a majority of the students would rather use a simple project to substitute for a complicated multimedia project if they had to spend a lot of time fixing the problem in the original multimedia artifact. In other words, when they encountered a problem that was not easy to solve, they would give up on that artifact. Therefore, the researchers recommend that accessible, ongoing technical assistance be provided to students. This would not only save students' time in dealing with technical problems, but also provide more opportunities for new learning.

This study showed that communicating, interacting, and collaborating with peers was vital in dealing with technical problems and design problems. One recommendation is to create an online forum on the topic of electronic portfolios. Students could post their problems online and get answers from different perspectives. Students indicated they liked to share their technical skills with their peers. Thus, an online forum would be especially helpful for those full-time teachers who live out of town and cannot come to the campus frequently.

This study confirmed the benefits of giving students an open-ended project that allows them to express their creativity and individuality rather than a software template in which all portfolios would have a similar design. The open-ended project motivated students to make their project unique and this motivation pushed most (not all) students to spend extra time on it and learn new skills. However, the freedom inherent in an open-ended project became an excuse for those students who did not want to put much effort into the development of their portfolios. Therefore, a rubric for evaluation of portfolios is needed. The rubric should have detailed criteria for pass and fail combined with rating scales. It should include the four major parts in developing electronic portfolios: (a) design, (b) artifacts, (c) reflection, and (d) presentation. The rubric would not only provide students with a criterion-based evaluation of their portfolios but also a detailed guideline for the portfolio design.

CONCLUSION

This chapter investigated students' learning experiences while developing electronic portfolios. It is apparent that developing an electronic portfolio encouraged students to be active, independent, and creative in their learning. The process also provided students with opportunities to review and synthesize the skills they had learned from different courses and motivated them to learn new applications. Electronic portfolios helped students assess their own learning and achievements, while encouraging collaborative learning and higher-order skills such as problem-solving, synthesis, and reflection.

The study has implications for learning environments that enhance or support collaborative, reflective, and meaningful learning. Viewing samples, collaborating with peers, and communicating and interacting with teachers are important for students in constructing their learning through the development of electronic portfolios. Learning how to learn is vital for students in order for them to become lifelong learners. Assembling electronic portfolios creates a technology-enriched learning environment for students to foster their knowledge and skills in the use of technology and achieve proficiency in the NETS·T.

This study focused on a unique group of graduate students who were majoring in computer education and technology, and thus they had developed considerable confidence and skills in using technology. In comparing the experiences of a similar group of graduate students in educational technology with non-technology undergraduates in teacher education, Bartlett and Sherry (2006) found that the graduate students had fewer technical difficulties and were more likely to view the portfolio assignment as a reflective process. Further research will be helpful in understanding the experiences of education students with limited experience with technology.

A final point is that the students in this study had considerable freedom to express their creativity in designing the structure and layout of their portfolios. They used generic tools such as Web authoring and multimedia authoring software to compile their artifacts. Since this study, however, commercial Web-based databases (such as LiveText and TaskStream) for compiling and storing student artifacts and reflections, linked to relevant national standards, have been implemented in a growing number of colleges of education (Wilhelm et al., 2006). Typically, students developing e-portfolios using these web-based systems are limited to a pre-designed template. Further research is needed to extend this study to teacher education students who use a pre-designed template. Undoubtedly, their learning experiences will be different.

REFERENCES

Barrett, H. (2001). *Electronic portfolio development strategies.* Presentation at the PT3 annual grantees meeting, Washington, DC.

Barrett, H. (2005). Research electronic portfolios in teacher education. *Society for Information Technology and Teacher Education 2005 Conference Proceedings*, 37.

Bartlett, A., & Sherry, A. (2006). Two views of electronic portfolios in teacher education: Non-technology undergraduates and technology graduate students. *International Journal of Instructional Media, 33*(3), 245-253.

Brown, J. O. (2002). Know thyself: the impact of portfolio development on adult learning. *Adult Education Quarterly, 52*(3), 228-245.

Bruner, J. (1996). *The culture of education.* Cambridge, MA: Harvard University Press.

Creswell, J. W. (1998). *Qualitative inquiry and research design.* Thousand Oaks, CA: Sage.

Dewey, J. (1963). *Experience and education.* New York: Macmillan.

Fiedler, R., & Baumbach, D. (2005). Portfolio as a comprehensive exam: Instigating change. *Society for Information Technology and Teacher Education 2005 Conference Proceedings* (pp. 27-32).

Foti, S. (2002). A comparison of two electronic portfolio models. *Society for Information Technology and Teacher Education 2002 Conference Proceedings*, 561.

Franklin, T. (2005). Seeking the perfect electronic portfolio solution: A case study. *Society for Information Technology and Teacher Education 2005 Conference Proceedings* (pp. 84-87).

Geier, C. (2002). Electronic portfolio assessment in teacher education. *Society for Information Technology and Teacher Education 2002 Conference Proceedings*, 564.

Guba, E. G., & Lincoln, Y. S. (1994). Competing paradigms in qualitative research. In N. K. Denzin & Y. S. Lincoln (Eds.), *Handbook of qualitative research* (pp. 105-107). Thousand Oaks, CA: Sage.

Hein, G. E. (1991). *Constructivist learning theory in the museum and the needs of people*. CECA (International Committee of Museum Educators) Conference, Jerusalem Israel (pp. 15-22).

Holstein, J. A., & Gubrium, J. F. (1995). *The active interview*. Thousand Oaks CA: Sage.

Jonassen, D. H., Peck, K. L., & Wilson, B. G. (1999). *Learning with technology: A constructivist perspective*. Upper Saddle River, NJ: Merrill Publishing.

Marshall, C., & Rossman, G. (1995). *Designing qualitative research* (2nd ed.). Thousand Oaks, CA: Sage.

McKinney, M. (1998). Preservice teachers' electronic portfolios: Integrating technology, self-assessment, and reflection. *Teacher Education Quarterly, 25*, 85-103.

Piaget, J. (1977). *Equilibration of cognitive structures*. New York: Viking Press.

Porter, C., & Cleland, J. (1995). *The portfolio as a learning strategy*. Portsmouth, NH: Boynton/Cook Publishers, Inc.

Punch, K. F. (2000). *Introduction to social research: Quantitative & qualitative approaches*. Thousands Oaks, CA: Sage Publications.

Strudler, N., & Wetzel, K. (2005). The diffusion of electronic portfolios in teacher education: Issues of initiation and implementation. *Journal of Research on Technology in Education, 37*(4), 411-433.

Vygotsky, L. (1978). *Mind in society: The development of higher psychological processes*. Cambridge, MA: Harvard University Press.

Wetzel, K., & Strudler, N. (2005). The diffusion of electronic portfolios in teacher education: Next steps and recommendations from accomplished users. *Journal of Research on Technology in Education, 38*(2), 231-243.

Wilhelm, L., Puckett, K., Beisser, S., Wishart, W., Merideth, E., & Sivakumaran, T. (2006). Lessons learned from the implementation of electronic portfolios at three universities. *TechTrends, 50*(4), 62-71.

Chapter XX
Strategic Planning for E-Learning in the Workplace

Zane L. Berge
University of Maryland, Baltimore County, USA

Lenora Giles
University of Baltimore, USA

ABSTRACT

New information and communication technology, specifically computer networked systems, create both a demand and an opportunity for businesses to approach training and knowledge management from new perspectives. These new training perspectives are driven by the need for businesses to provide the right training quickly and efficiently and to support knowledge systems that are current, accessible, and interactive. This article will discuss strategic planning in terms of the organizational elements and the e-learning program requirements that are necessary to build a framework in order to institutionalize and sustain e-learning as a core business process.

INTRODUCTION

The building blocks in a framework necessary to sustain e-learning and knowledge building begin with a foundation laid out by the strategic plan. The next two building blocks of the framework are the organizational structure and support processes, and the e-learning and knowledge management system requirements. The elements of the organizational framework include leadership, change management strategies, the technology infrastructure, and the organizational structure. The e-learning program requirements include instructional systems, roles and competencies of key staff people, and budgeting.

Figure 1. Building blocks for implementing and sustaining e-learning

E-LEARNING PROGRAM ELEMENTS
Instructional Systems
Roles and Competencies of Key Staff
Budgeting

ORGANIZATIONAL ELEMENTS
Leadership
Change Management Strategies
Technology Infrastructure

STRATEGIC PLAN
Mission Statement
Culture and Value System
Vision Statement

Strategic Plan

To implement and sustain e-learning in the workplace a strategic plan can serve as a dynamic blueprint to guide organizational practices based on the organization's strengths, values, and its mission (Schermerhorn & Chappell, 2000). The strategic plan provides a foundation that supports a learning culture by integrating learning and knowledge management with organizational business processes and business goals. Kilfoil (2003) defines strategic planning as a macro-level tool that involves change and focuses on the future by building a bridge between the organization's current position and its vision of the future based on evaluation of its internal and external environments. Strategic planning is:

- A disciplined, fact-based, decision making process
- Based on an analysis of internal and external contexts and data
- Related to choices on how to commit resources
- Compatible with the vision and mission
- Optimizes strengths and opportunities and minimizes weaknesses and threats (Kilfoil, 2003).

Strategic planning primarily involves two important components: the organizational mission and the vision for the future. Developing these two components requires the organization to analyze its current circumstances and to determine what strategy it needs to move forward and to thrive. According to Rosenberg (2001) vision statements are created through an organization-wide consensus-building activity and then refined by senior management.

The vision identifies how the organization will conduct business in the future. Rosenberg (2001) describes the mission statement as a "succinct, specific and powerful articulation of the steps the organization will take to reach its vision" (p. 297). The vision statement of an organization that intends to position itself as an e-learning organization of the future needs to determine how it will provide support and direction for the initiative.

Gap analysis and SWOT analysis are tools that can assist in identifying what an organization needs to do in order to implement and sustain e-learning as a business process (Rosenberg, 2001; Schreiber, 1998). The gap analysis identifies disparities in current e-learning status with those outlined in the vision statement. The SWOT analysis looks at the internal environment and identifies strengths and weaknesses while looking at the external environment to identify opportunities and threats (Rosenberg, 2001; Schermerhorn & Chappell, 2000).

Rosenberg (2001) explains that an organization can build a foundation for e-learning strategy that reinvents the training model. This model can encompass knowledge management, a learning architecture, the organization's technology infrastructure, a learning culture, and a sound business case (Rosenberg, 2001). These ingredients are key to sustaining e-learning over the long term because they institutionalize learning, support it with technology, and link learning to business goals. The blend of organizational learning programs linked to improved business goals as a strategic plan is the foundational building block in a framework to implement and sustain e-learning.

Organizational Issues

The second building block in a framework to move through the stages of technological maturity and to sustain distance training and knowledge management includes a commitment to strategically blend strong leadership, change management, a networked electronic technology infrastructure, and the organizational structure with the goals put forth in the mission and vision.

Training and knowledge management must be viewed as a core business process. According to Conner and Clawson (2003), in order to remain competitive, organizations need to adapt quickly to changing environmental factors. As a result, training and knowledge are critical to growth and survival. From an organizational perspective, this means developing a plan that includes training and knowledge management as integral system components that produce outcomes that are needed to reach business goals. Institutionalizing learning by gaining stakeholder buy-in is critical here. Ensuring access to learning systems, highlighting the personal benefits of e-learning, and illustrating improved business outcomes are methods that can be employed to gain stakeholder buy-in.

E-Learning Maturity Model

E-Learning can be defined as the use of computers, digital media, and communication and Internet technology to deliver learning or training solutions that enhance knowledge and performance (Berge & Kearsley, 2003; Rosenberg, 2001). According to Berge (2001), two primary benefits of e-learning are that it tailors learning to the individual needs of each learner by offering just-in-time and just-for-me learning, anytime and any place. This is a unique difference between e-learning models and traditional training as well as historical distance education models. Learning materials in traditional models are often outdated before they can be implemented into work functions (Rosenberg, 2001). Traditional delivery methods are often costly, synchronous events that halt workplace productivity and require travel expenses for learners and instructors (Rosenberg, 2001).

Schreiber (1998) provides a model of organizational technology maturity stages:

- **Stage 1:** The organization supports sporadic distance learning events.
- **Stage 2:** The organization has sufficient technological capability to support distance-learning events. When these events occur, they are replicated through an interdisciplinary team that responds to different staff/management inquiries and recommendations about distance learning.
- **Stage 3:** The organization has established a distance learning policy such that a stable and predictable process is in place to facilitate the identification and selection of technology to deliver distance training.
- **Stage 4:** Distance learning has been institutionalized in the organization. Distance learning policy, communication, and practice all are aligned in such a way that business objectives are being addressed. The organization has established a distance learning identity, and it conducts systematic assessment of distance training events within an organizational perspective.

These stages are designed to measure organizational maturity and capability in terms of maximizing the use of technology, institutionalizing e-learning, and linking learning outcomes to business goals (Schreiber, 1998).

Leadership

The transformation from traditional learning models to e-learning requires strong leadership. According to O'Rourke (1993), individuals in leadership positions might be senior administrators, top-level teaching staff, training managers, or human resource managers. Whatever the title, these individuals must have the ability to create and communicate the vision for change, implement change, and guide e-learning through its growth process. This includes conducting an environmental scan, securing funding, overcoming barriers, and recruiting and retaining key staff.

During the strategic planning process, leaders analyze external and internal environmental factors affecting the organization. Gap analysis and SWOT analysis of the current situation are highly effective tools here. Gap and SWOT analysis guide strategy planning designed to overcome barriers to building and sustaining distance training (Rosenberg, 2001; Schreiber, 1998). They aid in positioning distance training as a business process by identifying opportunities, clarifying goals, and highlighting strengths. Berge (2001) describes this process as developing an innovative roadmap that includes budgeting, funding support, infrastructure, communication, human development, and policies and procedures.

A crucial role of leadership is to gain support from top-level management in order to ensure proper funding for sustaining the program. One method of accomplishing this is to show how e-learning outcomes positively affect business. External issues concerning competition, the product market, and government activities are some primary considerations for top-level management. Leaders need to show that these issues also drive the training needs of the organization. From an internal perspective, leadership needs to promote a shared vision of where the organization wants to go and how it will conduct business in the future.

Leadership must develop strategies that overcome any barriers to implementing and sustaining the technology initiatives. According to a survey conducted by Berge, Muilenburg, and Haneghan (2002), resistance to e-learning is greatest during the early stages of organizational maturity, and their ranking changes as the organization progresses through the maturation process. The following list shows Berge and Kearsley's (2003) list of challenges to e-learning:

- Time and costs associated with the development of e-learning
- Demonstrating return on investment for e-learning
- Formalizing the processes associated with e-learning
- Keeping up with rapid changes in technology
- Finding and retaining e-learning staff
- Identifying what training needs can be met best by e-learning
- Creating and maintaining interest in e-learning
- Providing the technical support needed
- Misconceptions about e-learning that result in under-use or overuse
- Budget or resource limitations
- Inadequate bandwidth for complex applications
- Need for instructor acceptance of e-learning
- Getting employees to make time for e-learning
- Too much time spent on developing the technology and not enough on instruction
- Lack of consistent direction, support, or involvement from management or senior management.

To combat this resistance, leaders must communicate the benefits of e-learning and encourage involvement from all stakeholders. According to Conner and Clawson (2003), leaders that support and focus on institutionalizing learning can inspire ordinary people to accomplish extraordinary things. Another important task for leadership is to recruit, support, and retain a team of competent technology and instructional professionals and to have them work collaboratively in order to build and support the e-learning initiative.

According to Berge and Kearsley (2003), frequent personnel changes on the champion level can stifle the growth and development of the e-learning initiative. Troha (2002) advises leadership to clearly define the roles and responsibilities of team members to minimize resentment and overlapping of tasks.

Change Management

Four elements necessary to sustain e-learning in an organization are culture, champions, communication, and change (Rosenberg, 2001). In order to achieve a level of technological maturation, an organization can use a change management approach that builds a learning culture, identifies champions, and creates open communication channels to promote the initiative (Rosenberg, 2001). Change management strategy involves assessing the organization to determine its capability to transform into an organization that values learning and is willing to use technology to meet communication and learning goals.

Change management is first about people. It involves assessing the real levels of organizational support and resistance to e-learning. This support or resistance is influenced by knowledge of selected technology and the desire to change familiar behaviors (Snider, 2002). An assessment reveals administrators, managers, and other key players regarded as champions, who can be used to communicate the benefits of e-learning and to gain the trust of workers throughout the organization. It also assesses what actually needs to be taught and learned and what technology and methods would best deliver it and support users (Snider, 2002).

Critical to sustaining e-learning is whether it will be accepted into the organizational culture. Conner and Clawson (2003) define organizational culture as "the shared history, expectations, written and unwritten rules, values, relationships, and customs that affect everyone's behavior" (p.6). They further explain that "the organizational culture stands between the leader's intentions and the results the organization achieves" (p. 6).

A primary cultural barrier to e-learning is that e-learning methods do not feel like traditional training events. Equally important here is that people within the organization do not perceive training time as work time. Rosenberg (2001) suggests nine strategies that pull rather than push an organization toward becoming a learning culture and help to overcome barriers:

- Make direct manager accountable for learning
- Focus at the enterprise level
- Integrate learning directly into work
- Design well and certify where appropriate
- Pay for knowledge
- Everyone's a teacher
- Get rid of the training noise
- Eliminate the ability to pay as a gatekeeper
- Make access as easy as possible

Cross (2003) focuses on learner acceptance of new training methods and suggests that the failure of e-learning to take hold in many organizations is that it is not promoted properly. Cross (2003) contends that e-learning should be marketed internally as a consumer product in order to increase acceptance. Although this strategy is suggested for the learning audience, it could be equally effective in selling the idea to upper management and other stakeholders, because it applies proven marketing techniques such branding, positioning, segmenting, and promoting to increase acceptance.

Technology Infrastructure

The technology infrastructure entails more than just hardware and software solutions. McGraw (2001) defines the infrastructure as the foundation of e-learning that incorporates the organizational culture, values, activities, and structures. Bement (2007) describes the next revolution in organizational infrastructure as cyberinfrastructure and defines it as the "engine for change for the next revolution" in information technology and organizational infrastructure. Bement (2007) further explains that cyerinfrastructure is a user-centered, collaborative creation, dissemination, preservation and application of knowledge within an organization which facilitates creation of learner communities.

The infrastructure is supported by a shared vision, policy, and language that define the procedures and interpretations of e-learning (McGraw, 2001). Common language and governing principles work together to sustain e-learning. According to Rosenberg (2001) and McGraw (2001), there are practical guidelines that are critical to the e-learning organizational and technical infrastructure.

Most important is access to standardized technology hardware, software, and learning materials by all users anytime and anywhere (Bates, 1995; Rosenberg, 2001). The IT department must ensure that all users have a computer, intranet and Internet access to organizational knowledge stores through a web portal as well as access to training materials through a learning management system.

Another key ingredient is a collaborative relationship between the information technology (IT) department and the training department in order to ensure appropriate content that is interactive, consistent, individualized and linked to organizational policies and values (McGraw, 2001; Rosenberg, 2001). The IT department is responsible for building and maintaining the technical aspects of access, speed of connectivity, platform selection, integration, functionality and compatibility of the technology infrastructure. IT support is critical to e-learning, and all activities and decisions must be coordinated with the IT staff.. However, McGraw (2001) suggests that the infrastructure is the sum of business strategy, architecture, organizational legacies, and learner needs. Failure to not view infrastructure as more than technology and any lack of access to the infrastructure can cripple the e-learning effort.

Organizational Structure

The placement of an instructional design unit can greatly affect its success (Lent, 1990). The unit should be placed as closely as possible to its targeted audience. Lent (1990) advises that a training unit with a mandate to improve overall business should be placed highly in the organizational hierarchy, close to the power base, highly visible, and have access to key decision makers. Conner and Clawson (2003) advise that technology must be viewed as a tool playing a supporting role in enhancing learning and communication within the organization.

The social network of people within the organizational structure is the crucial factor in interpretation and application of the learning delivered via that technology. This social network component cannot and should not be automated (Conner & Clawson, 2003).

Berge and Schrum (1998) advise that institutions interested in distance education develop a strategic plan. Some steps they suggest to answer critical questions about current circumstances and desired outcome might be:

Resources Inventory

A first step is to take an inventory of resources such as available hardware, software, distance delivery technologies, technical and faculty support staff, and identify any technology-enhanced learning projects already functioning.

Financial and Market Assessment

A thorough review by the advisory committee of the strategic financial planning and opportunity costs should be made. Often, an outside consultant is hired to expedite this strategic assessment process.

Evaluation of Academic Standards and Roles

Establish a timetable for the roll-out of specific courses/programs and ensure equitable distribution of learning resources for

Program Requirements

The final building block in a framework to implement and sustain e-learning is the e-learning program. This involves implementing an enterprise learning system with a focus on instructional design processes that assess organizational business needs and links them to training outcomes. It also includes an administrative process that manages a team of specialists to facilitate a collaborative work environment. Merging the organization's technological infrastructure and learner access to instruction are key program considerations. Equally important are budgeting and costs justification functions.

Learning and Knowledge Management Systems

It is important to consider that not all training should be delivered online. The tasks analysis performed by an instructional design team can determine what should be delivered through e-learning and what should be delivered through traditional or other means. According to Waller (2003), organizations with goals to deliver training effectively and at lower cost can use e-learning as a component of an overall blended learning strategy. Snider (2002) suggests that all good solutions are blended and grounded in behavioral outcomes, not necessarily in content or pedagogy.

Bates' (1995) ACTIONS model provides further guidelines and considerations for selecting and implementing technology and learning formats:

- **Access:** How accessible is a particular technology for learners? How flexible is it for a particular target group?
- **Cost:** What is the cost structure of each technology? What is the unit cost per learner?
- **Teaching and learning:** What kinds of learning are needed? What instructional approaches will best meet these needs? What are the best technologies for supporting this teaching and learning?
- **Interactivity and user friendliness:** What kind of interaction does this technology enable? How easy is it to use?
- **Organizational issues:** What are the organizational requirements, and the barriers to be removed, before this technology can be used successfully? What changes in organization need to be made?
- **Novelty:** How new is this technology?
- **Speed:** How quickly can courses be mounted with this technology? How quickly can materials be changed?

Learning management systems (LMS) and Web portals are organization-wide components of the technology infrastructure that manage, monitor, and maintain electronic data and communication (Rosenberg, 2001). Although the technical responsibility for the system rests with the IT department, organizational learning is a combination of formal and informal activities that run horizontally and vertically through the entire organizational structure (Snook, 2003). According to Snook (2003), this means that the LMS and portals need to be integrated with all other business processes to support a learning culture and to benefit the organization. A high level of collaboration between the IT staff and the e-learning team is necessary during all stages of design, development, and implementation of learning and knowledge solutions.

Figure 2. Instructional design model for distance learning

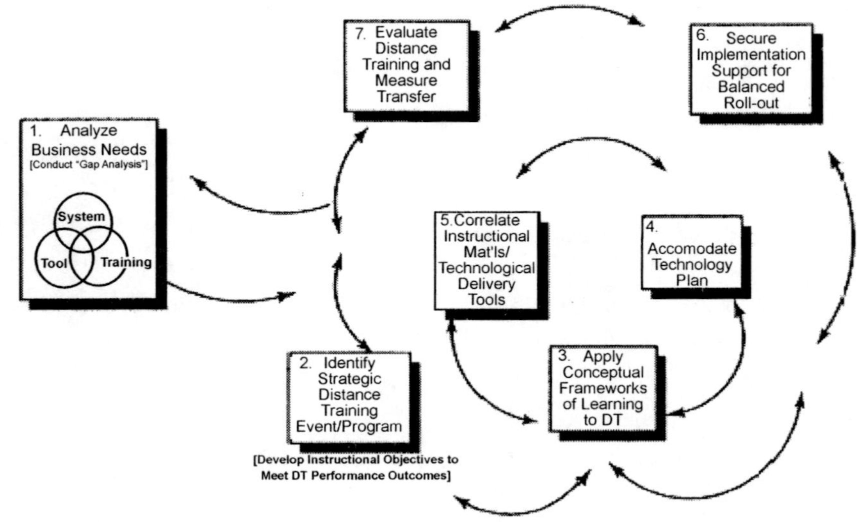

The goal of the proposed instructional design model for distance training is to maximize utilization of technology and institutionalize an organization's distance learning efforts. Instructional needs and performance outcomes of a distance learning event or program are defined by business goals and objectives. The impact of organizational culture, as well as internal corporate dynamics are also considered within the model.

LMS can be developed in-house or contracted out. There are advantages and disadvantages to either choice. Troha (2002) and Snider (2002) advise that selecting a provider is a challenging decision that should be planned carefully and that no single vendor can deliver all solutions. Troha (2002) suggest that organizations:

- Develop and confirm precise, comprehensive selection criteria before meeting prospective providers.
- Use a preliminary design document and selection criteria to interview prospective providers.
- If new to e-learning, start small by limiting the financial commitment to a small initiative.

Design, development, and technology delivery of the learning content is the main task of the training fepartment. Schreiber's (1998) instructional design model for distance training (IDM-DT) provides a reiterative systems processing model for developing and implementing distance training. This is a systems approach that bases performance outcomes and training needs on business goals and focuses on determining the most effective use of technology. It serves as a good model for organizations that are considering implementing and sustaining distance-learning systems.

Applications

New Web 2.0 applications, information and communication technologies such as simulations, gaming, blogs, wikis, podcasts, PDAs, forums and instant messaging offer new options for communication and interaction as well as management of learning (Madden & Fox, 2006; Willis, 2007) However, they all rely on four basic delivery options:

- **Voice:** Instructional audio tools include the interactive technologies of telephone, audioconferencing, and short-wave radio. Passive (i.e., one-way) audio tools include tapes and radio.

- **Video:** Instructional video tools include still images such as slides, pre-produced moving images (e.g., film, videotape), and real-time moving images combined with audioconferencing (one-way or two-way video with two-way audio).
- **Data:** Computers send and receive information electronically. For this reason, the term "data" is used to describe this broad category of instructional tools. Computer applications for distance education are varied and include:
  - **Computer-assisted instruction (CAI):** uses the computer as a self-contained teaching machine to present individual lessons.
  - **Computer-managed instruction (CMI):** uses the computer to organize instruction and track student records and progress. The instruction itself need not be delivered via a computer, although CAI is often combined with CMI.
  - **Computer-mediated education (CME):** describes computer applications that facilitate the delivery of instruction. Examples include electronic mail, fax, real-time computer conferencing, and World Wide Web applications.
- **Print:** is a foundational element of distance education programs and the basis from which all other delivery systems have evolved. Various print formats are available including: textbooks, study guides, workbooks, course syllabi, and case studies. (Willis, 2007)

These new technologies encourage learner centered and learner engaged training and knowledge management. They enhance the mandate of traditional distance education by providing real time interaction and communication to users who are mobile and connected through wireless technology. Organizations can utilize these technologies to provide access and training materials.

Staffing

An e-learning organization requires staff input from a variety of competency areas. Staff can belong to the organization or be external to it. O'Rourke's Roles and Competencies Report can serve as a guide for e-learning staffing needs and activities. According to O'Rourke (1993), staffing areas can be grouped by category according to the roles and the competencies they hold.

- **Leadership roles:** Administrators, managers, and senior teaching staff with vision and access to financial support.
- **Administrative roles:** Directors, managers, and project leaders who identify training needs, recruit staff, and handle finances.
- **Teaching and course development roles:** SME; instructional and graphic designers; media specialist with knowledge of technology, content, and learning theory, and may not have direct contact with learners.
- **Teaching, tutoring, and student support:** Mentors, facilitators, or teacher with direct contact to learners, materials, and delivery technology. Need interpersonal skills and ability to communicate the organization's perspective to learners.
- **Logistics and coordination:** This area would include IT and technology infrastructure and handles registering students and ensuring that materials and technology are accessible.
- **Research and evaluation:** Monitor, test, and review results of training evaluation.

When compared, traditional training and e-learning staffing needs differ in critical ways. These differences result from the fact that e-learning uses networked computers to deliver content and knowledge instead of the lecturer mode. This changes the role of the subject matter expert (SME) to a content developer who may or may not have direct contact with the learners. It also creates the need for a team of specialists that is familiar with adult learning theory, computer technology, and instructional design theory, among other areas.

Some staff services also can be outsourced to vendors. The benefits of outsourcing services are reduced cost in the areas of salaries for technical staff, development and delivery technology, overheads, and some content or training solutions (Rosenberg, 2001). Outsourcing allows an organization to devote its resources to developing staff in other areas. Organizations must have a solid knowledge of vendor products and services as well as an understanding of what solution the organization needs. Staff dedicated to researching, negotiating, and contracting with vendors is essential.

Budget and Cost Justification

This is one area that gets the full attention of upper-level management, because it requires a substantial monetary investment that must be justified by linking e-learning outcomes to business goals. Upper management will want verification that the program will show a return on investment (ROI) and reduce training cost, and that it is cost-effective and cost-efficient (Raths, 2001). The goal of leadership and champions is to promote training as an investment in order to secure funding support for distance training (Berge, 2001). According to Carliner (2000), champions may thoroughly understand the benefits of e-learning, but the costs and organizational disruptions associated with it have a sobering effect on executives.

Carliner (2000), therefore, suggests presenting the proposal as a business case or a request for project investment that identifies costs and returns and compares this with other potential investments. This might include:

- Research and compare relevant alternatives such as traditional methods.
- Show all component costs such as instructional design and authoring software.
- Present realistic return projections based on market rates and real enrollment data.
- Explain technical concepts in familiar terms.
- Recommend a course of action and outline benefits (Carliner, 2000).

Bates (1995, pp.1-2; 2000) cautions that student numbers and long term planning are essential to selecting technology. Cost are divided into fixed capital costs and variable operating costs which include technology infrastructure, administrative applications and academic applications costs (Bates, 2000, p.122). Bates (2000) further suggests that e-learning will initially require high fixed cost but variable cost will decrease as student enrollments increase (see Box 1).

Whalen and Wright (1999) indicate that the breakeven point, the point at which costs are recovered, and return on investment, which illustrates the economic gain or loss from having undertaken a project, are two common measures of financial performance that can be used in costing e-learning.

Box 1.

Bates' formula for calculating cost is: $\$= t / h \times n$
$: cost per student hour
t: total cost of materials
h: hours spent studying
n: number of students (Bates, 2000 pp.126)

Breakeven Number of Students

To offset the high fixed costs of Web-based courses, a certain number of students must be trained at a delivery cost per student of less than that of the delivery cost per student for classroom training. The number of students that offsets the fixed costs of Web-based training is the break-even point.

Return on Investment

The return on investment (ROI) is the percentage that represents the net gain or loss of using Web-based training instead of classroom delivery. For example, an ROI of 300 percent means that $3 was saved in reduced delivery costs for every $1 spent on Web-based training (Whalen & Wright, 1999).

Distance education literature has always noted economies of scale as a primary benefit of distance learning structures. Because the same materials can be delivered repeatedly to increasing numbers of students, distance education realized lower development cost as student numbers increased (Bates, 2000; Moore & Kearsley,1996). The demands of corporate training coupled with delivery via computer and Internet technology change that distance learning scenario (Rumble, 1992). Additionally, according to the Technology Costing Methodology Handbook (Jones, 2001) and Raths (2001), both higher education and traditional business training are notorious for their inability to classify and justify costs. Knowledge management and e-learning can allow systems to tailor information to specific learner needs. They also require constant updating of electronic information.

Consequently, e-learning cannot rely on traditional distance education economies of scale arguments to justify costs. Rosenberg (2001) uses Hammer and Champy's four success criteria for business performance—cost, quality, service, and speed—as a means to justify e-learning. According to these practitioners, the value of e-learning can be measured by how well these criteria enhance business performance. Justification of e-learning also can be shown in terms of gains in productivity hours or time saved and increased and better worker productivity.

According to Raths (2001), e-learning professionals are inventing bottom-line-oriented tactics to measure and justify e-learning. These include measures such as time to competency, achieved competency, and return on expectations. Kraack (2003) notes that lower direct costs, such as travel expenses, facility overheads, instructor fees, publishing costs, and lower program tuitions, are well-known ways that e-learning reduces training expenses.

Opportunity costs resulting from productivity gain is another area that results in reduced costs. According to Kraack (2003), industry standards indicate that one hour of e-learning is as effective as two hours of traditional training. E-learning workers spend less time away from work and receive training en masse, which results in more productivity time and faster application of learned material (Rosenberg, 2001).

CONCLUSION

The building blocks in a framework to implement and sustain e-learning and knowledge building begin with a strategic plan. The process of developing a strategic plan involves analyzing the internal and external environments in order to help the organization determine what the current situation is and how it sees itself doing business in the future. The two components that guide the future activities of the organization are the mission statement and its vision of the future. Once this strategic foundation is laid, the organization can go about the business of transforming itself into a learning culture that maximizes the use of technology and depends on the investment in learning to produce outcomes that further business processes and goals.

The next building block in the framework is the organizational elements. Strong leadership to oversee a change management program, the technology infrastructure, and the recruitment and support of key staff to champion and communicate about the e-learning initiative are key elements here. A primary function is to actively inform all stakeholders about the vision of becoming a learning organization of the future. This means involving them in the development, implementation, and future use of technology by outlining personal and organizational benefits. Support for this effort comes in the form of an interactive technology infrastructure with the role of supporting human communication networks.

The third building block is the distance training and knowledge management system. This is the merging of the organization-wide learning management system, the instructional system's design program, and the IT department. This combination of entities ensures organization-wide access to individualized information files, quality learning content, and support of business goals, all while utilizing technology to become a learning organization of the future. The reach of the learning program becomes global, with access to just-in-time and just-enough training and information available at all times. Instructional design of program content ensures that the right training is delivered via the correct media and method to the right people. The IT and LMS functions ensures that it is easy to use and allows collaboration between users. In order to successfully position itself as an organization of the future that values and supports learning, a business needs to see training as an investment and look to its outcomes to further it business goals.

REFERENCES

Bates, A. W. (Tony). (1995). *Technology, open learning and distance education* (pp. 1-2). New York: Routledge.

Bates, A.W. (2000). *Managing technological change: Strategies for college and university leaders*. San Francisco: Jossey-Bass.

Bement, A. (2007). Cyberinfrastructure: The second revolution. *The Chronicle of Higher Education, Information Technology, Section B*, 5. January 5, 2007.

Berge, Z. (Ed.). (2001). *Sustaining distance training: Integrating learning technologies into the fabric of the enterprise*. San Francisco: Jossey-Bass.

Berge, Z., & Kearsley, G. (2003). The sustainability of distance training: Follow-up to case studies. *The Technology Source*. Retrieved June 3, 2004, from http://ts.mivu.org/default.asp?show=article&id=2027

Berge, Z.L., Muilenburg, L.Y., & Haneghan, J. V. (2002). Barriers to distance education and training: Survey results. *The Quarterly Review of Distance Education, 3(4), 409-418*. Retrieved June 21, 2006, from http://www.emoderators.com/barriers/barriers2002.shtml

Berge, Z., & Schrum, L. (1998). Linking strategic planning with program implementation for distance education. *CAU SE/EFFECT journal, Volume 21 Number 3*. EDUCAUSE. Retrieved April 1, 2003 from http://www.educause.edu/ir/library/html/cem9836.html

Carliner, S. (2000, February). Build a business case for online learning projects. *Learning Circuits*. Retrieved August 6, 2004, from http://www.learningcircuits.org/2000/feb2000/feb2000_elearn.html

Conner, M.L., & Clawson, J.G. (2003). *Creating a learning culture*. Retrieved July 7, 2004 from http://www.darden.edu/batten/clc/Articles/clc.pdf

Cross, J. (2003). E-learning: You build it—Now promote it. *The e-Learning Developers' Journal*. Retrieved July 7, 2004, from http://www.internettime.com/Learning/articles/DevelopersJournal.pdf

Hammer, M., & Champy, J. (2001). *Reengineering the corporation. A manifesto for business revolution.* New York: HarperCollins.

Jones, D. (2001). Western Cooperative for Educational Telecommunications (WCET) and Western Interstate Commission for Higher Education. *Technology costing methodology handbook: Version1.0.* (WICHE). Retrieved July 9, 2003, from http://www.wcet.info/projects/tcm/TCM_Handbook_Final.pdf

Kilfoil, W. (2003). Strategic planning in distance education. *Distance Education and Training Council (DETC).* Retrieved July 7, 2004, from http://www.detc.org/downloads/Strategic%20Planning%20in%20Distance%20Education.pdf

Kraack, T. (2003). Managing cost: The goldilocks phenomenon. *Learning Circuits, 11*(3). Retrieved June 3, 2004, from, http://www.learningcircuits.org/2003/nov2003/kraack.htm

Lent, R. (1990). Building an instructional design organization. In K. Johnson & L. Foa (Eds.), *Instructional design: New alternatives for effective education and training* (pp. 121–136). NewYork: NUCEA.

Madden, M., & Fox, S. (2006). Riding the Waves of Web 2.0: More than a buzzword, but still not easily defined. *Pew Internet and American Life Project.* Retrieved January 15, 2007, from http://www.pewinternet.org/pdfs/PIP_Web_2.0.pdf

McGraw, K.L. (2001). E-learning strategy equals infrastructure. *Learning Circuits.* Retrieved August 6, 2004, from http://www.learningcircuits.org/2001/jun2001/McGraw.htm

Moore, M.G., & Kearsley, G. (1996). *Distance education: A systems view.* Belmont, CA: Wadsworth.

Newburger, E.C. (2000). Home computers and Internet use in the United States: August 2000. Current Population Reports. *U.S. Department of Commerce Economics and Statistics Administration. U.S. Census Bureau.* Retrieved on April 12th, 2003, from http://landview.census.gov

O'Rourke, J. (1993). *Roles and competencies in distance education.* Commonwealth of Learning. Retrieved July 9, 2004, from http://www.col.org/10th/about/images/Roles_Competencies.html

Raths, D. (2001). Measure of success. *Training Mag.com*, Retrieved July 7, 2004, from http://www.onlinelearningmag.com

Rosenberg, M. (2001). *E-Learning: Strategies for delivering knowledge in the digital age.* New York: McGraw-Hill.

Rumble, G. (1992). The comparative vulnerability of distance teaching universities. *Open Learning, 2*, 31-45. Retrieved August 21, 2005, from http://www.uni-oldenburg.de/zef/cde/econ/readings/rumble92.pdf

Schermerhorn, Jr., J. & Chappell, D.S. (2000). *Introducing Management!* New York: John Wiley and Sons.

Schreiber, D.A. (1998). Organizational technology and its impact on distance training. In D.A. Schreiber & Z.L. Berge (Eds.), *Distance training: How innovative organizations are using technology to maximize learning and meet business objectives* (pp. 3-18). San Francisco: Jossey-Bass Inc.

Snider, A. (2002). Find a winning blend. *Elearning Magazine.* Retrieved June 26, 2004, from http://www.elearningmag.com/ltimagazine/article/article-Detail.jsp?id=11688

Snook, A. (2003). Moving towards fully integrated systems for the learning organization. *The Training Foundation.* Retrieved August 6, 2004, from http://www.traininfoundation.com/articles/default.asp?PageID=1458

Troha, F. (2002). Ensuring e-learning success: Six simple tips for initiative leaders. *USDLA*

Journal 16(12). *United States Distance Learning Association.* Retrieved July 17, 2004, from http://www.usdla.org/html/journal/DEC02_Issue/article02.html

Waller, V. (2003). Implementing an e-learning strategy. *The Training Foundation*. Retrieved August 6, 2004, from http://www.trainingfoundation.com/articles/default.asp?PageID=1397

Willis, B. (2007). Distance Education at a Glance. Guides 1 - 7. Engineering Outreach at the University of Idaho Presents. Retrieved January 14, 2007, from http://www.uidaho.edu/eo/distglan.html

Whalen, T., & Wright, D. (1999). Methodology for cost-benefit analysis of Web-based tele-learning: Case study of the Bell Online Institute. *The American Journal of Distance Education, 13*(1), 25-43.

Compilation of References

AAUW. (2004). *Under the microscope: A decade of gender equity projects in the sciences*. Retrieved March 12, 2004, from http://www.aauw.org/research/underthe-microscope.pdf

Abdelraheem, A., & Almusawi, A.S. (2003). Instructional uses of internet services by Sultan Qaboo's University faculty members. *International Journal of Instruction Media, 30*(1), 45-59.

Abernethy, K., & Piegara, G. (2007). Teaching project management: An experiential approach. *Journal of Computing Sciences in Colleges, 22*(3), 198-205.

Aduwa-Ogiegbaen, S.E., & Iyamu, E.O.S. (2005). Using information and communication technology in secondary schools in Nigeria: Problems and prospects. *Educational Technology and Society, 8*(1), 104-112.

Agarwal, R., & Prasad, J. (1999). Are individual differences germance to the acceptance of new information technologies? *Decision Sciences, 30*(2), 361-391.

Airasian, P. (2001). *Classroom assessment: Concepts and applications* (4th Ed.). New York: McGraw-Hill.

Aladwani, A. M. (2002). IT project uncertainty planning and success: An empirical investigation from Kuwait. *Information Technology & People, 15*(3), 210-226.

Al-Khaldi, M. A., & Al-Jabri, I. M. (1998). The relationship of attitudes to computer utilization: New evidence from a developing nation. *Computers in Human Behavior, 14*(1), 23-42.

American Distance Education Consortium. (January, 2003). *ADEC Guiding Principles for Distance Teaching and Learning*. Retrieved January 14, 2004 from http://www.adec.edu/admin/papers/Distance-teaching_principles.html

Ananda, S. (2000, July). *Equipped for the Future Assessment Report: How Instructors Can Support Adult Learners Through Performance-Based Assessment.* (Equipped for the Future (EFF) Technical Report). Washington, DC: National Institute for Literacy. Retrieved August 19, 2002 from http://eff.cls.utk.edu/resources//ananda_eff.pdf

Anao, A.R. (2004). *Stewardship report of professor A.R. Anao, Vice-chancellor; University of Benin: February 10, 1999 — February 9, 2004*. Benin City: Ambik press, Ltd.

Anderson, J.R. (1982). Acquisition of Cognitive Skill. *Psychological Review, 89,* 369-406.

Anderson, L. W., Krathwohl, D. R., Airasian, P. W., Cruikshank, K. A., Mayer, R. E., Pintrich, P. R. Raths, J. & Wittrock, M. C. (Eds.). (2001). *Taxonomy for learning, teaching and assessing: A revision of Bloom's taxonomy of educational objectives*. New York: Addison Wesley Longman, Inc.

Andrews, P. (1997). Secondary teacher's perception of computer availability: A qualification study. *Journal of Information technology for Teacher Education, 6*(3), 321-335.

Anthony, R.N. (1965). *Planning and control systems: A framework for analysis*. Boston, MA: Harvard University Graduate School of Business Administration.

Argawal, R., & Karahana, E. (2000). Time flies when you're having fun: Cognitive absorption and beliefs about information usage. *MIS Quarterly, 24*(4), 665-694.

Armitage, J. (2004). Design: Are agile methods good for design? *ACM interactions, 10*(1), 14-23.

Association of African Universities. (2000). Technical experts meeting on the use and application of information and communication technologies in higher education in institutions in Africa. Retrieved January 22, 2007 from http://www.aau.org/english./documents.aau-ktreport-p2.htm.

Augustin L., Bressler D., & Smith, G. (2002). Accelerating software development through collaboration. In *Proceedings of the International Conference on Software Engineering*, (pp. 559-566). Orlando, Florida.

Ausubel, D.P. (1968). *Educational psychology: A cognitive view*. New York: Holt, Rinehart and Winston, Inc.

Awad, E.M. (2007). *Electronic commerce: From vision to fulfillment* (3rd Ed.). Upper Saddle River, NJ: Pearson Prentice Hall.

Badamas, M. A. (2001). Mobile computer systems—security considerations. *Information Management & Computer Security, 9*(2/3), 134-136.

Bailey, E., & Cotler, M. (1994). Teaching via internet. *Communication Education, 43*(2), 184-193.

Bailey, J. R., Langdana, F. K., Rotunda, P. D., & Ryan, J.C. (1997). A factor analytic study of teaching methods that influence retention among MBA alumni. *Journal of Education for Business, 72*(5), 297-302.

Bandura, A. (1965). Behavioral modifications through modeling procedures. In L. Krasner & L.P. Ullmann, L. (Eds.), *Research in behavior modification: New developments and implications* (pp. 310 - 340). Chicago: Holt, Rinehart and Winston, Inc.

Banerjee, S., & Lin, W. (2006). Essential entry-level skills for systems analysts. *Journal of Education for Business*, 81(5), 282-286.

Barak, M., Lipson, A., & Lerman, S. (2006). Wireless laptops as means for promoting active learning in large lecture halls. *Journal of Research on Technology in Education, 38*(3), 245-263.

Barak, M., Maymon T., & Harel, G. (1999). Teamwork in modern organizations: Implications for technology education. *International Journal of Technology and Design Education, 9*(1), 85-101.

Barrett, H. (2001). *Electronic portfolio development strategies*. Presentation at the PT3 annual grantees meeting, Washington, D.C.

Barrett, H. (2005). Research electronic portfolios in teacher education. *Society for Information Technology and Teacher Education 2005 Conference Proceedings*, 37.

Bartlett, A., & Sherry, A. (2006). Two views of electronic portfolios in teacher education: Non-technology undergraduates and technology graduate students. *International Journal of Instructional Media, 33*(3), 245-253.

Bates, A. W. (1995). *Technology, open learning and distance education* (pp. 1-2). New York: Routledge.

Bates, A.W. (2000). *Managing technological change: Strategies for college and university leaders*. San Francisco: Jossey-Bass.

Batra, D., & Satzinger, J. W. (2006). Contemporary approaches and techniques for the systems analyst. *Journal of Information Systems Education, 17*(3), 257-265.

Battle, A., & Wigfield, A. (2003). College women's value orientations toward family, career, and graduate school. *Journal of Vocational Behavior, 62*, 56-75.

Bazillion, R. J., & Braun, C. L. (2001). Classroom, library and campus culture in a networked environment. *Campus-Wide Information Systems, 18*(2), 61-67.

Becker, H. (2000). Internet use by teachers. Retrieved February 25, 2005 from http://www.crito.uci.edu/TLC/findings/internet-use/stArtspage.htm.

Bell, J. T., & Scott, H. F. (1995). The investigation and application of virtual reality as an educational tool. *Proceedings of the American Society for Engineering Education Annual Conference*, (pp. 1718-1728).

Bement, A. (2007). Cyberinfrastructure: The second revolution. *The Chronicle of Higher Education, Information Technology, Section B*, 5. January 5, 2007.

Bender, T. (2003). Discussion-based online teaching to enhance student learning: Theory, practice and assessment. VA: Stylus Publishing.

Berge, Z. (Ed.). (2001). *Sustaining distance training: Integrating learning technologies into the fabric of the enterprise*. San Francisco: Jossey-Bass.

Berge, Z., & Kearsley, G. (2003). The sustainability of distance training: Follow-up to case studies. *The Technology Source*. Retrieved June 3, 2004, from http://ts.mivu.org/default.asp?show=article&id=2027

Berge, Z., & Schrum, L. (1998). Linking strategic planning with program implementation for distance education. *CAU SE/EFFECT journal, Volume 21 Number 3*. EDUCAUSE. Retrieved April 1, 2003 from http://www.educause.edu/ir/library/html/cem9836.html by

Berge, Z.L., Muilenburg, L.Y., & Haneghan, J. V. (2002). Barriers to distance education and training: Survey results. *The Quarterly Review of Distance Education, 3(4), 409-418*. Retrieved June 21, 2006, from http://www.emoderators.com/barriers/barriers2002.shtml

Berkhout, A. L., Hendriksson-Larséna, K., & Bongers, P. (2004). The effect of using a laptop station compared to using a standard laptop PC on the cervical spine torque, perceived strain and productivity. *Applied Ergonomics, 35*(2), 147-152.

Bernauer, J. A. (1999). Emerging standards: Empowerment with purpose. *Kappa Delta Pi Record, 35*(2), 68-70.

BEST. (2002, April 2004). *BEST*. Retrieved September 28, 2004, from http://www.bestworkforce.org/

BEST. (2004, April 2004). *What it takes: Pre-K design principles to broaden participation in science, technology, engineering, and mathematics*. Retrieved February 11, 2007, from http://www.bestworkforce.org/PDFdocs/BESTPre-K-12Rep_part1_Apr2004.pdf

Biggs, J.B. (2003). *Teaching for quality at university: What the students does*. Society for Research into Higher Education Open University Press, Buckingham, Pennsylvania.

Bilyk, R. (September, 2005). E-learning that goes beyond text and graphics. *T.H.E. Journal, 33*(2), 36-37.

Bischoff, B. (2000). The elements of effective online teaching. In K. W. White & B. H. Weight (Eds.), *The online teaching guide* (pp. 57-72). Needham Heights, MA: Allyn and Bacon.

Blau, J. (2003). German Midsize Firms go for Linux. IDG News Service. Retrieved from http://www.infoworld.com.article/03/10/29/HNgermangoforlinux_1.html.

Blickenstaff, J. C. (2005). Women and science careers: Leaky pipeline or gender filter? *Gender and Education, 17* (4), 369-386.

Bloom, B.S., Engelhart, M.D., Furst, E.J. Hill, W.H., & Krathwohl, D.R. (1956). *Taxonomy of educational objectives: The classification of educational goals*. New York: Longmans, Green and Co.

Boards. The Creative Edge in commercial Production. (September, 2005). *Public Health Department, Belgium - Teenage Mum*. http://www.boardsmag.com/screeningroom/animation/1146/

Boaz, M. (1999). Effective methods of communication and student collaboration. In *Teaching at a distance: A handbook for instructors* (pp. 41-48). Mission Viejo, CA: League for Innovation in the Community College.

Boettcher, J. V., & Conrad, R. M. (1999). *Faculty guide for moving teaching and learning to the Web*. Mission Viejo, CA: League for Innovation in the Community College.

Bonk, C.J., & Graham, C.R. (Eds.). (2006). *The handbook of blended learning: Global perspectives, local designs*. San Francisco, CA: Pfeiffer Publishing.

Borja, R.R. (May, 2005) Cyber schools status. *Education Week, 24*(35), 22-23.

Bostrom, R.P., Olfman, L., & Sein, M. K. (1990). The importance of learning style(S)? In end-user training. *MIS Quarterly, 14*(1), 101-119.

Boucher, A. (1998). Information technology-based teaching and learning in higher education: A view of the economic issues. *Journal of Information Technology for Teacher Education, 7*(1), 87-111.

Boyles, D. (2002). *GIS means business* (Vol. 2). ESRI Press.

Brandao, C. (2002). Teaching online: Harnessing technology's power at Florida virtual school. *T.H.E. Journal, 29*, 1-6.

Bransford, J.D., Brown, A.L., & Cocking, R.R. (Eds.) (1999). *How people learn: Brain, mind, experience, and school.* Washington, DC: National Academy Press.

Breiman, L., Friedman, J. H., Olshen, R. A., & Stone, C. J. (1984). *Classification and re-gression trees*. Belmont, CA: Wadsworth.

Brewer, E., DeJonge, J., & Stout, V. (2001). *Moving to online: Making the transition from traditional instruction and communication strategies.* Thousand Oaks, CA: Corwin Press, Inc.

Brock, S., Hendricks, D., Linnell, S., & Smith, D. (2003). A balanced approach to IT project management. *Proceedings of the 2003 Annual Research Conference of the South African Institute of Computer Scientists and Information Technologists on Enablement through Technology*, (pp. 2-10).

Brookfield, S.D. (1995). *Becoming a critically reflective teacher.* San Francisco: Jossey-Bass Publishers.

Brown, D. G. (2003). *Ubiquitous computing: The universal use of computers on college campuses.* Bolton, MA: Anker Publishing Company.

Brown, D.G., & Petitto, K. R. (2003). The status of ubiquitous computing. *Educase Review, 38*, 25-33.

Brown, D.G., Burg, J.J., & Dominick, J.L. (1998). A strategic plan for ubiquitous laptop computing. *Communications of the ACM, 41*(1), 26-35.

Brown, G.F., & Read, A.R. (1984). Personnel and training policies — some lessons from western companies. *Long Range Planning, 17*(2), 48-57.

Brown, J. O. (2002). Know thyself: the impact of portfolio development on adult learning. *Adult Education Quarterly, 52*(3), 228-245.

Bruner, J. (1996). *The culture of education.* Cambridge, MA: Harvard University Press.

Bullock, L. D. (1997). Efficacy of a gender and ethnic equity in science education curriculum for preservice teachers. *Journal of Research in Science Teaching, 34*(10), 1019–1038.

Butler, M. B. (1999). Factors associated with students' intentions to engage in science learning activities. *Journal of Research in Science Teaching, 36*(4), 455–473.

Cairncross, S. (2002). *Interactive multimedia and learning: Realizing the benefits.* PhD Thesis. Napier University, Scotland.

Campbell, J.P. (1988). Training design for performance improvement. in J.P. Campbell, R.J. Campbell, & Associates (Eds.), *Productivity in organizations*, (pp. 177-216). San Francisco: Jossey-Bass.

Campbell, L., Campbell, B., & Dickinson, D. (2004). *Teaching and learning through multiple intelligences (3rd* Ed.). Boston: Pearson Education, Inc.

Campbell, R. (1999). The development and validation of an instructional design model for creating problem based learning (Doctoral dissertation, University of Pittsburgh, 1999). *Dissertation Abstracts International, 61*(01), 67.

Caramazza, A., McCloskey, M., & Green, B. (1981). Naive beliefs in 'sophisticated' subjects: Misconceptions about trajectories of motion. *Cognition, 9*, 117-123.

Carliner, S. (2000, February). Build a business case for online learning projects. *Learning Circuits.* Retrieved August 6, 2004, from http://www.learningcircuits.org/2000/feb2000/feb2000_elearn.html

Catlett, J. (1992). Peepholing: Choosing attributes effectively for megainduction. In *Proceedings of the 9th International Workshop on Machine Learning*, (pp. 49-54). Aberdeen Scotland.

Center for Charter Reform. (2002). Beyond brick and mortar: Cyber charters revolutionizing education. *CER Action Paper.* Retrieved on December 10, 2005 from http://edreform.com/index.cfm?fuseAction=document&documentID=1001

Champagne, A. L., Klopfer, J., & Anderson, M. (1980). Factors influencing the learning of classical mechanics. *American Journal of Physics, 48*, 1074-1079.

Changchit, C., Cutshall, R., & Elwood, S. (2006). Students' perceptions of the laptop program: what factors should be considered before implementing the program? *International Journal of Information & Communication Technology Education*, *2*(2), 53-61.

Chaptman, Dennis. (2004). Video-games in the classroom. *Wisconsin Technology Network*. Retrieved from http://www.wistechnology.com/article.php?id=513

Chau, P. Y. K., & Hu, P. J. H. (2001). Information technology acceptance by individual professionals: A model comparison approach. *Decision Sciences*, *32*(4), 699-719.

Chen, C. (2003). A constructivist approach to teaching: Implications in teaching computer Networking. Information Technology. *Learning and performance Journal, 21*(2), 17-27.

Christodoulopoulou, E. (2006). *Certification and validation of algorithms for decision trees*. Degree Thesis at Hellenic Open University, Patras, Greece (in Greek).

Christodoulou, S. (2001). *A development environment for programming libraries for decision trees*. Diploma Thesis at Computer Engineering and Informatics Department, University of Patras, Greece (in Greek).

Christodoulou, S., Hatzara, K., Kalles, D., & Papagelis, A. (2004). Building decision trees with components. In *Proceedings of the 4th Panhellenic Conference on Artificial Intelligence*, (pp. 42-51). Samos, Greece.

Chrysler, E., & Van Auken, S. (2002). Entry level value versus career value of MIS courses: Faculty expectations versus alumni perceptions. *Journal of Computer Information Systems,* 42(3), 38-43.

Chu, K., Heeter, C., Egidio, R., & Mishra, P. (2004). Girls and games literature review. *Michigan State University Mind Games Collaboratory*. Retrieved March 10, 2005 from http://spacepioneers.msu.edu/girls_and_games_lit_review.htm

Clark, D., Perrone, C., & Repenning, A. (2005). Webquest: Using WWW & Interactive Simulation Games in the Classroom. *First Monday*. Retrieved from http://www.firstmonday.dk/issues/issue5/perrone/

Clement, J. (1982). Students' preconceptions in introductory mechanics. *American Journal of Physics, 50*, 66-71.

Clewell, B. C., & Campbell, P. B. (2002). Taking stock: Where we've been, where we are, where we're going. *Journal of Women and Minorities in Science and Engineering, 8*(3&4), 255-284.

Cohen W., & D. Levinthal. (1990) Absorptive capacity: A new perspective on learning and innovation. *Administrative Science Quarterly, 35*(1), 128-152.

Cohen, A., & Nachmias, R. (2006). A quantitative cost effectiveness model for web-supported academic instruction. *The Internet and Higher Education, 9*(2), 81-90.

Cohen, J. (1995). *Computer mediated communication and publication productivity among faculty association of Jesuit colleges and universities (AJCU) Institution*. PhD Dissertation, SUNY, Buffalo.

Collis, B., Boer, W.D., & Slotman, K. (2001). Feedback for web-based assignments. *Journal of Computer Assisted Learning, 17*, 306-313.

Compeau, D.R., & Higgins, D.A. (1995). Computer self-efficacy: Development of a measure and initial test. *MIS Quarterly, 19*(2), 189-211.

Conner, M.L., & Clawson, J.G. (2003). *Creating a learning culture*. Retrieved July 7, 2004 from http://www.darden.edu/batten/clc/Articles/clc.pdf

Coppola, E.M. (2004). *Powering up: Learning to teach well with technology*. New York: Teachers College Press.

Cougar, J.D., G. B. Davis, D. G., Dologite, D. L. Feinstein, J. T. Gorgone, A. M. Jenkins, et al. (1995). IS '95: guidelines for undergraduate IS curriculum. *MIS Quarterly*, 9, 341-359

Council on Competitiveness. (2004). *Council on competitiveness website: Worldclass workforce initiatives.* Retrieved November 1, 2004, from http://www.compete.org/

Creswell, J. W. (1998). *Qualitative inquiry and research design.* Thousand Oaks, CA: Sage.

Cross, J. (2003). E-learning: You build it—Now promote it. *The e-Learning Developers' Journal.* Retrieved July 7, 2004, from http://www.internettime.com/Learning/articles/DevelopersJournal.pdf

Cutshall, R., Changchit, C., & Elwood, S. (2006). Campus laptops: Logistical and technological factors perceived critical to students. *Journal of Educational Technology & Society, 9*(3), 112-121.

Daft, R.L., Lengel, R.H., & Trevino, L.K. (1987). Message equivocality, media selection, and manager performance: Implications for information systems. *MIS Quarterly, 11*(3), 355-366.

Daloz, L.A. (1986). *Effective teaching and mentoring: Realizing the transformational power of adult learning* San Francisco, CA: Jossey-Bass, Inc., Publishers.

Dargan, C. P. (2003) *Designing online courses: Reflections from the cyber trenches* (pp. 1-2). (Graduate project, University Of Northern Iowa).

Darkwa, O., & Mazibiko, F. (2000). Creating virtual learning communities in Africa: challenges and prospects. *First Monday,* 5(5).

Davis, D. L., & Davis, D. F. (1990). The effect of training techniques and personal characteristics on training end users of information systems. *Journal of Management Information Systems, 7*(2), 93-110.

Davis, F. D. (1989). Perceived usefulness, perceived ease of use, and user acceptance of information technology. *MIS Quarterly, 13*(3), 318-339.

Davis, F. D., Bagozzi, R. P., & Warshaw, P. R. (1989). User acceptance of computer technology: A comparison of two theoretical models. *Management Science, 35*(8), 982-1003.

Davis, S. A., & Bostrom, R. P. (1983). Training end users: An experimental investigation of the roles of the computer interface and training methods. *MIS Quarterly, 17,* 61-85.

Dede, C. (1998, June 5-6). Future visions of information technology in math education: Design heuristics for vignettes embedded in the standards. Technology and NCTM Standards 2000 Conference, Arlington, VA.

Dede, C., Salzman, M., & Loftin, B. (1996). The development of a virtual world for learning newtonian mechanics. In P. Brusilovsky, P. Kommers, & N Streitz, (Eds), *Multimedia, hypermedia, and virtual reality.* Berlin: Springer (in press).

Demb, A., Erickson, D., & Hawkins-Wilding, S. (2004). The laptop alternative: student reactions and strategic implications. *Computers & Education, 43*(4), 383-401.

Dennis, G., & Espinosa, S. (1997). Models of internet use: Approaches from various disciplines. Retrieved February 14, 2005 from

Denzin, N. K., & Lincoln, Y. S. (1994). *Handbook of qualitative research.* Thousand Oaks, CA: Sage Publications.

Devaraj, S., & Kohli, R. (2003). Performance impacts of information technology: Is actual usage the missing link. *Management Science, 49*(3), 273-289.

Dewey, J. (1963). *Experience and education.* New York: Macmillan.

DiVesta, F. (1989). Applications of cognitive psychology to education. In M. C. Wittrock & F. Farley (Eds.), *The Future of Educational Psychology* (pp. 37-73). Hillsdale, NJ: Lawrence Erlbaum Associates.

Dodge, B. (2005). *The WebQuest Page.* Retrieved January 15, 2007, from http://webquest.sdsu.edu/

Donaldson, M. (1984). *Children's minds.* London: Fontana.

Dori, Y., Belcher, J., Bessette, M., Danziger, M., McKinney, A., & Hult, E. (2003). Technology for active learning. *Materials Today, 6*(12), 44-49.

Driver, M. (2002). Exploring student perceptions of group interaction and class satisfaction in the web-enhanced classroom. *The Internet and Higher Education, 5*, 35-45.

Drossos, L., Bassisliadis, B., Stefani, A., Xenos, M., Sakkopoulos, E., & Tsakalidis A., (2006). Introducing ICT in a traditional higher education environment: Background, design, and evaluation of a blended approach. *International Journal of Information and Communication Technology Education, 2*(1), 65-78.

Drossos, N., Papagelis, A., & Kalles, D. (2000). Decision tree toolkit: A component-based library of decision tree algorithms. *In proceedings of the 4th European Conference on Principles and Practice of Knowledge Discovery in Databases*. Lyon, France.

Du, S. M., Johnson, R. D., & Keil, M. (2004). Project management courses in IS graduate programs: What is being taught? *Journal of Information Systems Education, 15*(2), 181-187.

Duffy, T.M., & Jonassen, D.H. (1992). *Constructivism and the technology of instruction: A conversation*. Lawrence Erlbaum Associates.

Duhaney, D.C. (2001). Teacher education: Preparing teachers to integrate technology. *International Journal of Instructional Media, 28*(1), 23-30.

Duncan, D. (April, 2006). The cyber school option. *Valley News Dispatch*. Retrieved on December 6, 2006 from http://www.pittsburghlive.com/x/pittsburghtrib/s_449113.html

Earle, R.S. (2002). The integration of instructional technology into public education: Promises and challenges. *Educational Technology, 42*(1), 5-13.

EDC. (2004). *GSDL, FAQ*. Retrieved April 12, 2004, from http://gsdl.enc.org/ Education Development Center. (2004). *GSDL website*. Retrieved January 15, 2007, from http://gsdl.enc.org/

Edwards, S.H. (2003). Improving student performance by evaluating how well students test their own programs. *ACM Journal on Educational Resources in Computing, 3*(3). Retrieved from http://portal.acm.org/citation.cfm?doid=1029994.1029995

E-LeGI. (2007). *European Learning GRID Infrastructure project*. Retrieved from http://www.elegi.org

Ellen, N., & West, J. (2003). Classroom management of project management: A review of approaches to managing a student's information system project development. *Journal of American Academy of Business, 3*(1/2), 93-97.

Ellner, C.L., & Barnes, C.P. (Eds.). (1983). *Studies of college teaching: Experimental results, theoretical interpretations, and new perspectives*. Lexington, MA: D.C. Heath & Company.

Elwood, S., Changchit, C., & Cutshall, R. (2006). Investigating students' perceptions on laptop initiative in higher education: an extension of the technology acceptance model. *Campus-Wide Information Systems, 23*(5), 336-349.

Etheris, A.I., & Tan, S.C. (2004). Computer-supported collaborative problem solving and anchored instruction in a mathematics classroom: an exploratory study. *Int J Learning Technology, 1*(1), 16–39.

Evers, J. (2004). My daughter killed her brother in "The Sims." *IT World.Com*. http://www.itworld.com/App/4201/041221sibssims/

Falba, C.J., Strudler, N.B., & Boone, R. (1999). College of education faculty use of technology: A snapshot. In Price, J., Willis J., Willis, D., Jost M. & Boger-Mehall, S. (Eds.) *Proceedings of society for information technology and teacher education 10th annual conference*. Charlottesville, VA: Association for Advancement of Computing in Education.

Fall, K., & Varadhan, K. (2007). *The Ns Manual. The VINT Project*. Retrieved March 5, 2007, from http://www.isi.edu/nsnam/ns/

Fanning, E. 2006. The sims in therapy: An examination of feasibility and potential. Under review for publication by *The Journal of the American Art Therapy Association*.

Fenton, N., & Pfleeger, S. (1997) Software metrics a rigorous and practical approach. Thomson Computer Press.

Fiedler, R., & Baumbach, D. (2005). Portfolio as a comprehensive exam: Instigating change. *Society for Information Technology and Teacher Education 2005 Conference Proceedings*, (pp. 27 – 32).

Finley, L., & Hartman, D. (2004). Institutional charge and resistance. Teacher preparatory faculty and technology integration. *Journal of Technology and Teacher Education, 12*(3), 319-337.

Finn, S., & Inman, J. (2004). Digital unity and digital divide: Surveying alumni to study effects of a campus laptop initiative. *Journal of Research on Technology in Education, 36*(3), 297-317.

Fischer, C.G., & Grant, G.E. (1983). Intellectual levels in college classrooms. In C.L. Ellner & C.P. Barnes (Eds.). *Studies of college teaching: Experimental results, theoretical interpretations, and new perspectives* (pp. 47-60). Lexington, MA: D.C. Heath & Company.

Fisher, C., Dwyer, D.C., & Yocam, K. (Eds.) (1996). *Educational and technology: reflections on computing in classrooms.* San Francisco. Apple Computer, Inc., Jossey-Bass Publishers.

Flanagan, J. C. (1954, July). The Critical Incident Technique. *Psychological Bulletin, 51*(4), 327-358.

Fogarty, R. (1997). *Problem based learning and other curriculum models for the multiple intelligences classroom.* Arlington Heights, IL: IRI/Skylight Training and Publishing, Inc.

Fogler, H. S., Montgomery, S.M., & Zipp. R.P. (1992). Interactive computer modules for chemical engineering instruction. *Journal of Computer Applications in Engineering Education, 1*(1), 11-24.

Ford, J.K. (1997). Advances in training research and practice: A historical perspective. In J. K. Ford, & Associates (Eds.), *Improving training effectiveness in work organizations*, (pp. 1-16). Hillsdale, New Jersey: Lawrence Erlbaum Associates,.

Ford, L. (1969). *Using the case study in teaching and training.* Nashville: Broadman Press.

Forouzan, B. A. (2003). *TCP/IP Protocol Suite* (2nd Ed.): McGraw-Hill.

Fortugno, N. & Zimmerman, E.. (2005). Soapbox: Learning to play to Llearn — lessons in educational game design. *Gamesutra.* Retrieved from http://www.gamasutra.com/features/20050405/zimmerman_01.shtml

Foti, S. (2002). A comparison of two electronic portfolio models. *Society for Information Technology and Teacher Education 2002 Conference Proceedings*, 561.

Fouts, J., & Stuen, C. (1997). Copernicus Project: Learning with laptops: year 1 evaluation report. (ERIC No. ED 416 847)

Fox, E. (2005). Tracking U.S. trends. *Education Week, 24*(35), 40-42. Retrieved September 21, 2005, from edweek.org/ew/articles/2005/05//05/35tracking.h24.html.

Franklin, T. (2005). Seeking the perfect electronic portfolio solution: A case study. *Society for Information Technology and Teacher Education 2005 Conference Proceedings*,(pp. 84-87).

Fraser, A.B. (1999). Colleges should tap the pedagogical potential of the world-wide web. *Chronicle of Higher Education, 45*(48), B8.

Freedman, R. (2002). Adopting PMI standards in an IT environment. Retrieved October 28, 2005, from http://techrepublic.com.com/5100-6330-1038859.html

Freeman, C. E. (2004). *Trends in educational equity of girls & women: 2004* (No. NCES 2005-016). Washington, DC: U.S. Government Printing Office: U.S. Department of Education, National Center for Education Statistics.

Fry, F.L., & Stoner, C.R. (1985) Business plans: Two major types. *Journal of Small Business Management, 23*, 1-6.

Fulton, K., Yoon, I., & Lee, C., (2005, August). Induction into learning communities. National Comission on Teaching and America's Future. Retrieved from http://www.nctaf.org/documents/NCTAF_Induction_Paper_2005.pdf

Furuta, R., Shipman III, F., Hsieh, H., Francisco-Revilla, L., Karadkar, U., Rele, A., et al. (1999). *Using the Internet in the classroom: Variety in the use of Walden's Paths.* Paper presented at the ED-MEDIA 99: World Conference on Educational Multimedia, Hypermedia, and Telecommunications, Seattle,Washington.

Fusilier, M., & Durlabhji, S. (2003). No downturn here: Tracking e-business programs in higher dducation. *Decision Sciences Journal of Innovative Education, 1*(1), 73-98.

Gackowski, Z.J. (2003). Case/real-life problem-based learning with information systems projects. *Journal of Information Technology Education, 2,* 357-365.

Galbraith, M.W. (Ed.). (1998). *Adult learning methods: A guide for effective instruction.* (2nd Ed.). Malabar, FL: Krieger Publishing Company.

Gallivan, M. J., Spitler, V. K., & Koufaris, M. (2005). Does information technology training really matter? A social information processing analysis of coworkers' influence on IT usage in the workplace. *Journal of Management Information Systems, 22*(1), 153-192.

Garrison, D.R. & Kanuka, H. (2004). Blended learning: Uncovering its transformative potential in higher education. *The Internet and Higher Education, 7*(2), 95-105.

Gasen, J. B., Weistroffer, H. R., & Haynes, K. (1991-92). Development of an Instrument for Assessing MIS Majors. *Journal of Computer Information Systems,* Winter, 20-22.

Gaskell, P. J., Hepburn, G., & Robeck, E. (1998). Re/presenting a gender equity project: contrasting visions and versions. *Journal of Research in Science Teaching, 36*(8), 859–876.

Gee, J.P. (2003).What video games have to teach us about learning and literacy. New York: Palgrave Macmillan.

Geier, C. (2002). Electronic portfolio assessment in teacher education. *Society for Information Technology and Teacher Education 2002 Conference Proceedings,* 564.

Gilbert, S. D. (2001). *How to be a successful online student.* New York: McGraw-Hill.

Gipps, C.V. (1994). *Beyond testing: Towards a theory of educational assessment.* London, England: The Falmer Press.

GirlTech.org. (2004). *Girl Tech website.* Retrieved February 8, 2007, from http://www.girltech.org/

Gist, M. E., Schwoerer, C., & Rosen, B. (1987). Effects of alternative training methods of self efficacy and performance in computer software training. *Journal of Applied Psychology, 72*(2), 307-313.

Goetz, J.P., & LeCompte, M.D. (1984). *Ethnography and qualitative design in educational research.* Orlando, FL: Academic Press, Inc.

Goldberg, F.M., & Anderson, J.M. (1989). Student difficulties with graphical representations of negative values of velocity. *The Physics Teacher, 27,* 254-260.

Goldstein, I.L. (1992). *Training in organizations: Needs assessment, development, and evaluation* (3rd ed.). Pacific Grove, CA: Brooks/Cole.

Goold, A. (2003). Providing process for projects in capstone courses. *ACM SIGCSE Bulletin, 35*(3), 26-29.

Gorgone, J. T., Davis, G. B., Valacich, J. S., Topi, H., Feinstein, D. L., & Longenecker, H. E. (2002). IS 2002 model curriculum and guidelines for undergaduate degree programs in information systems. Retrieved February1, 2007, from www.acm.org/education/is2002.pdf

Gorilla Mask. (2005,September). *Trapped in the closet performed by the sims (Part 1).* Retrieved from http://gorillamask.net/rksims1.shtml

Gotsis, K. A., Goudos, S. K., & Sahalos, J. N. (2005). A test lab for the performance analysis of TCP over ethernet LAN on windows operating system. *IEEE Transactions on Education, 48*(2), 318-328.

Gottfried, J., & McFeely, M. (1998). Learning all over the place: Integrating laptop computers into the classroom. *Learning & Leading with Technology, 24*(4), 6-12.

Goulding, J. S., & Alshawi, M. (2002). Generic and specific IT training: A process protocol model for construction. *Construction Management and Economics, 20,* 493-505.

Goulding, J. S., & Alshawi, M. (2004). A process driven IT training model for construction: Core development issues. *Construction Innovation, 4*, 243-254.

Gowan, J. A., & Mathieu, R. G. (2003). The role of project methodology in large-scale IS project management. In *Proceedings of the Ninth Americas Conference on Information Systems, (pp. 1328-1332).*

Grabowski, B. & Koszalka, T. (1999). *Web-enhanced learning environment strategies for classroom teachers.* Retrieved April 12, 2002 from http://www.comms.uab.es/inet99/inet98/4e/4e_3.htm

Grabowski, B. (2004). Generative learning: Past, present, and future. In D.H. Jonassen (Ed.). *Handbook of research on educational communications and technology* (2nd Ed.), (pp. 719-743). Mahwah, NJ: Lawrence Erlbaum Associates, Publishers.

Graham, J., & Semich, G. (2006). A model for effectively integrating technology across the curriculum: A three-step staff development program for transforming practice. *International Journal of Information and Communication Technology Education,* 2(1), 1-10.

Gramoll. K. (2001). An internet portal for statics and dynamics engineering courses. *Proceedings of the International Conference of Engineering*, (pp. 1-6).

Green, H. (2004). Researchers and teachers push for games in schools. *The Institute of Education, University of London.* Retrieved March 10, 2005 from http://ioewebserver.ioe.ac.uk/ioe/cms/get.asp?cid=1397&1397_1=10817

Grimshaw, J.D. (1999). *Bringing geographical information systems into business.* Wiley, John & Sons, Incorporated.

Groccia, J.E., & Miller, J.E. (2005). *On becoming a productive university: Strategies for reducing cost.* Bolton, MA: Anker Publishing Company.

Grundy, J. C. (1997). A comparative analysis of design principles for project-based IT courses. *Proceedings of the 2nd Australasian Conference on Computer Science Education ACSE '97*, (pp. 170-177).

Guardian Newspaper (2003) Tuesday, October 12 2003 p. 39.

Guardian Newspapers (2003) Tuesday, December 2003, page 45.

Guardian Newspapers (2005) Tuesday, January 18 p. 41.

Guardian Newspapers Nigerian Limited. (2002). Nigeria's floppy skill development scheme worries manufacturers, July 23.

Guba, E. G., & Lincoln, Y. S. (1994). Competing paradigms in qualitative research. In N. K. Denzin & Y. S. Lincoln (Eds.), *Handbook of qualitative research,* (pp. 105-107). Thousand Oaks, CA: Sage.

Gunstone, R.F., & White, R. (1981). Understanding of gravity. *Science Education, 65*, 291-299.

Hackney, R., McMaster, T., & Harris, A. (2003). Using cases as a teaching tool in IS education. *Journal of Information Systems Education, 14*(3), 229-234.

Haddad, W.D. (2003). Is Instructional technology a must for learning? *Techknowlogia.* Jan-March, Oakton: Knowledge Enterprise Inc.

Hall, M., & Elliott, K. M. (2003). Diffusion of technology into the teaching process: strategies to encourage faculty members to embrace the laptop environment. *Journal of Education for Business*, *78*(6), 301.

Hammer, M., & Champy, J. (2001). *Reengineering the corporation. A manifesto for business revolution.* New York: HarperCollins.

Hanchey, Cindy M. An assessment model for surveying graduates (1995-96). *Journal of Computer Information Systems*, Winter, 67-75.

Hanson, K. (2002, October 3-6, 2002). *Space aliens? Women, ICTs, and gender-equitable electronic resources.* Paper presented at the Association of Women in Development(AWID) Conference, Guadalajara, Mexico.

Harasim, L., Hiltz, S.R., Teles, L., & Turoff, M. (1995). *Learning networks: A field guide to teaching and learning online*. Cambridge, MA: The MIT Press.

Harder, C. (1997). *GIS means business.* ESRI Press.

Harris, A. L., Lang, M., Oates, B., & Siau, K. (2006). Systems analysis and design: An essential part of IS education. *Journal of Information Systems Education,* 17(3), 241-248.

Harrison, A. W., & Rainer, R. K. (1992). The influence of individual differences on skill in end user computing. *Journal of Management Information Systems, 9*(1), 93-111.

Hassan, M., & Jain, R. (Eds.). (2004). *High performance TCP/IP networking — concepts, issues, and solutions.* New York: Pearson Prentice Hall.

Haywood, J., Anderson, C., Coyle, H., Day, K., Haywood, D., & Macleod, H. (2000). Learning technology in Scottish higher education — a survey of the views of senior managers, academic staff and experts. *ALT-J,* 8(2), 5-17.

Hedin G., Bendix L., & Magnusson, B. (2003). Introducing Software Engineering by means of Extreme Programming. In *Proceedings of the 25th International Conference on Software Engineering,* (pp. 586-593). Portland, Oregon.

Hein, G. E. (1991). *Constructivist learning theory in the museum and the needs of people.* CECA (International Committee of Museum Educators) Conference, Jerusalem Israel, (pp. 15-22).

Hibbeler, R. C. (2001). *Engineering mechanics: Dynamics* (9th Ed.) Prentice-Hall.

Higginbotham-Wheat, N.L. (1991). Use of generative learning strategies in computer-based instruction: Effects of learner-generated visual and verbal elaborations (Doctoral Dissertation, Memphis State University, 1991). *Dissertation Abstracts International, A(52/12),* 4205.

Hiltz, S.R. (1997). Impacts of college-level courses via asynchronous learning networks: Some preliminary results. *Journal of Asynchronous Learning Networks (JALN), 1*(2), 1-19.

Hiltz, S.R., & Turoff, M. (2002). What makes learning networks effective? *Communication of the ACM, 45*(4), 56-58.

Hirumi, A. (2003). Get a life: Six tactics for optimizing time spent online. In M. Corry & C. Tu (Eds.), *Distance education: What works well* (pp. 73-101). New York: The Haworth Press.

Hiss, A. (2000). Talking the talk: Humor and other forms of online communication. In K. W. White & B. H. Weight (Eds.), *The online teaching guide* (pp. 24-36). Needham Heights, MA: Allyn and Bacon.

Hoffman, B. (1996). What drives successful technology planning? *Journal of Information Technology for Teacher Education, 5(*1&2), 43-55.

Holt, T. (2004). How mods are really built. *Serious Games Summit DC.* Retrieved March 10, 2005 from http://www.cmpevents.com/GDe04/a.asp?option=C&V=11&SessID=3305&Mgt=0&RVid=0

Honey, M., Culp, K.M., & Spielvogel, R. (1999). *Using technology to improve student achievement* {Online}. Retrieved from www.ncrel.org/sdrs/areas/issues/methods/techlgy/te800.htm.

Hooper, S., & Rieber, L.P. (1995). Teaching with technology. In A.C. Ornstein (Ed.), *Teaching: Theory into practice*, (pp. 134-170). Needham Heights, MA: Allyn and Bacon.

Howard, W.G. (2006). Socrates and technology: A new millennium conversation. *International Journal of Instructional Media, 33*(2) 197-204.

http:/www.coe.uh.edu/insite/elec.pub/HTML1997/tg.denn.htm.

Huang, A.H. (2002). A three-tier technology training strategy in a dynamic business environment. *Journal of End User Computing, 14*(2), 30-38.

Huang, G., Taddese, N., & Walter, E. (2000). *Entry and persistence of women and minorities in college science and engineering* (No. NCES 2000–601,). Washington, DC: U.S. Department of Education, National Center for Education Statistics.

Hung, D., & Nichani, M. (2001). Constructivism and e-learning: Balancing between the individual and social levels of cognition. *Educational Technology, 41*(2), 40-44.

Hussey, D.E. (1985). Implementing corporate strategy: Using management education and training. *Long Range Planning, 18*(5), 28-37.

IEEE Learning Technology Standards Committee. (2002). The learning object metadata standard (WG12). *IEEE Learning Technology Standards Committee (ILTSC),* Piscataway, NJ: IEEE. Retrieved December 29, 2006, from ieeeltsc.org/wg12LOM/lomDescription

International Society for Technology in Education. (2000). *Educational Technology Standards and Performance Indicators for All Teachers.* Retrieved January 2005 from: http://cnet.iste.org/teachers/t_stands.html

International Technology Education Association. (2002). (ITEA) is the professional organization of technology teachers. Retrieved January 2005 from: http://www.iteaconnect.org/TAA/TAA.html

Ismail, J. (2001). The design of an e-learning system: Beyond the hype. *The Internet and Higher Education, 4*(3-4), 329-345.

ISO (1999). Information technology — evaluation of software — quality characteristics and guides for their use. *International Standard, ISO/IEC 9126.*

Jackson, S. (2004). Ahead of the curve: Future shifts in higher education. *Educause Review, 39*(1), 10-18.

Jackson, S. A. (2004). *The quite crisis: Falling short in producing American scientific and technical talent.* Retrieved February 10, 2007, from http://www.bestworkforce.org/PDFdocs/Quiet_Crisis.pdf

Jackson, S.E., & Schuler, R.S. (1990). Human resource planning: Challenges for industrial/organizational psychologists. *American Psychologist, 45*, 223-239.

James, M., & Voigt, M. (2001). Tips from the trenches: Delivering online courses effectively. *Business Education Forum, February*, 56-60.

Janson, J. L. (1992). Computer-based training helps firms trim budgets. *PC Week, January 27,* 90-93.

Jarmon, C. (1999). Strategies for developing effective distance learning experience. In *Teaching at a distance: A handbook for instructors* (pp. 1-14). Mission Viejo, CA: League for Innovation in the Community College.

Jefferies, A., Thornton, M., Alltree, J., & Jones, I. (2004). Introducing web-based learning: An investigation into its impact on university lecturers and their pedagogy. *Journal of Information Technology Impact, 4*(2), 91-98.

Jeffries, C., & Maeder, D.W. (2004). *Using vignettes to build and assess teacher understanding of instructional strategies.* Unpublished manuscript, University of California at Los Angeles.

Jih, W. (2003). Simulating real world experience using accumulative system development projects. *Journal of Information Systems Education, 14*(2), 181-191.

Johari, A. (2006). Creating a knowledge-based society through e-learning in Korea. *Educational Technology Research and Development, 54*(5), 529-540.

Johnson, A.P. (2000). *Up and out: Using creative and critical thinking skills to enhance learning.* Needham Heights: Allyn & Bacon.

Johnson, D.L., & Lui, L. (2000). First steps toward a statistically generated information technology model. *Computers in the Schools, 16*(2), 3-12.

Johnson, J. L. (2003). *Distance education: The complete guide to design, delivery, and improvement.* New York: Teachers College Press.

Johnson, J., Chapman, C., & Dyer, J. (2006). Pedagogy and innovation in education with digital technologies. In A Mendez-Vilas, A. Solano Martin, J. A. Mesa Gonzalez, & J. Mesa Gonzalez (Eds.) *Current developments in technology-assisted education* (pp. 135-139). Badajoz, Spain: Formatex.

Jonassen, D. (Ed). (1988). *Instructional designs for microcomputer courseware101.* Hillsdale, NJ: Lawrence Erlbaum.

Jonassen, D. H., Peck, K. L., & Wilson, B. G. (1999). *Learning with technology: A constructivist perspective.* Upper Saddle River, NJ: Merrill Publishing.

Jonassen, D.H. (Ed.). (2004). *Handbook of research on educational communications and technology* (2nd Ed.). Mahwah, NJ: Lawrence Erlbaum Associates, Publishers.

Jonassen, D.H., Howland, J., Moore, J., & Marra, R.M. (2003). *Learning to solve problems with technology.* Pearson Education.

Jones, B.F., Valdez, G., Nowakowski, J., & Rassmusen, C. (1995). Plugging in: Choosing and using educational technology. *Washington, DC: Council for Educational Development and Research, and North Central Regional Educational Laboratory.* Retrieved from www.ncrel.org/sdrs/edtalk/toc.htm

Jones, D. (2001). Western Cooperative for Educational Telecommunications (WCET) and Western Interstate Commission for Higher Education. *Technology costing methodology handbook: Version1.0.* (WICHE). Retrieved July 9, 2003, from http://www.wcet.info/projects/tcm/TCM_Handbook_Final.pdf

Jones, M. G., Howe, A., & Rua, M. J. (2000). Gender differences in students' experiences, interests and attitudes toward science and scientists. *Science Education, 84*(2), 180-192.

Judson, E. (2006). How teachers integrate technology and their beliefs about learning: Is there a connection? *Journal of Technology and Teacher Education. Association for the Advancement of Computing in Education (AACE), 14*(3), .581-597.

Julia, M., Niall, S., & Su, W. (2002). Report on the British Council seminar and Asian association of open universities exhibition in Malaysia. Retrieved from http://cvu.strath.ac.uk/niall/malaysia/

Kahn, T. M., (1992). Multimedia literacy at Rowland: A good story, well told. *Technological Horizons in Education. 19*(7), 77-83.

Kalles, D. (2007). Students working for students on programming courses. *Computers and Education (in press).*

Kalles, D., & Papagelis, A. (2006). Managing the decision tree life-cycle with components. *International Journal of Information and Communications Technology Education, 2*(3),1 – 13.

Kanaya, T., & Light, D. (2005). *Duration and relevance of a professional development program: Using Intel Teach to the Future to illuminate successful programmatic features.* Presented at Society for Information Technology and Teacher Education International Conference 2005. Norfolk, VA: AACE.

Kang, D., & Santhanam, R. (2003). A longitudinal of training practices in a collaborative application environment. *Journal of Management Information Systems, 20*(3), 257-281.

Kauchak, D. P., & Eggen, P.D. (1998). *Teaching & learning: Research-based methods.* (3rd Ed.). Boston: Allyn and Bacon.

Keil, M., Cule, R. E., Lyytinen, K., & Schmidt, R. C. (1998). A framework for identifying software project risks. *Communications of the ACM, 41*(11), 76-83.

Kent, A. (1996). Technology: Moving faster. *Australian Accountant,* 66(1), 26.

Kerka, S. (2001). *Capstone experiences in career and technical education.* (ERIC Publications 071). Washington, DC: Office of Educational Research and Improvement (ED). (ERIC Reproduction Service No. ED456333).

Kilfoil, W. (2003). Strategic planning in distance education. *Distance Education and Training Council (DETC).* Retrieved July 7, 2004, from http://www.detc.org/downloads/Strategic%20Planning%20in%20Distance%20Education.pdf

Kim, J., Heon, P. S., Tae, Lee, H., & Yuk, K. C. (2001). Virtual reality simulations in physics education. *Interactive Multimedia Electronic Journal of Computer-Enhanced Learning, 3*,(2). Retrieved from http://imej.wfu.edu/articles/2001/2/02/index.asp

Kim, K.A. (1999). A comparison of academic achievement between the predominantly adult classes and the predominantly traditional age student classes and their teachers' perceptions of teaching (Doctoral Dissertation, North Carolina State University at Raleigh, 1999). *Dissertation Abstracts International, A60/09,* 3235.

Kim, L.S. (1995). Creative games for the language class. *Forum,* 1(33), 35.

King, B. (2003). Educators turn to games for help. *Wired News.* Retrieved March 10, 2005 from http://www.wired.com/news/games/0,2101,59855,00.html ().

Kirkpatrick, D.L. (1967). Evaluation of training. in R.L. Craig & L.R. Bittel (Eds.), *Training and development handbook*, (pp. 87-112). New York: McGraw-Hill,.

Kirkwood, A., & Price, L. (2006). Adaptation for changing environment: Developing learning and teaching with information and communication technologies. Retrieved May 10, 2006 from http://www.irrodl.org/index. php/irrodl/article/viewArticle/294/624.

Kish, M.H.Z. (2004). *Using vignettes to develop higher order thinking and academic achievement in adult learners in an online environment (Doctoral Dissertation,* Duquesne University, Pittsburgh, 2004).

Knight, C., Knight, B.A., & Teghe, D. (2006). Releasing the pedagogical power of information and communication technology for learners: A case study. Retrieved February 5, 2007 from http://ijedick.dec.uwi.edu/viewarticle/php?id=16T&layout=html.

Knowles, M. (1978). *The adult learner: A neglected species* (2nd Ed.). Houston: Gulf Publishing Company.

Knowles, M. (1995*). Designs for adult learning: Practical resources, exercises, and course outlines from the father of adult learning.* Alexandria, VA: American Society for Training and Development.

Knox, A.B. (1986). *Helping adults learn: A guide to planning, implementing, and conducting programs.* San Francisco: Jossey-Bass Inc., Publishers.

Ko, S., & Rossen, S. (2004). *Teaching online: A practical guide* (2nd Ed). Boston: Houghton Mifflin.

Kohavi, R., John, G., Long, R., Manley, D., & Pfleger, K. (1994). MLC++: A Machine Learning Library in C++. *Proceedings of IEEE Conference on Tools with Artificial Intelligence* (pp. 740-743). New Orleans.

Komives, S. R. (2001). Learning leadership: As individuals and in communities of practice. *Concepts and Connections, 9(*3),11.

Kozlowski, S.W.J. (1987). Technological Innovation and Strategic HRM: Facing the challenge of change. *Human Resource Planning, 10*, 69-79.

Kraack, T. (2003). Managing cost: The goldilocks phenomenon. *Learning Circuits, 11*(3). Retrieved June 3, 2004, from, http://www.learningcircuits.org/2003/nov2003/kraack.htm

Krasner, L., & Ullmann, L.P. (Eds.). (1965). *Research in behavior modification: New developments and implications.* Chicago: Holt, Rinehart and Winston, Inc.

Kraut, A.I., Pedigo, P.R., McKenna, D.D., & Dunnette, M.D. (1989). The role of manager: What's really important in different managerial jobs. *Academy of Management Executive, 3*, 286-293.

Kulik, J. A. (2002). School mathmatics and science programs benefit from instructional technology (InfoBrief, NSF 03-301). Washington DC: National Science Foundation: Retrieved September 21, 2005 from http://dwbrr.unl.edu/iTech/TEAC859/Read/KulikTech.pdf

Kung, M., Yang, S. C., & Zhang, Y. (2006). The changing information systems (IS) curriculum: A survey of undergraduate programs in the United States. *Journal of Education for Business*, 81(6), 291-299.

Kuperstein, J., Gentile, C.K. & Zwier, J. (1999). The connected learning community technology roadmap. Retrieved from www.microsoft.com/Education/vision/roadmap/default.asp

Kurose, J. F., & Ross, K. W. (2005). *Computer networking: A top-down approach featuring the internet* (3rd Ed.). New York: Addison Wesley.

Laat, M., & Lally, V. (2005). Investigating group structure in CSCL: Some new approaches. *Information Systems Frontiers, 7*(1), 13 – 25.

Laudon, K.C., & Traver, C.G. (2007). E-commerce: Business, technology, society (3rd Ed.). Upper Saddle River, NJ: Pearson Prentice Hall.

Laurillard, D. (2002). *Rethinking university teaching: A conversational framework for the effective use of learning technologies* (2nd Ed.), London: Routledge Falmer.

Lee, D.M.S., Trauth, E. M. & Farwell, D. (1995). Critical skills and knowledge requirements of IS professionals: A joint academic/industry investigation. *MIS Quarterly*, 19, 313-340.

Lehner, F., Nosekabel, H., & Lehmann, H. (2003). Wireless e-learning and communication environment: welcome at the University of Rosenburg. *E-Service Journal*, *2*(3) 23-41.

Lelliot, A. (2000). Online education in africa: Promises and pitfalls. Retrieved January 22, 2007 from http://generalupdate.rau.ac.za/infosci/conf/ Wednesday/lelliot.htm.

Lent, R. (1990). Building an instructional design organization. In K. Johnson, & L. Foa (Eds.), *Instructional design: New alternatives for effective education and training* (pp. 121–136). NewYork: NUCEA.

LePore, S. J., Kling, R., Iacono, S., & George, J. R. (1989). Desktop computerization and quality of work life: The role of implementation process and infrastructure. *In J.I. DeGross, J.C. Henderson, & B.R. Konsynski (Eds.) Proceedings of the Tenth International Conference on Information Systems,* (pp. 223-235). Atlanta: Association for Information Systems.

Levine, M. (2005). *Ready or not, here life comes.* New York: Simon & Schuster.

Liaw, S. (2002). An internet survey for perceptions of computer and the world wide web: Relationship, prediction, and difference. *Computers in Human Behavior, 18*(1), 17-35.

Linn, R.L., Baker, E.L., & Dunbar, S.B. (1991). Complex, performance-based assessment: expectations and validation criteria. *Educational Researcher*, *20*(8), 15-21.

Liu, L., Maddux, C., &Johnson, L. (2004). Computer attitude and achievement; is time an intermediate variable? Retrieved February 5, 2007 from http://www.accessmylibrary.com/coms2/summary_0286-18279752_ITM

Liu, X., Liu, L.C., Lu, J., & Koong, K. S. (2003). An examination of job skills posted on internet databases: Implications for information systems degree programs. *Journal of Education for Business*, 78(4), 191.

London, M. (1989). *Managing the training enterprise.* San Francisco: Jossey-Bass.

Longley, P., &Clarke, G. (1996).*GIS for business and service planning.* John Wiley & Sons.

Louca S., Constantinides, C., & Ioannou, A. (2004). Quality assurance and control model for e-learning. *In Proceedings of Computers and Advanced Technology in Education*, (pp. 468-472).

Lowe, G. S. (2000). Preparing students for the workforce. *Proceedings of the Australasian Conference on Computing Education ACSE '00*, (pp. 163-169).

Luarn, P., & Lin, H.H. (2005). Toward an understanding of the behavioral intention to use mobile banking. *Computers in Human Behavior*, *21*(6), 873-891.

Lucas, H. C., Jr. (1975). Performance and the use of an information system. *Management Science, 21*(8), 908-919.

Macpherson, C. (1998). Virtual reality: What is the state of play in education? *Australian Journal of Educational Technology, 14*(1), 60-74.

Madden, M., & Fox, S. (2006). Riding the Waves of Web 2.0: More than a buzzword, but still not easily defined. *Pew Internet and American Life Project.* Retrieved January 15, 2007, from http://www.pewinternet.org/pdfs/PIP_Web_2.0.pdf

Maier, J.L., & Gambill, S. (1996). CIS/MIS curriculums in AACSB-accredited colleges of business. *Journal of Education for Business*, 71(6), 329-333.

Majchrzak, A. (1988). *The human side of factory automation.* San Francisco: Jossey-Bass.

Man, C. L. T., Hasegawa, G., & Murata, M. (2006). ImTCP: TCP with an inline measurement mechanism for available bandwidth. *Computer Communications, 29*(10), 1614-1626.

Manjit Sidhu, S., Ramesh, S., & Selvanathan, N. (2003). Using multimedia to minimize computational effort in engineering. *Proceedings of the Malaysian Scientific & Technology Congress (MSTC) (*pp. 811-815).

Manseur, R. (2005). Virtual reality in science and engineering education. In proceedings of the 35th ASEE/IEEE Frontiers in Education Conference (pp. 8-13).

Mansfield, N. (2003). *Practical TCP/IP — Designing, using, and troubleshooting TCP/IP networks on Linux and Windows.* Addison-Wesley.

March, T. (1997, June 2001). *Working the web for education.* Retrieved February 10, 2007, from http://www.ozline.com/learning/theory.html

Marriott, M. (2003). Games made for remaking. *The New York Times.* Retrieved March 10, 2005 from http://www.nytimes.com/2003/12/04/technology/circuits/04modd.html?ex=1135227600&en=0fc160f73ad53e42&ei=5070

Marshall, C., & Rossman, G. (1995). *Designing qualitative research* (2nd Ed.). Thousand Oaks, CA: Sage.

Marshall, J.M. (2002). *Learning with technology: Evidence that technology can and does, support learning.* San Diego, CA: Cable in the Classroom.

Marsick, V. J. (1998). Case study. In M.W. Galbraith (Ed.), *Adult learning methods: A guide for effective instruction* (2nd Ed.), (pp. 197-218). Malabar, FL: Krieger Publishing Company.

Maslin-Ostrowski, P., & Ackerman, R.H. (1997, March). *A Case for Stories: Toward Further Understanding of Situated Knowledge and Practice.* Paper presented at the Annual Meeting of the American Educational Research Association, Chicago, IL. (ERIC Reproduction Service Document No. ED408265).

Maslin-Ostrowski, P., & Ackerman, R.H. (1998). Case Story. In M.W. Galbraith (Ed.), *Adult learning methods: A guide for effective instruction* (2nd Ed.), (pp. 303-316). Malabar, FL: Krieger Publishing Company.

McBride, N.K. (2005). A student driven approach to teaching e-commerce. *Journal of Information Systems Education, 16*(1), 75-83.

McCormack, C., & Jones, D. (1998). *Building a Web-based education system.* New York: John Wiley & Sons, Inc.

McDermott, L.C., Rosenquist, M.L., &. Van, Zee, E. H. (1987). Student difficulties in connecting graphs and physics: Examples from kinematics. *American Journal of Physics, 55*, 503-513.

McDonnell, F. (2006). Why so few choose physics; An alternative explanation for the leaky pipeline. *American Journal of Physics, 73* (7), 583-586).

McGhee, W., & Thayer, P.W. (1961). *Training in business and industry.* New York: Wiley.

McGraw, K.L. (2001). E-learning strategy equals infrastructure. *Learning Circuits.* Retrieved August 6, 2004, from http://www.learningcircuits.org/2001/jun2001/McGraw.htm

McGuire, K. (2003). *Some definitions and guidelines for distance education.* Retrieved February 3, 2003 from www.uscampus.com/get_help/library/library_edurelated/article_distance.htm

McIsaac, M., & Craft, E. (2003). Faculty development: Using distance education effectively in the classroom. In M. Corry & C. Tu (Eds.), *Distance education: What works well* (pp. 73-101). New York: The Haworth Press.

McIsaac, M.S., & Gunawardena, C.N. (2004). Distance education. In D.H. Jonassen (Ed.), *Handbook of research for educational communications and technology* (2nd Ed.), (pp. 355-395). New Jersey: Lawrence Erlbaum Associates, Publishers.

McKeachie, W.J. (1999). Introduction. In W.J. McKeachie (Ed.), *McKeachie's teaching tips: Strategies, research, and theory for college and university teachers* (10th Ed.), (pp. 1-8). Boston: Houghton Mifflin Company.

McKenna, J.F., & Oritt, P.L. (1981). Growth planning for small business. *American Journal of Small Business, 5*(4), 19-29.

McKenzie, J. (2001). *Planning good change with technology and literacy*. Bellingham, WA: FNO Press.

McKinney, M. (1998). Preservice teachers' electronic portfolios: Integrating technology, self-assessment, and reflection. *Teacher Education Quarterly, 25*, 85-103.

Mead, S. (2006). *The truth about boys and girls.* Washington, DC: Education Sector

MentorNet. (2004). *MentorNet Website*. Retrieved January 27, 2007, from http://www.mentornet.net/

Merriam, S.B. (1988). *Case study research in education: A qualitative approach.* San Francisco: Jossey-Bass Publishers.

Midkiff, S. F. (2005). An experiential course in wireless networks and mobile systems. *IEEE Pervasive Computing, 4*(1), 9-13.

Millar, R., & Shevlin, M. (2003). Predicting career information-seeking behavior of school pupils using the theory of planned behavior. *Journal of Vocational Behavior, 62*, 26-42.

Miller, F., Mangold, W.G., & Holmes, T. (2006). Integrating geographic information systems (GIS) applications into business courses using online business geographics modules. *Journal of Education for Business, 82*(2), 74-79.

Mingers, J. (1987). Experts Systems—Rule induction with statistical data. *Journal of the Operational Research Society, 38*, 39-47.

Mingers, J. (1989). An empirical comparison of selection measures for decision tree induction. *Machine Learning, 3*, 319-342.

Miron, G., & Nelson, C. (October, 2000). *Autonomy in exchange for accountability: An initial study of Pennsylvania charter schools.* Executive Summary. The Evaluation Center: Western Michigan University. Retrieved on December 10, 2005 from http://www.wmich.edu/evalctr/charter/pa_reports/pa_final_rpt.pdf

Miron, G., Nelson, C., & Risley, J. (October, 2002). *Strengthening Pennsylvania's charter school reform: Findings from the statewide evaluation and discussion of relevant policy issues.* The Evaluation Center: Western Michigan University. Retrieved on December 10, 2005 from http://www.wmich.edu/evalctr/charter/pa_5year/

Mitchell, T. (1997). *Machine Learning.* McGraw Hill.

Mitra, A. (1998). Categories of computer use and their relationship with attitudes toward computers. *Journal of Research on Computing in Education, 30*(3), 281-295.

Mitra, A., & Hullet, C. R. (1997). Toward evaluating computer aided instruction: Attitudes, demographics, context. *Evaluation and Program Planning, 20*(4), 379-391.

Mitra, A., & Steffensmeier, T. (2000). Changes in student attitudes and student computer use in a computer-enriched environment. *Journal of Research on Computing in Education, 32*(3), 192-222.

Molenda, M. (2003). In search of the elusive ADDIE model. *Performance Improvement, 42*(5), 34-36.

Moore, G., Winograd, K., & Lange, D. (2001). *You can teach online*. New York: McGraw-Hill Higher Education.

Moore, G.C., & Benbasat, I. (1991). Development of instrument to measure the perceptions of adopting an information technology innovation. *Information Systems Research, 2*(3), 192-222.

Moore, M.G., & Kearsley, G. (1996). *Distance education: A systems view*. Belmont, CA: Wadsworth.

Moore, R.C., & Bogotch, I.E. (1993, November). *No longer neglected: Adult learners in graduate learning programs*. Paper presented at the Annual Meeting of the Mid-South Educational Research Association, New

Orleans, LA. (ERIC Document Reproduction Service No. ED367905).

Morgan, C., Isaac, J. D., & Sansone, C. (2001). The role of interest in understanding the career choices of female and male college students. *Sex Roles, 44*(5-6), 295-320.

Mpekou, P. (2006). *Algorithmic library for decision trees: Development of educational content for decision trees and for object oriented designing and programming.* Degree Thesis at Hellenic Open University, Patras, Greece (in Greek).

Mueller, J. (2006). Authentic assessment toolbox. *Authentic Assessment.* Retrieved December 30, 2006, from http:/jonathan.mueller.faculty.noctrl.edu/toolbox/index.htm

Musick, R., Catlett, J., & Russel, S. (1993). Decision theoretic subsampling for induction on large databases. In proceedings of the 10th International Conference on Machine Learning, (pp. 212-219). Amherst, MA..

Nair, P. (2004). Learning without boundaries — Educational technology planning for the new millennium. retrieved from www.fetc.org/fetcon/1200/nair.htm

Nance, W. D. (1998). Experiences with an innovative approach for improving information systems students' teamwork and project management capabilities. *Proceedings of the 1998 ACM SIGCPR Conference on Computer Personnel Research*, 145-151.

Napier, H.A., Rivers, O., Wagner, S., & Napier, J.B. (2006). Creating a winning e-business. Boston, MA: Thomson Course Technology.

Naquin, D. (2000). The integration of technology at Loudon Campus, Northern Virginal community college; A faculty survey. *Inquiry,* 5(1), 58-65.

NASA. (2004). *Women of NASA website*. Retrieved February 10, 2007, from http://quest.arc.nasa.gov/women/intro.html

National Education Association. (2000). *A Survey Of Traditional and Distance Learning Higher Education Members*. NEA, Washington, DC. Retrieved November 22, 2002 from: http://www.nea.org/he

National Science Foundation. (2003). *New formulas for America's workforce: Girls in science and engineering* (No. nsf03207). Washington, D.C.: National Science Foundation.

Nelson, R. R., & Cheney, P. H. (1987) Training end users: An exploratory study. *MIS Quarterly, 11*(4), 547-559.

Nelson, R.R. (1991). Educational needs as perceived by IS and end user personnel: A survey of knowledge and skill requirements. *MIS Quarterly*, 15, 503-525.

Nelson, R.R., Whitener, E.M., & Philcox, H.H. (1995). The assessment of end user training needs. *Communications of the ACM, 38*(7), 27-39.

Newburger, E.C. (2000). Home computers and Internet use in the United States: August 2000. Current Population Reports. *U.S. Department of Commerce Economics and Statistics Administration. U.S. Census Bureau.* Retrieved on April 12th, 2003, from http://landview.census.gov.

Nielsen, J. (2000). *Designing web usability: The practice of simplicity.* Indianapolis, IN: New Riders Publishing.

Noe, R.A., & Ford, J.K. (1992). Emerging issues and new directions for training research. *Research in Personnel and Human Resource Management, 10*, 345-384.

Norhayati, A. M., Halimah, B. Z., & Siew, P. H. (2001). Courseware development to motivate life-long reading habits. Multimedia at work. *IEEE, 8*(4) 76-81.

Nuldén, U., & Schepers, H. (2002). Increasing student interaction in learning activities: Using a simulation to learn about project failure and escalation. *Journal of Information Systems Education, 12*(4), 223-232.

O'Rourke, J. (1993). *Roles and competencies in distance education.* Commonwealth of Learning. Retrieved July 9, 2004, from http://www.col.org/10th/about/images/Roles_Competencies.html

Odom, E. (1998, Fall). Critical thinking skills. *Word's worth,* (pp. 5,10). (ERIC Reproduction Service No. ED434227).

Ogle, D. (1986). K-W-L: A teaching model that develops active reading of expository text. *The Reading Teacher, 39,* 564-570.

Okebukola, P. (1997). Old, new and current technologies in education. *UNESCO Africa 14*(15), 7-18.

Okojie, M., & Olinzock, A. (2006). Developing a positive mind-set toward the use of technology for classroom instruction. *International Journal of Instructional Media, 33*(1), 33-41.

Olfman, L., & Mandviwalla, M. (1994). Conceptual versus procedural software training for graphical user interfaces: A longitudinal field experiment. *MIS Quarterly, 18*(1), 405-426.

Olgren, C. H. (1998). Improving learning outcomes: The effects of learning strategies and motivation. In C. C. Gibson (Ed.), *Distance learners in higher education* (pp. 77-96). Madison, WI: Atwood.

Olson, G.M., Malone, T.W., & Smith, J.B. (Eds.). (2001). *Coordination theory and collaboration technology.* Mawhah, NJ: Lawrence Erlbaum Associates.

OPNET Technologies. (2007). Retrieved February 20, 2007, from www.opnet.com

Oppenheimer, T. (2003). *The flickering mind: The false promise of technology in the classroom and how learning can be saved.* New York: Random House.

Ostroff, C., & Ford, J.K. (1989). Assessing training needs: Critical levels of analysis, In I. L. Goldstein (Ed.), *Training and development in organizations* (pp. 25-62). San Francisco: Jossey-Bass,.

Packard, B. W.-L., & Nguyen, D. (2003). Science career-related possible selves of adolescent girls: A longitudinal study. *Journal of Career Development, 29*(4), 251-263.

Palloff, R. M., & Pratt, K. (1999). *Building learning communities in cyberspace: Effective strategies for the classroom.* San Francisco: Jossey-Bass.

Palloff, R. M., & Pratt, K. (2001). *Lessons from the cyberspace classroom: The realities of online teaching.* San Francisco: Jossey-Bass.

Palmer, S. (2000). On-and off-campus computer usage in engineering education. Computers and education. *Elsevier Science, 33*(3), 141-154.

Panettieri, J.C. (2007). Brave new world-advance teaching technologies. *Campus Technology, 209*(4), 27-33.

Papagelis A., Drosos N. (1999). *A decision trees components library.* Diploma Thesis at Computer Engineering and Informatics Department, University of Patras, Greece (in Greek).

Pastore, M. (2001). Companies, universities moving toward e-learning. *Cyberatlas*.com. Retrieved October 12, 2002 from http://cyberatlas.internet.com/markets/education/article

Patton, M. Q. (2002). *Qualitative research and evaluation methods.* Thousand Oaks, CA: Sage Publications.

Pepicello, B., & Tice, E. (2000). Reshaping and teaching and learning: The role of liberal arts in online education. In K.W. White & B.H. Weight (Eds.), *The online teaching guide: A handbook of attitudes, strategies, and techniques for the virtual classroom* (pp. 45-56). Needham Heights, MA: Allyn & Bacon.

Petrova, K., & Claxton, G. (2005). Building student skills and capabilities in information technology and e-business: A moving target. *Journal of Information Systems Education*, 16(1), 27-41.

Phillips, R., Fairholme, E., & Luca, J. (1998). Using e-mail to improve the experience of students doing a group-based project unit. *Showing Best Practice at ECU. The Proceedings of the 1998 Best Practice Showcase* Retrieved September 30, 2004, 2004, from http://www.cowan.edu.au/eddev/98case/phillips.html

Piaget, J. (1977). *Equilibration of cognitive structures.* New York: Viking Press.

Picciano, A. G. (2006). *Educational leadership and planning for technology.* NJ: Pearson Prentice Hall.

Pinto, J. K., & Slevin, D. P. (1988). Critical success factors across the project life cycle. *Project Management Journal, 19*(3), 67-75.

Pittinsky, M.S. (2002). The wired tower: Perspectives on the impact of the internet on higher education. Financial Times/Prentice Hall.

Pohjolainen, S., Hautakangas, S., Ranta, P., Levasma J., & Pesonen, K. (2003). A learning experiment in mathematics using A&O-learning environment. *Int. J. Cont. Engineering Education and Lifelong Learning, 13*(1&2), 57-74.

Porter, C., & Cleland, J. (1995). *The portfolio as a learning strategy*. Portsmouth, NH: Boynton/Cook Publishers, Inc.

Posnick-Goodwin, S. (2005, March). In the march toward gender equity, girls surge forward, but boys fall back. *California Educator, 9*.

Pournaghshband, H. (1990). The students' problems in courses with team projects. *Proceedings of the 21st SIGCSE Technical Symposium on Computer Science Education*, (pp. 44-47).

Prensky, M. (2001). *Digital game-based learning.* New York: McGraw Hill.

Prensky, M. (2003.) "Modding"—the newest authoring tool. *Marc Prensky.* Retrieved from http://www.marcprensky.com/writing/Prensky%20-%20Modding%20-%20The%20Newest%20Authoring%20Tool.pdf

Price, R.W. (2004). Roadmap to entrepreneurial success: Powerful strategies for building a high-profit business. New York: American Management Association.

Project Management Institute. (2004). *A guide to the project management body of knowledge (PMBOK Guide)* (Third Ed.). PA: Project Management Institute Inc.

Punch, K. F. (2000). *Introduction to social research: Quantitative & qualitative approaches.* Thousands Oaks, CA: Sage Publications.

Quinlan, J. R. (1983). Learning efficient classification procedures and their application on chess endgames (p.469). In R. S. Michalski, J. R. Carbonell, & T. M. Mitchell (Eds.), *Machine learning: An artificial intelligence approach* (pp. 463-82). Los Altos, CA: Morgan Kaufmann.

Quinlan, J. R. (1986). Induction of decision trees. *Machine Learning, 1*(1), 81-106.

Quinlan, J. R. (1987). Simplifying decision trees. *International Journal of Man-Machine Studies, 27,* 221-234.

Quinlan, J. R. (1993). *C4.5: Programs for machine learning.* San Mateo, CA: Morgan Kaufmann Publishers.

Quinlan, J. R., & Rivest, R. L. (1989). Inferring decision trees using the minimum description length principle. *Information and Computation, 80*(3), 227-248.

Ratan K., & Mitty, P. (1997). Using contemporary tools to teach dynamics in engineering technology. *International Journal of Engineering Education,* 13(6), 407-411.

Raths, D. (2001). Measure of success. *Training Mag. com*, Retrieved July 7, 2004, from http://www.onlinelearningmag.com

Raymond, R. P. (2005). *Business data networks and telecommunications* (5th Ed.). Englewood Cliffs, NJ: Prentice Hall.

Reif, F., & Sue, A. (1992). Cognition for interpreting scientific concepts: A study of acceleration. *Cognition and Instruction, 9,* 1-44.

Reif, H. L., & Mitri, M. (2005). How university professors teach project management for information systems. *Communications of the ACM, 48*(8), 134-136.

Richards, B., & Waisbrot, N. (2002, June 24-28). *Illustrating networking concepts with wireless handheld devices.* Paper presented at the 7th Annual SIGCSE Conference on Innovation and Technology in Computer Science Education (ITiCSE'02), (pp. 29-33). Aarhus, Denmark.

Ringstaff, C., & Kelley, L. (2002) The learning return on our educational technology Investment. San Francisco: WestEd. Retrieved from www.wested.org/cs/we.view/rs/619

Rodrguez, D., Sicilia, M. A., Cuadrado-Gallego, J.J., & Pfahl, D. (2007). e-learning in project management using simulation models: A case study based on the replication of an experiment. *IEEE Transactions on Education 49*(4), 451-463.

Rola, M. (2002). Building IT into lesson plans. *Computing Canada, 28*(22), 28.

Romano, J., Wallace, T.L., Helmick, I.J, Carey L.M. & Adkins L. (2005). Study procrastination, achievement, and academic motivation in web-based and blended distance learning. *The Internet and Higher Education, 8*(4), 299-305.

Rosenberg, M. (2001). *E-Learning: Strategies for delivering knowledge in the digital age*. New York: McGraw-Hill.

Rosser, S. V. (1997). *Re-engineering female friendly science.* New York: Teacher's College Press.

RPP International. (2001). *Challenge and opportunity: The impact of charter schools on school districts.* Washington, DC: Office of Educational Research and Improvement, U.S. Department of Education.

Rumble, G. (1992). The comparative vulnerability of distance teaching universities. *Open Learning, 2*, 31–45. Retrieved August 21, 2005, from http://www.uni-oldenburg.de/zef/cde/econ/readings/rumble92.pdf

Rumble, G. (1999). Costs of networked learning: What have we learned? *In Proceedings of the FLISH'99. Conference on Flexible Learning on the Information Superhighway.* Sheffield, England. Retrieved from http://www.shu.ac.uk/flish/rumblep.htm.

Ryder, M. (1998, June 7-11). *The world wide web and the dialects of consciousness.* Paper presented at the Fourth Congress of the International Society for Cultural Research and Activity Theory, Aahus, Denmark. Retrieved October 15, 2003 from http://carbon.cudenver.edu/~mryder/iscrat_98.html

Sage, S.M. (2000). A natural fit: Problem-based learning and technology standards. *Learning & Leading with Technology, 28*(1), 6-12.

Salmon, G. (2002). *Developing e-tivities: The key to active online learning.* London: Kogan Page, Ltd.

Sandholtz, J.H., Ringstaff, C., & Dwyer, D.C. (1997). *Teaching with technology: Creating student-centred classrooms.* New York: Teachers College Press.

Sarkar, N. I. (Ed.). (2006). *Tools for teaching computer networking and hardware concepts* (Vol. 1). Hershey, PA: Information Science Publishing.

Sarkar, N. I., & Craig, T. M. (2006). Teaching wireless communication and networking fundamentals using Wi-Fi projects. *IEEE Transactions on Education, 49*(1), 98-104.

Sarkar, N., & Clear, T. (2000, June 30 - July 3). *Developing a new course for the software development pathway on the AUT Bachelor of applied science programme.* Paper presented at the 13th Annual Conference of the National Advisory Committee on Computing Qualifications (NACCQ), (pp. 303-309).Wellington, New Zealand.

Sarkar, N., & Petrova, K. (2001, July 2-5). *Teaching computer networking & telecommunications: a network analysis and software development approach.* Paper presented at the 14th annual conference of the national advisory committee on computing qualifications (NACCQ), (pp. 379-384).Napier, New Zealand.

Satterly, D. (1987). Piaget and education. In R. L. Gregory (Ed.). *The oxford companion to the mind.* Oxford: Oxford University Press.

Saunders, G., & Klemming, F. (2003). Integrating technology into a traditional learning environment. *Active Learning in Higher Education, 4*(1), 74-86.

Scardamalia, M., & Bereiter, C. (1986). Research on written composition. In M.C. Wittrock (Ed.), *Handbook of research on teaching* (3rd Ed.), (pp. 778-803). New York: Macmillan Publishing Company.

Scheinder, B., & Konz, A. (1989). Strategic job analysis. *Human Resource Management, 28*, 51-63.

Schermerhorn, Jr., J. & Chappell, D.S. (2000). *Introducing Management!* New York: John Wiley and Sons.

Schofield, J.W. (1995). *Computers and classroom culture.* Cambridge NY: Cambridge University Press.

Schön, D.A. (1987). Educating the reflective practitioner: Toward a new design for teaching and learning in the professions. San Francisco, CA: Jossey-Bass.

Schreiber, D.A. (1998). Organizational technology and its impact on distance training. In D.A. Schreiber, & Z.L. Berge (Eds.), *Distance training: How innovative organizations are using technology to maximize learning and meet business objectives* (pp.3–18). San Francisco: Jossey-Bass Inc.

Schrum, L., & Dehoney, J. (1998). Meeting the future: A teacher education program joins the information age. *Journal of Technology and Teacher Education, 6*(1), 23-27.

Schuler, R.S., & Walker, J.W. (1990). Human resources strategy: Focusing on issues and actions. *Organizational Dynamics, 19*(1), 5-19.

Schwalbe, K. (2004). Introduction to project management. In J. Locke & A. Poirier (Eds.), *Information Technology Project Management* (Third Ed.) (pp. 1-29). Cambridge, MA: Course Technology.

Schwartz, F., & White, K. (2000). Making sense of it all: Giving and getting online course feedback. In K. W. White & B. H. Weight (Eds.), *The online teaching guide* (pp. 167-182). Needham Heights, MA: Allyn and Bacon.

Schwartz, P. (1991). *The art of the long view*. New York: Double Day.

Schweizer, H. (1999). *Designing and teaching an on-line course: Spinning your Web classroom*. Upper Saddle River, NJ: Prentice Hall.

Schwier, R.A. (2004). Virtual learning communities. In G. Anglin (Ed.), *Critical issues in instructional technology*. Portsmouth, NH: Teacher Ideas Press.

Scott, N. W. (1996). *A study of the introduction of educational technology into a course in engineering dynamics: classroom environment and learning outcomes*. PhD. thesis.

Sein, M.K., & Bostrom, R.P. (1989). Individual differences and conceptual models in training novice users. *Human Computer Interaction, 4,* 197-229

Sept, J. (2004). The stone age in information age. In W.E. Becker and M.L. Andrews (Eds.), *The scholarship of teaching and learning in higher education,* (pp 47-80) Bloomington, IN: Indiana University Press.

Sexton, D.L., & Bowman-Upton, N.B. (1991). Entrepreneurship: Creativity and growth. New York: McMillan Publishing Company.

Seymour, E. (1998). *The role of socialization in shaping the career-related choices of undergraduate women in science, mathematics, and engineering majors.* Paper presented at the Choices and Successes: Women in Science and Engineering, New York.

Shelton, M., & Jones, M. (1996). Staff development that works! A tale of four T's. *NASSP Bulletin, 80*(582), 99-105.

Sivin-Kachala, J., & Bialo, E. (2000). *2000 Research report on the effectiveness of technology in schools.* (7th Ed.). Washington, DC: Software and Information Industry Association.

Sloan, J. D. (2004). *A remotely accessible network laboratory— TCP/IP Laboratories. http://webs.wofford.edu/sloanjd/netlab/IPLabs/labs.htm*. Retrieved December 27, 2004, from http://webs.wofford.edu/sloanjd/netlab/IPLabs/labs.htm

Smaldino, S.E., Russell, J. D., Heinich, R., & Molenda, M. (2005). *Instructional technology and media for learning.* NJ: Merrill Prentice Hall.

Snider, A. (2002). Find a winning blend. *Elearning Magazine*. Retrieved June 26, 2004, from http://www.elearningmag.com/ltimagazine/article/article-Detail.jsp?id=11688

Snook, A. (2003). Moving towards fully integrated systems for the learning organization. *The Training Foundation*. Retrieved August 6, 2004, from http://www.traininfoundation.com/articles/default.asp?PageID=1458

Sorgo, A. (2003). Searching for information on the internet — What if your students cannot speak English? *International Journal of Instructional Media, 30*(3), 315-319.

Squire, K. (2001). Games to teach project. *Education Arcade.* Retrieved from http://www.educationarcade.org/gtt/

Squire, K. (2002). Cultural framing of computer/video games. *Game Studies*, 2(3).Retrieved from the internet September 14, 2007 at http://www.gamestudies.org/0102/squire/

Squire, K. (2002). Replaying history: Learning world history through playing civilization III. Retrieved from http://website.education.wisc.edu/kdsquire

Squire, K. (2004). What happens when games go into any classroom situation? *Serious Games Summit DC.* Retrieved March 10, 2005 from http://www.cmpevents.com/GDe04/a.asp?option=C&V=11&SessID=3250

Squires, R. G. Andersen, P. K. Reklaitis, G. V. Jayakumar, S. & Carmichael, D. S. (1992). Multimedia-based applications of computer simulations of chemical engineering processes. *Journal of Computer Applications in Engineering Education, 1*(1), 25-30.

Srivastava, S.C. (2001). IT training needs for mechanical department over Indian railways. *In proceedings of Seminar on Training Needs for Indian Railway Officials, Indian Railways Institute of Mechanical & Electrical Engineering (IRIMEE),* (pp. 8-16). Jamalpur.

Srivastava, S.C., & Teo, T.S.H. (2004). IT training as a strategy for business productivity: A framework for small and medium-sized enterprises in Asia. *In proceedings of the 8th Pacific Asia Conference on Information Systems, (PACIS 2004),* Shanghai, China.

Stager, G.S. (1995). Laptop schools lead the way in professional development. *Educational Leadership, 53*(2), 78-81.

Stallings, W. (2007). *Data and coputer communications* (8 Ed.). NJ: Prentice Hall.

Starr, P. (1996). Computing our way to educational reform. *The American Prospect,*27 (July-August), 50-60.

Stefani, A., Vassiliadis, B., & Xenos, M., (2006). On the quality assessment of advanced e-learning services, Interactive Technology & Smart Education, *3*(3), 237-250.

Steif, P. S., & Naples, L. M. (2003). Design and evaluation of problem solving courseware modules for mechanics of materials. *Journal of Engineering Education*. Retrieved from http://findarticles.com/p/articles/mi_qa3886/is_200307

Stein, M. V. (2002). Using large vs. small group projects in capstone and software engineering courses. *The Journal of Computing in Small Colleges, 17*(4), 1-6.

Stonier, T., & Conlin, C. (1985). *The Cs: Children computers and communication.* London: Wiley.

Strudler, N., & Wetzel, K. (2005). The diffusion of electronic portfolios in teacher education: Issues of initiation and implementation. *Journal of Research on Technology in Education, 37*(4), 411-433.

Strudler, N., Archambault, L., Bendixen, L., Anderson D., & Weiss, R. (2003). Project THREAD: Technology helping restructure educational areas and delivery. *Educational Technology Research and Development, 51*(1), 41-56.

Sucher, Y. (2003, March 2003). *Digital libraries: Effective access survey of STEM educators, preliminary summary of survey results.* Retrieved January 27, 2007, from http://www2.edc.org/GDI/homepage_SR/Prelim_Report_March2003.doc

Swartz, N. (2006). Enterprise-wide records training: Key to compliance, success. *Information Management Journal, 40*(5), 35-44.

Tabor, S.W. (2005). Achieving significant learning in e-commerce education through small business consulting project. *Journal of Information Systems Education, 16*(1), 19-26.

Tannenbaum, S., & Yukl, G. (1992). Training and development in work organizations. *Annual Review of Psychology, 43*(1), 399-441.

Teale, P. (2003). ICT implementation; what makes the difference? *British Journal of Educational technology, 34*(5), 567-583.

Teo, T.S.H. (1999). Managing information technology at the housing and development board in Singapore: A holistic approach for the public sector. *Journal of Information Technology Cases and Applications (JITCA), 1*(3), 26-44.

Teo, T.S.H., & Ranganathan, C. (2003). Leveraging IT resources and capabilities at the housing and development board. *Journal of Strategic Information Systems, 12*, 229-249.

The Douglas Project. (2003). *GirlsTech: Girls, science, and technology*. Retrieved January 27, 2004, from http://girlstech.douglass.rutgers.edu/

The Standish Group. (2001). Extreme CHAOS. Retrieved October 28, 2005, 2005, from http://www.standishgroup.com/sample_research/index.php

The SUCCESS Program/ (2004). An unpublished manuscript by Melissa B. Tomei, Duquesne University.

Thomas, L.G. "The nets project." *International Society for Technology Standards*. Retrieved November 21, 2004, from enets.iste.org/nets_overview.html

Thomas, R.M. (1985). Educational radio and television. In R.M. Thomas and V.N. Kobayashi (Eds), *Educational technology — its creation, development and gross cultural transfer*, (pp 105-123). Oxford: pergamon press.

Thompson, L. (1997, May). *Education graduates' post-degree learning needs study*. (SIDRU Research Report No.19). Saskatchewan, Canada: Saskatchewan Instructional Development and Research Unit, Regina. (ERIC Document Reproduction Service No. ED434081).

Thorburn, D. (2004). Technology integration and educational change: Is it possible? *Educational Communications and Technology*, February, 2004. Retrieved from www.usask.ca/education/coursework/802papers/thorburn/ind.

Tiene, D., & Luff, P. (2001). Teaching in a technology rich classroom. *Educational Technology, 41*(4), 23-31.

Timmons, J.A. (1980). A business plan is more than a financing device. *Harvard Business Review, 58(*2), 28-34.

Tomei, L., & Wilcox, B. (1999). *Professional portfolios for teachers: A guide for learners, experts, and scholars*. Norwood, MA: Christopher-Gordon Publishers, Inc.

Tomek, I., & Muldner, T. (1999). Acadia advantage—evolution and experiences. *Interactive Learning Environments, 7*(2-3), 175-194.

Tomlinson, R. (2003). *Thinking about GIS: Geographic information system planning for managers*. ESRI Press.

Troha, F. (2002). Ensuring e-learning success: Six simple tips for initiative leaders. *USDLA Journal 16*(12). *United States Distance Learning Association*. Retrieved July 17, 2004, from http://www.usdla.org/html/journal/DEC02_Issue/article02.html

Turban, E., King, D., Lee, J., & Viehland D. (2006). Electronic commerce: A managerial perspective (2006 Ed.). Upper Saddle River, NJ: Pearson Prentice Hall.

U.S. Department of Commerce. (2000). Falling through the net: Toward digital inclusion. A report on American's access to technology tools. Retrieved September 21, 2005 from search.ntia.doc.gov/pdf/fttn00.pdf/

Vallim, B. R. (2006). Practicing engineering in a freshman engineering course. *IEEE Transactions on Education, 49*(1), 74-79.

Van Auken, S. (1991, April). Outcomes assessment: Implications for AACSB accredited business schools and marketing departments. *Western Marketing Educators' Association Conference Proceedings*, (pp.34-37).

Van der Wal, R.J., & Van der Wal, R. (2003). Assessing life skills in young working adults — Part 1: The development of an alternative instrument. *Education and Training, 45*(3), 139-151.

Van der Wende, M., & Van de Ven, M. (2003). The use of ICT in higher education: A mirror of europe. Utrecht, Lemma Publishers.

Van Weert, T.J., & Pilot, A. (2003). Task-based team learning with ICT, design and development of new learning. *Education and Information Technologies, 8*(2), 195 - 214.

VanSickle, J. (2003). *Making the transition to teaching online: Strategies and methods for the first-time, online instructor*. Morehead, KY: Morehead State University. (ERIC Document Reproduction Service No. ED479882)

Varvel Jr., V. E. & Thurston, C. (2002). Perceptions of a wireless network. *Journal of research on Technology in Education, 34*(4), 487-501.

Vaughan, J., & Burnes, B. (2002). Bringing them in and checking then out: laptop use in the modern academic library. *Information Technology and Libraries, 21*(2), 52-62.

VClab. (2007).Retrieved from http://www.esr.uni-bochum.de/VCLab/

Vella, J. (1994). *Learning to listen, learning to teach: The power of dialogue in educating adults*. San Francisco: Jossey-Bass, Inc., Publishers.

Vodanovich, S., & Piotrowski, C. (1999). Views of academic 1-0 psychologists toward internet-based instruction. Retrieved from http:/www.siop.org/tip/backissues/TIP July 99/7 vodanovich.htm.

Vonk, R. (2006). Six ways to link *training* with business performance. *Strategic HR Review, 6*(1), 20-23.

Vygotsky, L. (1978). *Mind in society: The development of higher psychological processes*. Cambridge, MA: Harvard University Press.

Waite, S. J., Wheeler, S., Bromfield, S. (2007). Our flexible friend: The implications of individual differences for information technology teaching. *Computers & Education, 48,* 80-99.

Waller, V. (2003). Implementing an e-learning strategy. *The Training Foundation*. Retrieved August 6, 2004, from http://www.trainingfoundation.com/articles/default.asp?PageID=1397

Warschauer, M. (2003). *Technology and social inclusion: Rethinking the digital divide*. The MIT Press.

Watt, H. M. G., Eccles, J. S., & Durik, A. M. (2006). The leaky pipeline for girls; A motivational analysis of high school enrolments in Australia and the USA. *Equal Opportunities International, 25,* 642-663.

Waxman, H. C., Connell, M. L., & Gray, J. (2002). *A quantitative synthesis of recent research on the effects of teaching and learning with technology on studtent outcomes*. Naperville, IL: North Central Regional Educational Laboratory. Retrieved September 21, 2005 from http://www.ncrel.org/tech/effects/effects.pdf

Wayne, R. (2002, March). Location, location, location. *SDMAGAZINE*.

Weiser, M. (1998). The future of ubiquitous computing on campus. *Communications of the ACM, 41*, 41-42.

Wetzel, K., & Strudler, N. (2005). The diffusion of electronic portfolios in teacher education: Next steps and recommendations from accomplished users. *Journal of Research on Technology in Education, 38*(2), 231-243.

Wetzel, K., Zambo, R., & Buss, R. (1996). Innovations in integrating technology into student teaching experiences. *Journal of Research on Computing in Education, 29*(Winter), 196-214.

Whalen, T., & Wright, David (1999). Methodology for Cost-Benefit Analysis of Web-Based Tele-Learning: Case Study of the Bell Online Institute. *The American Journal of Distance Education,* 13(1), 25-43.

Whitaker, B.E. (2001). Association of African Universities charts goals for 21st century. *International Higher Education, 23(spring 2001)*.

White, K. (2000). Face to face in the online classroom. In K. W. White & B. H. Weight (Eds.), *The online teaching guide* (pp. 1-12). Needham Heights, MA: Allyn and Bacon.

White, K.W., & Weight, B.H. (2000). *The online teaching guide: A handbook of attitudes, strategies, and techniques for the virtual classroom*. Needham Heights, MA: Allyn & Bacon.

Wieland, A. (2005). Evaluation of the use of electronic learning environments: perceptions of lecturers and students. *Recent research developments in learning technologies*. Rotterdam: FORMATEX.

Wiggins, G. (1987, Winter). Creating a thought-provoking curriculum. *American Educator,* 11(4), 10-17.

Wiggins, G. (1998). *Educative assessment: Designing assessments to inform and improve student performance.* San Francisco: Jossey-Bass Publishers.

Wilhelm, L., Puckett, K., Beisser, S., Wishart, W., Merideth, E., & Sivakumaran, T. (2006). Lessons learned from the implementation of electronic portfolios at three universities. *TechTrends, 50*(4), 62-71.

Willis, B. (2007). Distance Education at a Glance. Guides 1 - 7. Engineering Outreach at the University of Idaho Presents. Retrieved January 14, 2007, from http://www.uidaho.edu/eo/distglan.html

Wired. (2003). *Every sims picture tells a story.* Retrieved October 9, 2006 from http://www.bmedia.org/archives/00000330.php

Wiske, S. (2004). Using technology to dig for meaning. *Educational Leadership, 62*(1), 46-50.

WITI. (2004). *Women in Technology International website.* Retrieved February 10, 2007, from http://www.witi.com/

Witten, I., & Frank, E. (2000). Data mining: Practical machine learning tools and techniques with java implementations. San Mateo, CA: Morgan Kaufmann Publishers.

Wittrock, M.C. (1974). A generative model of mathematics learning, *5*(4), 181-196.

Wittrock, M.C. (1986). Students' thought processes. In M. C. Wittrock (Ed.), *Handbook of research on teaching* (3rd Ed.), (pp. 297-314). New York: Macmillan Publishing Company.

Wittrock, M.C. (1990). Generative processes of comprehension. *Educational Psychologist. 24*(4), 345-376.

Wittrock, M.C. (1991). Generative teaching of comprehension. *The Elementary School Journal, 92*(2), 169-184.

Wittrock, M.C. (1994). *Metacognition.* Unpublished manuscript, University of California at Los Angeles.

Wittrock, M.C. (2000, Winter and Spring). Knowledge acquisition and education. *The Journal of Mind and Behavior, 21*(1 and 2), 205-212.

Wittrock, M.C. (2001). The Knowledge Dimension. In L.W. Anderson, D.R. Krathwohl, P.W. Airasian, K.A. Cruikshank, R.E. Mayer, P.R. Pintrich, J. Raths, & M.C. Wittrock (Eds.), A Taxonomy for Learning, Teaching, and Assessing: A Revision of Bloom's Taxonomy of Educational Objectives (pp. 38-62). New York: Addison Wesley Longman, Inc.

Wittrock, M.C. (2001). The Cognitive Process Dimension. In L.W. Anderson, D.R. Krathwohl, P.W. Airasian, K.A. Cruikshank, R.E. Mayer, P.R. Pintrich, J. Raths, & M.C. Wittrock (Eds.), A Taxonomy for Learning, Teaching, and Assessing: A Revision of Bloom's Taxonomy of Educational Objectives (pp. 63-92). New York: Addison Wesley Longman, Inc.

Wittrock, M.C. (Ed.). (1986). *Handbook of research on teaching* (3rd Ed.). New York: Macmillan Publishing Company.

Wittrock, M.C. (Nov-Dec, 1987). Teaching and student thinking. *Journal of Teacher Education, 38*(6), 30-33.

Wlodkowski, R.J. (1999). *Enhancing adult motivation to learn (Revised Edition): A comprehensive guide for teaching all adults* (2nd Ed.). San Francisco: Jossey-Bass Inc., Publishers.

Wojnar, L. (2000). *Instructional design and implementation of a best practice model of online teaching and learning.* Unpublished doctoral dissertation, Duquesne University, USA.

Wojnar, L. (2002, August). Research summary of a best practice model of online teaching and learning. *English Leadership Quarterly, 25(1)*, 2-9.

Xenos M., Pierrakeas C., & Pintelas, P. (2002). Survey on student dropout rates and dropout causes concerning the students in the course of informatics of the Hellenic Open University. *Computers & Education, 39* (4), 361-377.

Xin, F., & Jamalipour, A. (2006). TCP performance in wireless networks with delay spike and different initial congestion window sizes. *Computer Communications, 29*(8), 926-933.

Yidirim, S. (2000). Effects of an educational computing course on preservice and in-service teachers: A discussion and analysis of attitudes and use. *Journal of Research on Computing in Education, 32*(4), 479-495.

Zeng, W. G., & Trajkovic, L. (2005, August 22-24). *TCP packet control for wireless networks.* Paper presented at the IEEE International Conference on Wireless And Mobile Computing, Networking And Communications (WiMob'2005) (pp. 196-203).

Zeng, X., Bagrodia, R., & Gerla, M. (1998). *GloMoSim: a library for parallel simulation of large-scale wireless networks.* Paper presented at the Twelfth Workshop on Parallel and Distributed Simulation, (pp. 154-161).

Zhao, Y., & Cizko, G.A. (2001). Teacher adoption of technology: A perceptual control theory perspective. *Journal of Technology and Teacher Education, 9*(11), 5-30.

Zinkoski, P. (August, 2001). The many benefits of writing. *Partnership for Learning.* Retrieved January 11, 2004 from http://www.partnershipforlearning.org/article.asp?ArticleID=319

About the Contributors

Lawrence A. Tomei is the associate vice president of academic affairs and associate professor of education at Robert Morris University. Born in Akron, Ohio, he earned a BSBA from the University of Akron (1972) and entered the U.S. Air Force, serving until his retirement as a lieutenant colonel in 1994. Dr. Tomei completed his MPA and MEd at the University of Oklahoma (1975, 1978) and EdD from USC (1983). His articles and books on instructional technology include: "Professional Portfolios for Teachers" (1999); "Teaching Digitally: Integrating Technology into the Classroom" (2001); "Technology Façade"(2002); "Challenges of Teaching with Technology Across the Curriculum" (2003); and "Taxonomy for the Technology Domain" (2005).

***

Lindsay Adams earned her teaching certification in elementary and special education at Geneva College. Her experience includes a variety of settings including a school for students with autism, emotional support classrooms, and currently, cyber education. Along with Dr. Hipsky, she recently presented a workshop on strategies for students with special needs in cyber schools at the Pennsylvania Educational Technology Expo and Conference. Adams is currently a special education teacher and instructional supervisor at The Pennsylvania Cyber Charter School.

S. E. Aduwa-Ogiegbaen is a senior lecturer in the Department of Educational Psychology and Curriculum Studies, University of Benin, Benin City-Nigeria. He teaches courses in instructional technology. His research interests include curriculum implementation, use of media for instructional purpose, gender studies, and studies related with problems of teaching the English language. He has been in the service of University of Benin for about 20 years.

Zane Berge is currently associate professor in the instructional systems development graduate program at the University of Maryland (UMBC campus). He served as director of the program from 1995 until 2001. Immediately prior to this, he was the founding director of the Center for Teaching and Technology and assistant director for Training Services, Academic Computer Center, Georgetown University. Berge has consulted in the areas of ISD, multimedia design and distance education with clients in higher education, government, special education and both large and small corporate organizations. His scholarship in the field of computer-mediated communication and distance education includes numerous articles, chapters, workshops, and presentations. His more recent books are "Sustaining Distance

Training: Integrating learning technologies into the fabric of the enterprise" (Jossey-Bass, 2001) and "Virtual Schools: Planning for Success." He consults internationally in distance education.

Chuleeporn Changchit is an associate professor of management information systems at Texas A&M University-Corpus Christi. Drawing from experiences from working for over five years as a senior consultant in industry, she provides her students with practical and comprehensive instructions, shares ideas and techniques enthusiastically, and enjoys substantial successes as an educator. Dr. Changchit is also actively engaged in the scholarly activities. Her articles have been published in several journals, such as decision support systems, information systems journal, expert systems with applications, and international journal of intelligent systems in accounting, finance, and management. She also serves as an associate editor for the *Journal of Electronic Commerce in Organizations*, as well as on the editorial review board for several journals, such as *Information Resources Management Journal, the Journal of Global Information Technology,* and *the International Journal of E-Business Research.*

Earl Chrysler has been associated with the MIS field for over 30 years. During that time he has been a systems analyst for Ford Motor Company; a systems consultant for Laventhol and Horwath, CPAs, an independent consultant serving national and international clients; chair of the Computer Information Systems Department at Quinnipiac University; and professor of MIS at California State University, Chico, where he designed the initial MIS program. He is currently a professor of MIS at Black Hills State University. He has published or presented over 40 papers in the MIS area and written chapters for two computer-related books.

Robert Cutshall is an assistant professor of management information systems at Texas A&M University-Corpus Christi. Dr. Cutshall's current interests are in the areas of technology in higher education, electronic commerce design, decision support systems, and expert systems. His publications have appeared in journals such as *Journal of Computer Information Systems* and *Industrial Management & Data Systems.* In addition, he has made numerous presentations at national and international conferences.

Arnold Depickere is the executive dean for the Division of Arts at Murdoch University. He has an MBA (information systems) from Maastricht School of Management in the Netherlands. Prior to joining Murdoch University, Professor Depickere had a combination of 30 years of experience in both the IT industry and IT education; nine years of which was in Europe followed by 21 years in the Far East. For four of these years he held a project management position in IT quality assurance and transfer of technology at the Kuala Lumpur International Airport. Over the last 15 years he has acted as a consultant to several multinational companies and has held senior executive appointments.

Lambros Drossos is an associate professor in the Applied Informatics in Management and Economy Department of the Technological Educational Institute of Messolonghi He has participated in over 10 research and development projects in the area of software engineering and educational technologies. His current research interest includes complex network constructions (LAN, MAN, WAN, Satellite) and programming in networks (Internet), Web development, Application development in UNIX, Software development, information retrieval and information systems, application of non-linear dynamics to particle accelerator problems, Symplectic maps and periodic orbits, numerical simulation of ODEs, PDEs, complex systems and singularity analysis, Fractals, non-linear time series analysis, and non-linear waves in biological systems. He has authored or co-authored two books in Greek and over 60 papers in international journals and conferences. Drossos serves as a reviewer for numerous journals and conferences.

Susan Elwood is currently an assistant professor of educational technology in the College of Education at Texas A&M University Corpus Christi. She has approximately 20 years experience in public and private, international, and national teaching, as well as multiple grants related to pedagogical applications of basic integrated technologies. Dr. Elwood's research interests include higher-order thinking skills applications of basic integrated technologies within portable computing environments, especially related to reality-based learning. She has made several presentations at state, national, and international conferences. Some of her publications have appeared in the *Encyclopedia of Information Technology Curriculum Integration, Educational Technology & Society, Campus-Wide Information Systems, International Journal of Information and Communication Technology Education,* and *The Clearing House.*

Elizabeth Fanning is a PhD candidate at the University of Virginia Curry School of Education, Department of Instructional Technologies. She teaches courses in instructional design and game-based learning. She earned her MA in educational technologies from San Francisco State University and a BA in English from the University of Michigan. For the past 15 years, she has consulted as an instructional designer, distance learning specialist, and multimedia producer. Her work includes educational games and simulations for adult learning audiences, including the educational CDROM entitled "The Wall: A Living Memorial," distinguished by awards from Multimedia Producer, AV Producer's Award, and the New York Festival Gold Medal. Currently, Ms. Fanning is with the University's School of Continuing and Professional Studies. Current projects and research focus on virtual learning spaces, game-based learning, and facilitating the development and conversion of existing instruction to online and blended deliveries.

David Gadish is a faculty member at California State University Los Angeles. He teaches technology management in the CIS, business, and MBA programs. Dr. Gadish's research spans geographic information systems, location-based services, and online marketing and public relations. Dr. Gadish has spent over 15 years in government and industry, and consults to organizations in a variety of business sectors in these areas.

Lenora Giles holds a Master of Distance Education from the University of Maryland (2004) and a BS in Corporate Communication from the University of Baltimore (2001). Lenora is the e-learning counselor for the admissions office at the University of Baltimore in Baltimore, Maryland managing email and web recruitment of prospective students for online undergraduate and graduate degree programs and Web marketing. Prior to this, she served as a designated school official handling admission and immigration concerns regarding the university admissions policy, U.S. Citizenship and Immigration Services (USCIS) regulations, Student and Exchange Visitor Program (SEVP) and Student and Exchange Visitor Information System (SEVIS) processes for nonimmigrant F-1 students and J-1 scholars.

John E. Graham is dean of the School of Education and Social Sciences at Robert Morris University and professor of education. Previously at Robert Morris University, Dr. Graham served as acting dean of various academic schools in his role as associate vice president of academic affairs, and was head of both the Education and Administrative Office Management Departments. Dr. Graham actively participates in the PA Association of Teacher Educators and frequently attends the Association for Curriculum Development and SITE national conferences. His research interests focus on technology and curriculum integration.

Shellie Hipsky's career includes teaching students from kindergarten to graduate school in the United States as well as in Rome, Italy. She presented at an international conference on educational leadership at Oxford University. A few of the journals that her over 40 publications are published in are: *Early Childhood Education Journal, Curriculum Review, The Northam Centre for Leadership Studies Monograph, Educational Review, The Qualitative Report,* and *Kappa Delta Pi Record.* Hipsky's books are: "The Drama Discovery Curriculum: Bibliotherapy" and "Theater Games for Students with Emotional and Behavioral Challenges and Lincoln Interactive Arts Alive Textbook." Shellie is a frequently requested speaker by educational organizations, conferences, and school districts on: differentiated instruction, educational leadership, the arts, and special education in the classroom. As a recent assistant principal in charge of curriculum and supervision at a school for students with emotional or behavioral disabilities, she is acutely aware of teacher and student needs. Dr. Shellie Hipsky is currently a contracted author for Prentice Hall, an educational consultant for the Tri-State Study Council at the University of Pittsburgh, and an assistant professor of education at Robert Morris University.

C. Derrick Huang is an assistant professor in the Department of Information Technology & Operations Management in the College of Business at the Florida Atlantic University. Previously, as a practitioner, his experience was in strategic planning and marketing in information technology as an executive in a number of high-tech companies. His research interest lies in the business value and strategic impact of information technology in organizations, and the current focus is on the economics of investments in information security. He holds PhD from Harvard University.

Dimitris Kalles was educated in Greece (Diploma) and in the UK (MSc,PhD). He sits on the board of directors of a software development company and is a researcher and tutor with the Hellenic Open University. Before that he worked as a research manager for a research institute. He has been teaching courses, authoring coursework and supervising diploma theses on AI, complexity, programming and software engineering for several years, at an undergraduate and postgraduate level. He has published about 30 papers in journals and conferences.

Maria H. Z. Kish is an adjunct instructor for the Instructional Technology Program in the Education Department at Duquesne University in Pittsburgh, Pennsylvania. Previously, she edited and wrote technical manuals at the corporate level and taught adult learners mathematics and software applications in colleges, business institutes, and companies. Her research interest lies in developing teaching strategies and learning techniques, such as vignettes, to work with adult learners, especially in an online environment. She developed and teaches classes to show teachers how to use various software packages to create digital images, illustrations, animations, and Websites. She received secondary certification to teach English and math from Seton Hill College, earned an MS in multimedia technology from Duquesne University, and holds an EdD from Duquesne University.

Tanya McGill is a senior lecturer in the School of Information Technology at Murdoch University in Western Australia. She has a PhD from Murdoch University. Her major research interests include end user computing and information technology education. Her work has appeared in various journals including *the Journal of Research on Computing in Education, European Journal of Psychology of Education, Information Resources Management Journal* and *the Journal of Organizational and End User Computing.*

Athanasios Papagelis graduated from the Department of Computer Engineering and Informatics (CEID with a MSC in computation from the University of Patras, Greece (1999). He has an extensive technical and research background. He has worked two years for the Computer Technology Institute, participating in five national and European projects. He has co-authored a book chapter, three journal papers, six conference papers and four technical reports.

Singh Ramesh is currently a senior lecturer in the Department of Mechanical Engineering at the University Tenaga Nasional. He earned his BEng (Honors, 1st class) in mechanical engineering (1994) and his PhD in advanced ceramics (1997), both from the University of Sunderland. He has published extensively and has filed six Malaysian patents. Dr. Singh is a registered chartered engineer with the Engineering Council UK, a registered professional engineer with the Board of Engineers Malaysia, and a corporate member of the following: The Institutions of Mechanical Engineers UK, Institutions of Engineers Malaysia, Institute of Materials Malaysia, and Electron Microscopy Society Malaysia. His research interests include processing of engineering ceramics, biomaterials, failure analysis of engineering components, and multimedia technology in engineering education.

Maria Delia Rojas is a tutor in the School of Information Technology at Murdoch University in Western Australia. She has an honors degree in information systems from Murdoch University. Her research interests include project management and how to use the practices to obtain a good software product.

George Saltsman is director of educational technology for the Adams Center of Teaching Excellence at Abilene Christian University and serves as an adjunct part-time instructor for the Department of Journalism and Mass Communications. He co-authored the recently released book, *An Administrator's Guide of Online Education,* as well as other journal articles related to online education. His education includes a BS in computer science and an MS in organizational and human resource development. Mr. Saltsman has managed ACU's distance education efforts for more than eight years, helping establish the initial strategic planning elements and the first online courses.

Nurul I. Sarkar is a senior lecturer in the School of Computing and Mathematical Sciences at AUT University, Auckland, New Zealand. He has more than 12 years of teaching experience in universities at both undergraduate and post-graduate levels and has taught a range of subjects, including computer networking, data communications, computer hardware, and e- commerce. His first edited book, entitled "Tools for Teaching Computer Networking and Hardware Concepts," was published by Information Science Publishing in 2006. He has published in international journals and conferences, including the *IEEE Transactions on Education, the International Journal of Electrical Engineering Education, the International Journal of Information and Communication Technology Education,* and *the International*

Journal of Business Data Communications and Networking. Sarkar's research interests include wireless communication networks, simulation and modeling of computer and data communication networks, and tools to enhance methods for teaching and learning computer networking and hardware concepts. He is currently serving as chairman of the IEEE NZ Communications Society Chapter; executive peer-reviewer of the SSCI indexed Journal of Educational Technology & Society; member of editorial review board of the International Journal of Information & Communication Technology Education; and a member on the IASTED Technical Committee on Computers and Advanced Technology in Education.

George W. Semich works in the Department of Secondary Education and Graduates at Robert Morris University and is an assistant professor of education. Currently, Dr. Semich is also the program administrator for the new doctoral program, the PhD in instructional management and leadership which is housed in the School of Education and Social Sciences. Dr. Semich also served as department head of communications at Robert Morris University where he helped to develop and implement a new BFA degree in media arts. His research interests are in technology integration and leadership studies.

Kaye Shelton is the dean of online education for Dallas Baptist University whose online education program now offers 22 fully online degree programs and maintains a 92 percent course completion rate. Ms. Shelton is a certified online instructor and winner of the 2005 Blackboard Bbionic course contest and the Instructional Technology Council's (ITC) E-Learning 2006 Outstanding Online Course award. She co-authored the recently released book, *An Administrator's Guide of Online Education*, as well as other journal articles related to online education and also practices as an online education consultant. Her education includes a BAS in management of information systems, an MS in education emphasizing online teaching and learning, and she is currently pursuing a PhD at University of Nebraska in educational leadership with a focus on distance learning administration.

Manjit Singh Sidhu is currently a senior lecturer in the College of Information Technology in the Deptartment of Informatics at the University Tenaga Nasional. He earned his BSc (Honors) in computer science from the University of Wolverhampton (1997) and an MS in information technology from University Putra Malaysia (2000). He completed his PhD in computer science from University Malaya (2007). He is a member of the Institute of Electrical and Electronics Engineers (IEEE), Computer and Communications Society, Malaysian National Computer Confederation (MNCC), and Malaysian Scientific Association (MSA). His research interests include patterns of interactions, user interface design approaches in multimedia and virtual reality applications, visualization, graphics, and computer simulations and animations.

Shirish C. Srivastava is a doctoral candidate in information systems and e-business at the School of Business, National University of Singapore. His research has been published in international refereed journals and books, including *MIS Quarterly Executive, International Journal of Information & Communication Technology Education (IJICTE), IIMB Management Review, Vikalpa: The Journal of Decision Makers,* and *Encyclopedia of Information Science and Technology.* He has also presented his research in international refereed conferences, like Academy of Management (AOM), Academy of International Business (AIB), International Conference on Information Systems (ICIS), Institute for Operations Research and the Management Sciences (INFORMS), Americas Conference on Information

Systems (AMCIS), and Asia Pacific Conference on Information Systems (PACIS). His research interests include IS and e-business strategy, IT offshoring, and e-government.

Antonia Stefani graduated from the University of Patras, Department of Mathematics in 1999 and received an MSc degree in computer science in 2001. She is currently a computer science Phd student at the Hellenic Open University. Her research interests include e-commerce and quality assessment.

Thompson S.H. Teo is information systems area coordinator and associate professor in the Department of Decision Sciences at the NUS Business School, National University of Singapore. His research interests include strategic use of IT, e-commerce, adoption and diffusion of IT, and strategic IT management and planning. He has published more than 80 papers in international refereed journals such as *Communications of the ACM, Database, Decision Sciences, Decision Support Systems, Information and Management, International Journal of Electronic Commerce, Journal of Management Information Systems* and *MIS Quarterly Executive*. He has also edited four books on IT and e-commerce, and is on the editorial board of several international refereed journals.

Sandra Turner is a professor in instructional technology at Ohio University. Four themes have characterized Dr. Turner's research and scholarship — middle school students as multimedia authors, constructivist learning environments, gender issues in education, and qualitative research methodology. She has worked extensively with teachers and their students in implementing multimedia technologies, project-based learning activities, and alternative assessment in a student-centered learning environment.

Raymond Uwameiye is a senior lecturer in the Department of Vocational and Technical Education, University of Benin, Benin City-Nigeria. He teaches courses in professional vocational and technical education. His research interests include studies related to teaching and learning of vocational and technical education subjects, use of media in teaching/learning of prevocational and vocational education subjects, gender studies, studies related to implementation of prevocational and vocational education curriculum, and studies in entrepreneurship education. He has been in the service of the University of Benin for about 14 years.

Stuart Van Auken holds the position of Alico chair in market analysis and strategy and the rank of eminent scholar at Florida Gulf Coast University. He is a former dean (California State University, Chico) and a former department chair at the University of Louisville. His research interests encompass cognitive aging, cross-cultural research, across-class associational advertising, and pedagogical issues. He has published in the *Journal of the Academy of Marketing Science, the Journal of Consumer Psychology, the Journal of Advertising Research, the Journal of Advertising, Decision Sciences, and Psychology and Marketing,* among others. During his career, he has received the title of distinguished teaching professor and the award of master teacher. He has also published numerous articles in pedagogically-oriented journals, including the *Journal of Marketing Education* and *the Journal of Business Education.*

Bill Vassiliadis obtained his diploma and Phd from the Department of Computer Engineering and Informatics of the University of Patras in 1995 and 2004 respectively. He is currently working as a post-doctoral research fellow at the Digital Systems and Media Computing Lab of the Hellenic Open

University. His current research interests include Information Retrieval and Information Systems. He has published more than 50 papers in international conferences and journals.

Shuyan Wang is an assistant professor in the Department of Technology Education at the University of Southern Mississippi. Her research interests include distance education and online course development, assessment and evaluation, and issues related to technology integration into language teaching and learning.

Mara H. Wasburn is an associate professor in the Department of Organizational Leadership in the College of Technology, Purdue University. Prior to joining the faculty, her experience was in fundraising and publicity/public relations. Her research and consulting focus on mentoring, with an emphasis on women in technology. She recently developed a team mentoring model, which is in the process of being trademarked. She holds a PhD from Purdue University.

Karen White is an intellectual property manager with Purdue Research Foundation, managing a portfolio of patents, educational technology, and technical educational materials produced by Purdue University. This work includes supporting community outreach of Purdue University for both educational programs and economic development, and facilitating commercial and non-profit dissemination of educational materials.

Michalis Xenos is an assistant professor in the Computer Science Department of the School of Sciences and Technology of the Hellenic Open University. He has participated in over 20 research and development projects in the area of software engineering and educational technologies. His current research interests include, inter alia, software quality, software metrics, e-commerce systems quality, and educational technologies. He has authored or co-authored six books in Greek and over 70 papers in international journals and conferences. Dr. Xenos serves as a reviewer for numerous journals and conferences. He is a member of the IEEE and member of the Technical Chamber of Greece.

Index

E

F

G

I

J

K

V

W